Air Quality 2018

Transportation Research Record: Journal of the Transportation Research Board

Volume 2672, Issue 25

ISBN: 978-1-5108-7829-7

Printed from e-media with permission by:

Curran Associates, Inc.
57 Morehouse Lane
Red Hook, NY 12571

Some format issues inherent in the e-media version may also appear in this print version.

Published by Sage Publications USA

ISBN (Print) 978-1-5108-7829-7
ISBN 2018 Printed Set (All Issues) 978-1-5108-7735-1

Printed by Curran Associates, Inc. (2019)

Additional copies of this publication are available from:

Curran Associates, Inc.
57 Morehouse Lane
Red Hook, NY 12571 USA
Phone: 845-758-0400
Fax: 845-758-2633
Email: curran@proceedings.com
Web: www.proceedings.com

Volume 2672 Number 25 December 2018

TRR
JOURNAL OF THE TRANSPORTATION RESEARCH BOARD

Contents

Air Quality

Articles

TRB

TRANSPORTATION RESEARCH BOARD

CORE SPONSORS

Transportation Departments of the 50 States and the District of Columbia

Federal Government

U.S. Department of Transportation
- Federal Aviation Administration (FAA)
- Federal Highway Administration
- Federal Motor Carrier Safety Administration (FMCSA)
- Federal Railroad Administration (FRA)
- Federal Transit Administration
- National Highway Traffic Safety Administration (NHTSA)
- Office of the Assistant Secretary of Transportation for Research and Technology (OST-R)

U.S. Air Force Civil Engineer Center (AFCEC)
U.S. Coast Guard (USCG)
U.S. Department of Energy
U.S. Department of the Interior (DOI)
U.S. Environmental Protection Agency (EPA)

Nongovernmental Organizations

American Public Transportation Association (APTA)
Association of American Railroads (AAR)
California Air Resources Board

The National Academies of
SCIENCES · ENGINEERING · MEDICINE

The **National Academy of Sciences** was established in 1863 by an Act of Congress, signed by President Lincoln, as a private, nongovernmental institution to advise the nation on issues related to science and technology. Members are elected by their peers for outstanding contributions to research. Dr. Marcia McNutt is president.

The **National Academy of Engineering** was established in 1964 under the charter of the National Academy of Sciences to bring the practices of engineering to advising the nation. Members are elected by their peers for extraordinary contributions to engineering. Dr. C. D. Mote, Jr., is president.

The **National Academy of Medicine** (formerly the Institute of Medicine) was established in 1970 under the charter of the National Academy of Sciences to advise the nation on medical and health issues. Members are elected by their peers for distinguished contributions to medicine and health. Dr. Victor J. Dzau is president.

The three Academies work together as the National Academies of Sciences, Engineering, and Medicine to provide independent, objective analysis and advice to the nation and conduct other activities to solve complex problems and inform public policy decisions. The Academies also encourage education and research, recognize outstanding contributions to knowledge, and increase public understanding in matters of science, engineering, and medicine.

Learn more about the National Academies of Sciences, Engineering, and Medicine at **www.national-academies.org.**

The **Transportation Research Board** is one of seven major programs of the National Academies of Sciences, Engineering, and Medicine. The mission of the Transportation Research Board is to increase the benefits that transportation contributes to society by providing leadership in transportation innovation and progress through research and information exchange, conducted within a setting that is objective, interdisciplinary, and multimodal. The Board's varied activities annually engage about 7,000 engineers, scientists, and other transportation researchers and practitioners from the public and private sectors and academia, all of whom contribute their expertise in the public interest. The program is supported by state transportation departments, federal agencies including the component administrations of the U.S. Department of Transportation, and other organizations and individuals interested in the development of transportation.

Learn more about the Transportation Research Board at **www.TRB.org**.

Article

TRR
JOURNAL OF THE TRANSPORTATION RESEARCH BOARD

Transportation Research Record
2018, Vol. 2672(25) 1–10
© National Academy of Sciences:
Transportation Research Board 2018
Article reuse guidelines:
sagepub.com/journals-permissions
DOI: 10.1177/0361198118769901
journals.sagepub.com/home/trr
SAGE

Effectiveness of Idle Reduction Technologies in Reducing Driver Exposure to Diesel Emissions

Josias Zietsman[1], Jeremy Johnson[1], Tara Ramani[1], Reza Farzaneh[1], Michael Rodgers[2], Alexander Samoylov[2], Yanzhi "Ann" Xu[2], and Amy Moore[2]

Abstract

The purpose of this study was to investigate the effectiveness of idle reduction technologies (IRTs) in reducing driver exposure to diesel exhaust, and to study the cost effectiveness of these technologies. IRTs are devices that provide heating and cooling to the cabin of a truck without idling the truck engine. The focus was on diesel-powered IRTs (auxiliary power units and fuel-operated heaters), and their impact on particulate matter (PM2.5) exposure of drivers sleeping or resting in the truck's cabin. The focus was on diesel-powered IRTs as these devices generate their own emissions, potentially exacerbating in-cab exposure levels. The project involved initial field data collection at truck stops in the states of Georgia, Texas and California. This was followed by laboratory testing in an environmentally controlled test chamber on a sample of trucks, with and without the use of IRTs. The study findings showed that the use of IRTs resulted in a significant reduction of PM2.5 levels in the truck cabin when compared with the baseline scenario of a truck parked with the engine off and the doors and windows closed. Idling the truck engine and running the air conditioning system was also found to reduce in-cabin PM2.5 levels relative to the baseline. The study supports the premise that IRTs reduce driver exposure to diesel exhaust. Additionally, it was found that these devices are cost effective in that they all have payback periods of less than five years, making them viable alternatives to idling the truck engine during long-duration rest periods.

Long-haul truck drivers usually spend extended periods of time in the cabin of a truck, both while driving and during driver rest periods. Often, drivers rest in the truck cabin with the truck engine idling to provide air conditioning or heating. The exposure of drivers to pollutants in the truck cabin during idling/rest periods has emerged as an area of concern, due to short-term impacts such as drowsiness and fatigue with an increased risk of crashes, as well as long-term health impacts such as lung cancer and heart disease.

Further, idle reduction is also being promoted as an emissions reduction strategy and fuel saving measure for trucks. Several idle reduction technologies (IRTs) are commercially available in the market. They provide heating and cooling in the truck cabin and eliminate the need to idle the truck's main engine during driver rest periods. While studies have indicated that IRTs have the potential to reduce overall emissions and fuel consumption, there is limited knowledge on the impact of IRTs on driver exposure to diesel exhaust.

This study investigated the effectiveness of various IRTs in reducing driver exposure to diesel exhaust and the cost effectiveness of these technologies. The focus was on diesel-powered IRTs and their impact on drivers' exposure to particulate matter (PM) of 2.5 micrometers or less (i.e., PM2.5) when sleeping or resting in the truck's cabin. The study involved a literature review, development of a testing protocol, field testing, laboratory testing, and a cost analysis, as described in the remainder of this paper.

Literature Review

Truck drivers, especially those on long-haul operations, are often exposed to vehicle exhaust, especially during driver rest periods at truck stops or rest areas. Diesel exhaust, specifically diesel particulate matter/PM2.5 is a known health hazard, with both short-term and long-term effects (*1*).

[1]Environment and Air Quality Division, Texas A&M Transportation Institute, College Station, TX
[2]School of Civil and Environmental Engineering, Georgia Institute of Technology, Atlanta, GA

Corresponding Author:
Address correspondence to Tara Ramani: t-ramani@tti.tamu.edu

Table 1. Technologies for Reducing Truck Idling or Idling Impacts

Technology	Description	Exposure reduction mechanism	Estimated cost
Automatic shut-down/start-up systems	The system is built into the engine's electronic control software and enables the engine to automatically stop and restart to maintain cab temperature.	Reduces unnecessary engine idling and reduces fuel consumptions and emissions.	\$1,000–\$2,500
Auxiliary power unit (APU)	A device, powered by a small (12–20HP) diesel engine, on a vehicle that provides energy for functions other than propulsion.	APUs consumes less fuel, produce less emissions while idling.	\$5,000–\$12,000
Battery air conditioner systems (BAC)	A system that uses batteries to power an independent electric cooling system.	Eliminates engine idling fuel consumptions and emissions.	\$3,000–\$8,000
Electrified parking spaces (EPS)	The system provides power to necessary systems, such as heating, air conditioning, or appliances, without idling engine.	Engine idling is eliminated at the EPS, so no emissions or fuel consumption.	\$12,000–\$22,000
Filtering devices	Devices designed to remove pollutants from the exhaust gas of engine.	Physically filters out pollutants from exhaust gas before it is emitted to environment.	\$5,000–\$7,000
Cabin air filters	HVAC filters located at the outside of air intake.	Removes pollutants before they enter the cabin.	Less than \$100
Fuel operated heater (FOH)	A system provides heat to the cabin.	This technology works by burning a small amount of fuel to produce heat for the cabin.	\$1,000–\$3,000
Shore connection systems	This system connects a power outlet with a truck refrigeration unit to provide energy for operation.	Eliminates using truck engine to operate truck refrigeration unit.	\$5,000–\$9,000
Thermal storage system (TSS)	The system uses energy created as a truck is driven to provide heating or cooling while the truck is off.	The stored heat or cooling energy could be used to cool the cabin and eliminate engine idling.	\$2,000–\$6,500

However, the Occupational Safety and Health Administration (OSHA) does not currently have a permissible exposure limit (PEL) for PM2.5 applicable to workers in general industry. Recent in-cab air quality studies (*2, 3*) have indicated that in-cab PM2.5 concentrations were close to the Environmental Protection Agency's (EPA) National Ambient Air Quality Standards (NAAQS), which were used for comparison in the absence of PELs set by OSHA. Additionally, occupational epidemiology literature indicates elevated mortality rates from lung cancer and heart disease in the unionized U.S. trucking industry (*4, 5*). This is especially the case among truck drivers, indicating a possible correlation to diesel exposure, most notably PM2.5. Therefore, control strategies that reduce driver exposure to diesel exhaust are likely to benefit the long-term health and well-being of truck drivers. Further, driver exposure to diesel exhaust is also a concern from a driver fatigue and alertness perspective (*6, 7*).

Prior studies have indicated that in-cab PM concentrations are higher during truck idling than other times (*2, 8*). Heavy-duty trucks spend, on average, 30% of their operational time idling (*9*) or between 6 and 16 hours daily (*10*). In-cab pollutant exposure is a function of cabin pollutant concentrations and time spent in the cabin. The concentrations found in the cabin are a function of pollutant concentrations in the ambient environment surrounding the truck, the rate of air transfer from external to internal environment, and infiltration of emissions directly from the truck. In some cases, in-cab air quality can be negatively affected by leaks in the engine exhaust system or from an open crankcase. The ambient concentrations near the truck are related to background concentrations (which can be quite high in areas such as truck stops where many other trucks are idling) and the contribution of pollutants generated by the truck itself.

While idling is seen to be significant from a driver exposure perspective, idle reduction has generally been promoted as an emissions reduction strategy and fuel saving measure for trucks. Several IRTs are available on the market, which allows drivers resting in the truck cabin to turn off the main truck engine and maintain in-cab heating and cooling for comfort. Table 1 summarizes major available IRTs, including a description and other relevant information.

As mentioned previously, the focus of this study is specifically on diesel-powered IRTs, namely auxiliary power units (APUs) and fuel operated heaters (FOHs), due to their potential for increasing driver exposures from the emissions they generate.

Development of Test Protocol

Test protocols were developed for measuring driver exposures and assessing the effectiveness of IRTs in reducing driver exposures to diesel exhaust, especially PM2.5, during

Table 2. Summary of Ambient PM2.5 Concentration Data

State	Type	Number of data points	Concentrations - μg/m^3 mean	Concentrations - μg/m^3 maximum	Concentrations - μg/m^3 minimum
GA	Highway	1,670	9.16	39.9	4.7
GA	Truck stops	1,438	10.23	62.6	4.4
CA	Highway	5,618	6.36	37.8	0.3
CA	Truck stops	3,039	6.42	285.4	0.3
TX	Highway	3,835	4.70	48.4	1.4
TX	Truck stops	1,896	5.90	105.3	1.5

long duration idling events. These events generally refer to periods where the driver is resting within the truck cab for an extended period, often up to 10 consecutive hours, as required by the Federal Motor Carrier Safety Regulations (*11*).

Driver exposure to PM2.5 and other pollutants is affected by complex interactions of: a) personal exposure, which is a function of occupational exposure and smoking status; b) occupational exposure in general, and idle-related exposure in particular; and c) background exposure (*12*). For the laboratory testing in this study, no contribution to cabin interior pollutant levels from driver behavior (such as smoking) was considered. The testing focused on exposure due to the ambient conditions surrounding the test vehicle and the impact of the idling truck and/or IRT under consideration.

The testing was divided into two components:

- Field Testing—These measurements were conducted at truck stops, to determine the ambient conditions drivers are exposed to in the field during long-duration idling events. The findings from the field measurements at truck stops were used to develop laboratory testing conditions representative of what was seen in the field.
- Laboratory Testing—This testing was conducted in a controlled laboratory environment to determine the effects that IRTs have on driver exposure, using a representative sample of test trucks. The field data were used to provide ambient testing conditions representative of conditions observed at truck stops.

The following sections describe the field testing protocol and results, laboratory testing protocol, and laboratory testing results in further detail.

Field Testing Protocol and Results

The main purpose of conducting the field testing was to establish a range of realistic ambient concentrations to be replicated during the laboratory testing task. In order to establish this range, the project team conducted real-time measurements at three geographically diverse locations across the United States, covering 24 truck stops in Georgia (near Calhoun), California (near Bakersfield), and Texas (near Houston). The data from the field measurements were based on a collection of ambient concentrations at the truck stops, regional background concentrations, meteorology, and PM2.5 source activities. To collect the ambient PM2.5 data a GRIMM 1.109 aerosol spectrometer with an isokinetic sampling pipe was used. The units were calibrated prior to the project and have a measuring range from 0.25 μm to 32 μm with a reproducibility of 5%. Measurements at each stop were conducted during overnight hours when there were the highest numbers of trucks parked at the truck stops. The overall sampling period ranged between 8 and 10 hours to cover truck activity throughout the night.

The summary of the field testing results is shown in Table 2. At each test location measurements were taken at both the truck stop itself, as well as on the highway near the stop. The data was collected as 6 second averages for the duration of the testing. Overall, while truck stops were seen to have higher ambient PM2.5 concentrations compared with adjacent highways (with statistically significant differences in the Texas and Georgia datasets), the spatial analysis of the data also indicated variations between truck stops in a region, and instances of elevated PM2.5 levels outside of trucks stops as well. A temporal analysis of concentrations at truck stops measured over time also showed little correlation to the number of idling trucks or APUs at a location. However, local meteorological conditions were found to affect the observed concentrations.

Based on a review of the data collected and a consideration of several factors, a range of 35–55 μg/m^3 was determined by the research team as a reasonable representation of the higher end of the ambient concentrations observed at truck stops, for use in the laboratory testing. It represented a pollutant level range that could plausibly be experienced at a fairly polluted truck stop. Although a majority of the data points in the field observations were below this range, it was selected to ensure representation of the higher end of the spectrum of observed ambient readings. The lower end of the range was double that of the 99th percentile of the field measurements and also coincided with the current NAAQS for PM2.5 (*13*). The upper limit was selected based on the higher readings that were seen in the field.

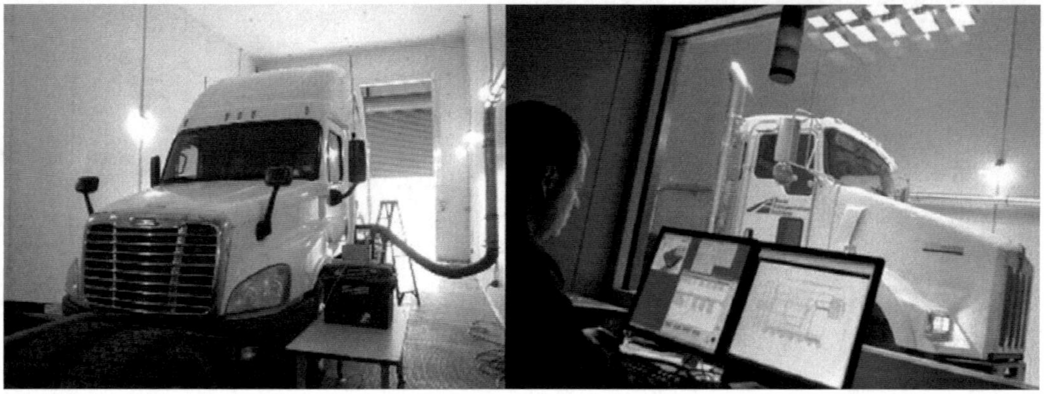

Figure 1. Test facility and test vehicle.

Table 3. Test Scenario Descriptions and Tested Operational Modes

Test scenario	Operational mode: Heating (Ambient temperature 32°F)	Operational mode: Cooling (Ambient temperature 100°F)
Truck baseline (no idle) – engine off	Tested	Not tested for this study
Truck idle (recirculation)	Tested, depending on truck	Not tested for this study
Truck idle (no recirculation)	Tested, depending on truck	Not tested for this study
Test truck with APU	Tested depending on APU	Tested
Test truck with FOH	Tested	Not applicable (no cooling capabilities)

Laboratory Testing Protocol

The purpose of the laboratory testing was to allow for multiple, repeatable, standardized tests to determine the effects of various IRTs on driver exposure to diesel exhaust emissions. The field data collected were used to identify representative ambient conditions that were replicated in the laboratory setting. This allowed for multiple test runs to be conducted under similar ambient conditions, with the added benefits of a controlled environment. Thus, multiple vehicle and IRT combinations were tested to determine the impact IRT devices had on the PM2.5 exposure levels of a driver resting in the truck cab.

Test Facility and Equipment

The laboratory tests for this study were conducted under environmentally controlled conditions at the Texas A&M Transportation Institute's Environmental and Emissions Research Facility (EERF). The EERF has a fully environmentally controllable test chamber capable of controlling the temperature between −15 and 130°F (−26 to 55°C). At 75-ft long and 23-ft wide, the chamber is large enough to fit an entire tractor trailer inside for testing, though for this test no trailer was required. Figure 1 shows one of the test vehicles in the chamber during a test, along with the chamber control center.

Just as was done in the field testing, the laboratory testing used a GRIMM 1.109 aerosol spectrometer to measure the ambient (inside the test chamber) and the in-cab PM2.5 levels. Since both the in-cab and ambient PM2.5 levels were required during laboratory testing, two separate GRIMM units were used. For all laboratory testing, the GRIMM units were set to report one-minute averages of the PM levels.

Testing Approach

The testing was designed to cover the various possible modes (termed as test scenarios) that a truck would operate under at a truck stop, which includes: parked with the engine turned off (defined as the baseline for testing purposes), idling the main truck engine, or operating an IRT (APU or FOH). Depending on ambient conditions, the truck engine or IRT would provide heating or cooling to the truck cab. Thus, some of the test scenarios may also have an associated operational mode, a cooling mode (during ambient temperatures of 100°F), and/or a heating mode (during ambient temperatures of 32°F). All baseline tests (truck parked with the engine off) and idling tests were conducted at 32°F. The idling tests included two scenarios, with and without recirculation of air in the cabin, to assess whether these affected in-cab concentrations. Table 3 summarizes the test scenarios and operational modes used for all the test trucks. As noted, the operational mode tested for each scenario was determined by the capabilities of the IRT; for example, FOHs were not tested for cooling operations since they do not have cooling capabilities.

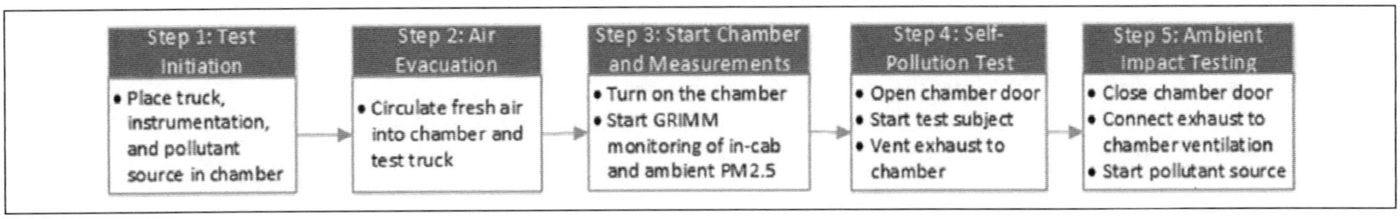

Figure 2. Laboratory testing steps.

Table 4. Test Truck and IRT Information

Truck number	Truck make	Model year	Truck model	Engine make	Certified clean idle	APU make	FOH make
1	Freightliner	2006	Columbia CL120	Caterpillar	No	Comfort Pro	Espar Airtronic D2
2	Freightliner	2008	Cascadia CA-125	Detroit Diesel DD15	No	ThermoKing TK270M	Espar Airtronic D2
3	Freightliner	2012	Coronado CC-132	Detroit Diesel DD15	Yes	ThermoKing TK270M	Espar Airtronic D2
4	Freightliner	2013	Cascadia CA-125SLP	Detroit Diesel DD15	Yes	Comfort Pro	Espar Airtronic D2
5	Peterbilt	2014	Peterbilt 279	PACCAR Diesel	Yes	Dynosys 2TNV70-KBP	NA[a]

[a]The 2014 truck was not equipped with a FOH and one could not be installed in the truck due to truck owner request.

Each truck was tested for each test scenario, under applicable operational modes to assess how the use of IRTs affects in-cab exposure to PM2.5. The broad approach to the testing involved the truck operating in an environmentally controlled chamber, with a pollutant source (secondary diesel engine) used to simulate ambient PM2.5 pollution levels observed at truck stops during the field data collection task. Each test run, for a specific test scenario and operational mode, involved the steps shown in Figure 2. Further details on the test steps can be found in the project report (*14*).

Test Truck Selection

A total of five trucks were tested in the laboratory, each equipped with a preinstalled APU and FOH (i.e., the diesel-powered IRTs of relevance to this study). The trucks were selected for testing based on their model year (MY). In order to be representative of the long-haul fleet and the emissions reduction devices that were available at the time the trucks were manufactured, the trucks were selected based on the following proposed MY ranges:

- One truck pre-2007
- One truck from MY 2008–2010
- One truck from MY 2011–2012
- Two trucks from MY 2013–2016

The MYs were selected to ensure they captured the introduction of new technologies, specifically diesel particulate filters (DPF) in MYs 2008–2010 and selective catalytic reduction (SCR) in MYs 2011–2012. The pre-2007 vehicle was selected since it would have no DPF or SCR. Table 4 outlines the details of the trucks and the IRT devices that were used in the testing. In terms of applicable test scenarios and operational modes, APUs on trucks 1, 4, and 5 were capable of both heating and cooling, and were subjected to both heating and cooling tests. The ThermoKing APUs only had cooling capabilities and were only tested under cooling conditions.

Laboratory Testing Results

Data Analysis and Reporting Approach

The laboratory testing involved two different tests, self-pollution and ambient impact, and the data analysis and reporting approach was slightly different for each. For the ambient impact testing, all results are reported as a ratio of the minute-by-minute in-cab readings to the average ambient reading for the duration of the test, to better account for daily variations in the PM2.5 measurements, and to allow for comparisons across the technologies used. The ratio was calculated on a per-minute basis by dividing the one-minute reading of the in-cab PM2.5 levels by the average ambient PM2.5 level for the duration of the test. The average of the ambient levels was used due to the cyclical nature of the ambient PM levels, which were caused by turning the pollutant source on and off to keep the ambient levels within the desired range. Each one-minute ratio was then averaged together over the duration of the test to come up with the overall ratio for each test that was conducted. A ratio of 1 would indicate that the average ambient and in-cab PM2.5 levels were equal, while a ratio below 1 shows that the in-cab level is lower than the average ambient level (and vice versa for ratios above 1).

Results

Self-Pollution Testing. The self-pollution testing was conducted to determine if the idling truck or IRT being tested

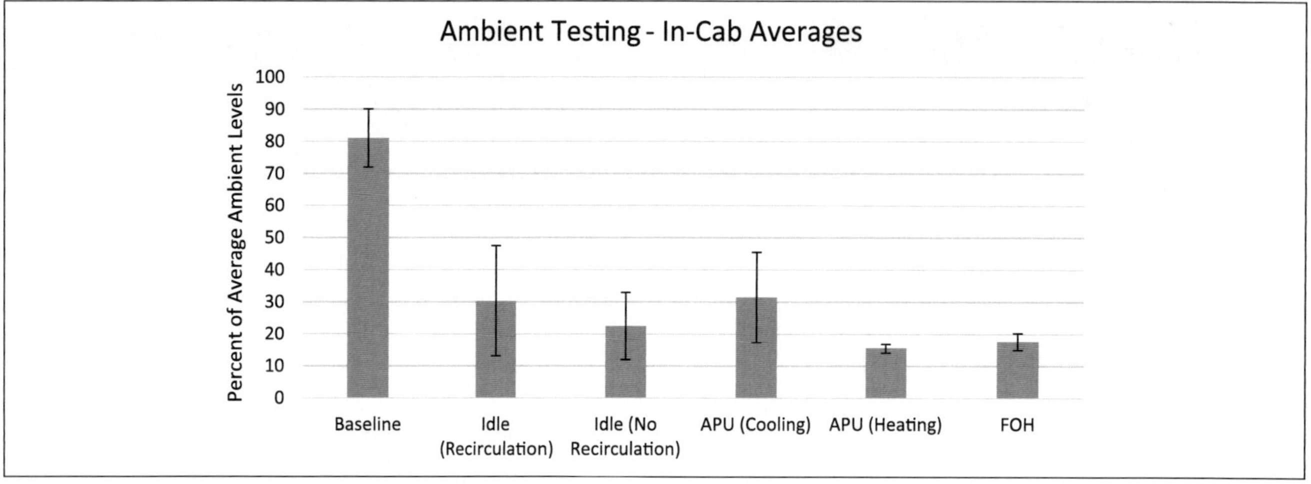

Figure 3. Ambient testing results by test scenario.

were introducing PM2.5 emissions into the cab. The results showed that only one of the test trucks self-polluted the cab, the 2006 truck while idling, which occurred during both the recirculation and no-recirculation tests. The 2006 truck saw in-cab PM levels of 45.4 $\mu g/m^3$ for the recirculation mode and 26.8 $\mu g/m^3$ for the non-recirculation mode. These were much higher than the starting in-cab levels of 4.7 and 3.6 $\mu g/m^3$ respectively. No other test scenario had an in-cab PM level higher than 10.2 $\mu g/m^3$ for the self-polluting test. The self-polluting from the 2006 truck is likely due to the crankcase exhaust from the truck, since, per EPA regulations, closed crankcase ventilation systems were required only for engines meeting the 2007 standards onward.

Ambient Impact Results. The main purpose of the laboratory testing was to determine the effect that each IRT device had on the exposure a driver would have during long-duration idle events. The following section details the results obtained from each of the test scenarios that were conducted. Figure 3 shows the results of each test scenario during the ambient impact testing. The error bars shown are the standard deviation of the measured data. As the graph shows, all of the IRT devices and the truck engine idling outperformed the baseline testing (truck parked with engine and IRT off), which saw an average in-cab reading of 80% of the ambient levels. The highest average reading from each of the other test scenarios was the APU device when cooling the cabin. These results are skewed slightly by one reading (from the MY 2008 truck), which had a reading of just under 59%, 30% higher than any of the other APUs. If this outlier is not considered, the average APU cooling results drop from 31% to below 25%.

Technology Comparison. The previous section outlines the average results from each of the test scenarios. The following section compares the results of each of the individual trucks under the different test scenarios. Since no tests (other than the MY 2006 truck idle test) showed any self-pollution, this section focuses on the results from the ambient impact testing. For all the graphs in this section, the results are presented as a ratio, comparing the in-cab values to the average ambient values for the duration of the test.

Baseline Comparison. Figure 4 shows the results from the baseline tests. The graph shows that the baseline tests for each truck were very similar, with each test ending with an in-cab PM2.5 level above the average ambient PM2.5 level. Due to the ambient levels being controlled within a range, it was possible to have levels above the average used to calculate the ratios in the graphs. By the end of the test, the in-cab levels were approaching the upper limit of the ambient PM2.5 range being used, so the ratios were above 1. Two of the trucks (MY 2006 and MY 2008) took slightly longer to reach their ending levels, and the MY 2012 truck started with slightly elevated in-cab readings, but had approximately the same rate of change as the other tested vehicles.

Truck Idling Comparison. Figure 5 shows the results from the six truck idling tests that were conducted. As the graph shows, the in-cab ratio levels fluctuated during the tests, especially for the MY 2012 test. These fluctuations were seen when the pollutant source in the chamber was turned on/off.

APU Comparison. Figure 6 shows the results from the APU tests. A total of 8 APU tests were conducted, with most ratios being in the 0.15–0.30 range for the duration of the testing, with the exception of the MY 2008 truck test, which had a much higher in-cab to ambient ratio close to 0.6 for most of the test. The average across all tests for the APU technologies was 0.255. Unlike the idling tests, which had

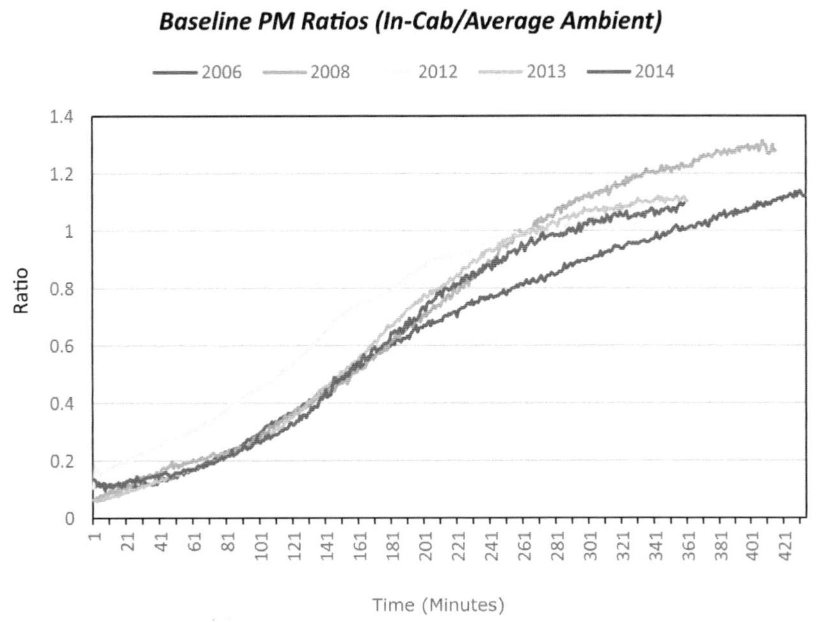

Figure 4. Baseline PM results.

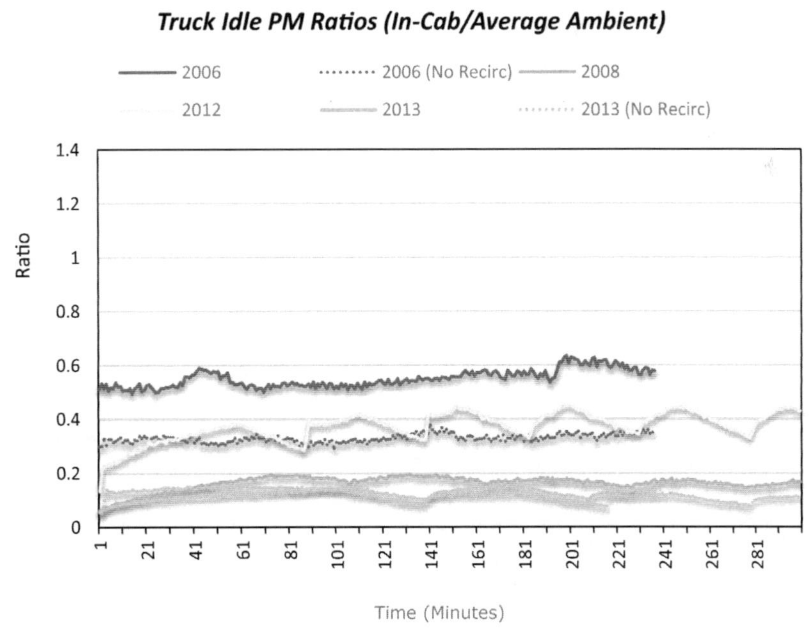

Figure 5. Truck idle PM test results.

the exhaust venting in different locations, each of the APUs' exhausts was vented in the same manner, just off the side and underneath the APU itself.

FOH Results. Figure 7 shows results from the testing of the FOHs. As the graph shows, the results from the FOH tests are much more consistent than the tests from the other technologies, with the difference between the highest and lowest ratio only 0.138 (high of 0.328 and low of 0.19). This could be attributed to the fact that each of the FOHs tested are the exact same make and model (and the exact same unit was used on 3 of the 4 tests). All the FOHs had the same exhaust location, with the tubing extruding from the bottom of the cab.

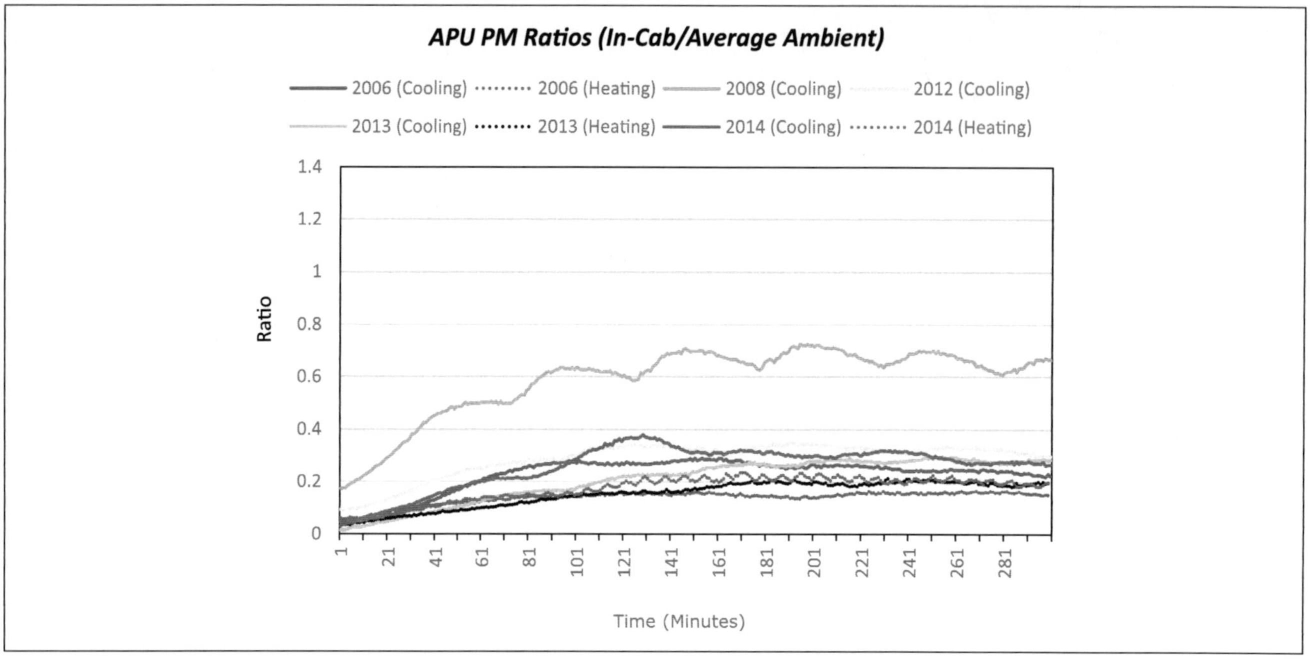

Figure 6. APU PM test results.

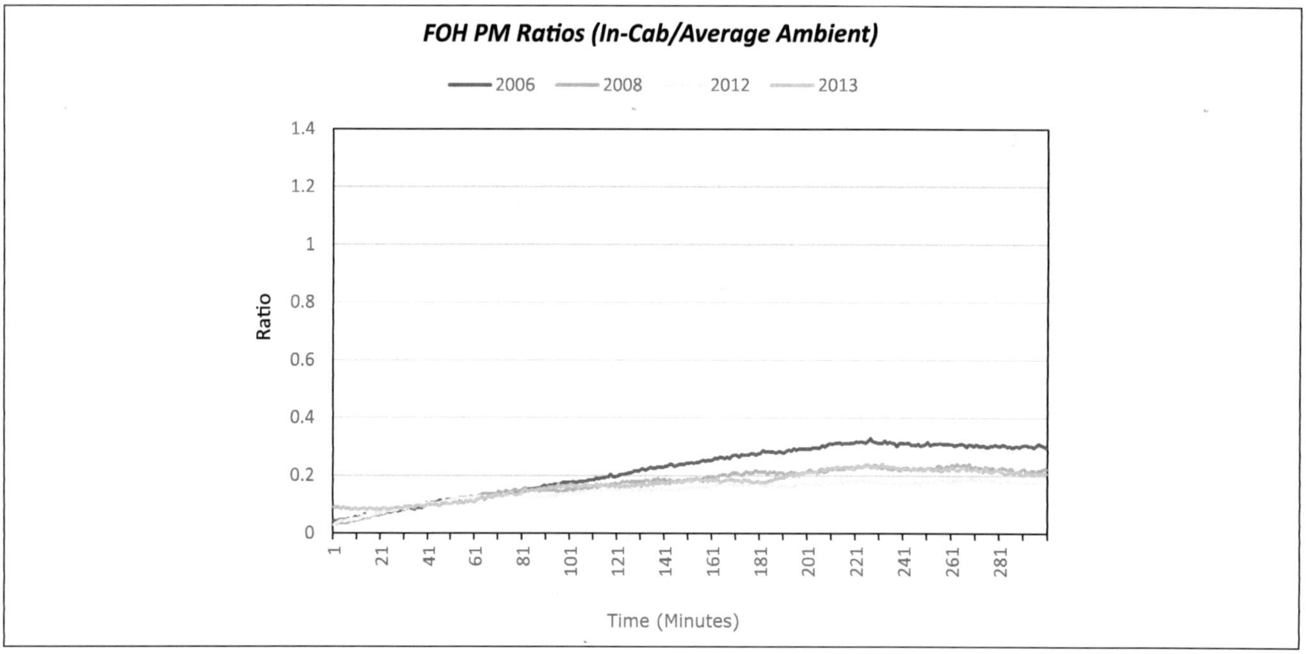

Figure 7. FOH PM test results.

Cost Analysis

A cost analysis was conducted using the Idling Reduction Savings Calculator (IRSC) developed by Argonne National Laboratory (ANL) to assess fuel savings against capital and maintenance costs for each IRT. Five different IRTs, including APUs and FOHs were analyzed, using data and assumptions from past studies, including those documented in the literature review. The cost analysis compared not only APUs and FOHs, but other types of non-diesel powered IRTs. Overall, FOHs and APUs were found to be on the opposite ends of the spectrum in terms of cost effectiveness, with FOHs having the lowest payback period among the studied IRTs, and APUs having the highest. Among the other options investigated, battery air conditioning (BAC) was found to have a payback period similar to APUs.

Electrified parking spaces (EPSs) and thermal storage systems (TSSs) had intermediate payback periods, being lower than the APU and BAC options, but higher than FOHs. However, all IRTs analyzed had a payback period of 2.1 years or less, under an assumption of low fuel costs ($2.71 per gallon) and with idling rates at 75% of rest time. This indicates their overall financial viability under fairly conservative assumptions. At the highest assumed fuel costs of $5.00 per gallon, the payback period lowers to 1.1 year or less at a 75% idling time.

Conclusion

Key findings from the investigation are as follows:

- There were indications that prevailing ambient PM2.5 concentrations at truck stops were linked more to location and meteorological and geographic factors than the number of idling trucks or APUs in the truck stop. The average ambient concentrations at truck stops were higher than the averages measured on adjacent roadways in all three states, with the difference being statistically significant in the cases of Texas and Georgia. This indicates that truck stops potentially show elevated levels of PM2.5 concentrations relative to adjacent roadways.
- Laboratory testing indicated that the exhaust location on the truck did not affect in-cab concentrations of PM2.5, and self-pollution of trucks was not seen with the exception of one test truck, a 2006 Freightliner. The self-pollution seen in the case of this particular truck could be attributed to the open crankcase design. Crankcase emissions were required to be measured starting in 2007, so most manufacturers closed the crankcase as a result, eliminating this issue with self-pollution.
- It was found that IRTs, on all tested modes, resulted in reduced PM2.5 levels in the truck cab compared with the baseline scenario of the engine turned off and the door and windows closed. Idling the truck engine, with and without air recirculation also showed lower PM2.5 concentrations compared with the baseline scenario. The range of PM2.5 levels found for all tests, except the baseline, was between 11.9 and 58.9% of the ambient average. Based on these numbers, in order for the in-cab levels to be over the NAAQS limit of 35 $\mu g/m^3$ the ambient levels would need to be at least 59.4 $\mu g/m^3$.
- The cost effectiveness analysis indicated that all IRTs were financially viable from a payback period perspective, making them viable options for reducing driver exposure to diesel exhaust.

This study supports the premise that IRTs can reduce driver exposure to diesel exhaust at truck stops. Even if the extent of the benefits of IRTs compared with idling a truck's engine are not as clear, a case for IRTs can be made due to overall emissions and fuel consumption benefits.

Further investigation can help draw more conclusive findings on the benefits of specific IRTs and the impact of other factors outside of those considered in the study. This includes understanding how driver exposure to diesel exhaust at truck stops itself varies, based on location at the truck stop (whether the truck is near another idling truck or APU) and the geography and meteorology of the truck stop area itself. Further research is also needed into truck characteristics, which may include factors such as insulation and filtration systems present. Thus, extending this study to cover a wider spectrum of truck models and ages can promote better understanding of the impacts across the current heavy-duty truck fleet. As alternative fueled trucks are increasingly prevalent in the market, their impacts on in-cab exposure also warrant further investigation. Finally, since this study was focused on diesel-powered IRTs, there is potential to expand the research to include a broader range of IRT options, such as electric-powered alternatives.

Acknowledgments

This work was conducted as part of a project funded by the Federal Motor Carrier Safety Administration (FMCSA), titled "Study of Alternatives to Reduce Driver Exposure to Diesel Exhaust" (Contract Number DTMC-75-14-D-00010). The authors wish to thank Quon Kwan of FMCSA for his oversight, guidance and input over the course of this study. The project team would also like to thank the project review panel members, Linda Gaines, John Storey, and Dana Lowell for their invaluable input on this project.

References

1. United States Department of Labor, Occupational Safety and health Administration, Diesel Exhaust/Diesel Particulate Matter. https://www.osha.gov/dts/hazardalerts/diesel_exhaust_hazard_alert.html. Accessed October 2016.
2. Fu, J. S., J. A. Calcagno, W. T. Davis, and A. Alvarez. Evaluation of Noise Level, Whole-Body Vibration, and Air Quality Inside Cabs of Heavy-Duty Diesel Vehicles. *Transportation Research Record: Journal of the Transportation Research Board*, 2010. 2194: 29–36.
3. Zhu, Y., T. J. Smith, M. E. Davis, J. I. Levy, R. Herrick, and H. Jiang. Comparing Gravimetric and Real-Time Sampling of PM2.5 Concentrations Inside Truck Cabins. *Journal of Occupational and Environmental Hygiene*, Vol. 8, No. 11, 2011, pp. 662–672.
4. Garshick, E., F. Laden, J. E. Hart, B. A. Rosner, M. E. David, E. A. Eisen, and T. J. Smith. Lung Cancer and Vehicle Exhaust in Trucking Industry Workers. *Environmental Health Perspectives*, Vol. 116, No. 10, 2008, pp. 1327–1332.
5. Laden, F., J. E. Hart, T. J. Smith, M. E. Davis, and E. Garshick. Cause-Specific Mortality in the Unionized U.S. Trucking Industry. *Environmental Health Perspectives*, Vol. 115, No. 8, 2007, pp. 1192–1196.

6. Richard, T., Z. Tan, and X. Cheng. Relationship Between Perceived In-cab Air Quality and Truck Drivers' Self-Reported Health and Alertness. *Journal of the Transportation Research Forum*, 2008. 47: 115–133.
7. Kilburn, K. H. Effects of Diesel Exhaust on Neurobehavioral and Pulmonary Functions. *Archives of Environmental Health: An International Journal*, Vol. 55, No. 1, 2000, pp. 11–17.
8. Doraiswamy, P., W. T. Davis, T. L. Miller, N. Lam, and P. Bubbosh. Air Quality Measurements Inside Diesel Truck Cabs During Long-term Idling. *Transportation Research Record: Journal of the Transportation Research Board*, 2006. 1987: 82–91.
9. Huai, T., S. D. Shah, J. W. Miller, T. Younglove, D. J. Chernich, and A. Ayala. Analysis of Heavy-duty Diesel Truck Activity and Emissions Data. *Atmospheric Environment*, Vol. 40, 2006, pp. 2333–2344.
10. Grupp, D., M. Forrest, P. Mader, C. J. Brodrick, M. Miller, and H. Dwyer. Development of a Retrofit Fuel Cell Auxiliary Power Unit for Truck Idle Reduction. *SAE Transactions*, Vol. 113, 2004, pp. 237–247.
11. Federal Motor Carrier Safety Administration. Interstate Truck Driver's Guide to Hours of Service. https://www.fmcsa.dot.gov/sites/fmcsa.dot.gov/files/docs/Drivers%20Guide%20to%20HOS%202015_508.pdf. Accessed September 2015.
12. Davis, M. E., T. J. Smith, F. Laden, J. E. Hart, L. M. Ryan, and E. Garshick (2006). Modeling particle exposure in U.S. trucking terminals. *Environmental Science & Technology*, Vol. 40, No. 13, 2006, pp. 4226–4232.
13. United States Environmental Protection Agency, Federal Register. https://www.gpo.gov/fdsys/pkg/FR-2013-01-15/pdf/2012-30946.pdf. Accessed August 2016.
14. Zietsman, J., J. Johnson, T. Ramani, F. Farzaneh, Y. Xu, M. Rodgers, A. Samoylov, and A. Moore. *Study of Alternatives to Reduce Driver Exposure to Diesel Exhaust*. https://www.fmcsa.dot.gov/research-and-analysis/research/study-alternatives-reduce-driver-exposure-diesel-exhaust. Accessed 2016.

The Standing Committee on Transportation and Air Quality (ADC20) peer-reviewed this paper (18-03013).

The contents of this paper reflect the views of the authors, who are responsible for the accuracy of the data presented. The contents do not necessarily reflect the official policy of the USDOT.

Article

TRR
JOURNAL OF THE TRANSPORTATION RESEARCH BOARD

Transportation Research Record
2018, Vol. 2672(25) 11–20
© National Academy of Sciences:
Transportation Research Board 2018
Article reuse guidelines:
sagepub.com/journals-permissions
DOI: 10.1177/0361198118755714
journals.sagepub.com/home/trr
SAGE

Reducing Transportation Greenhouse Gas Emissions Through the Development of Policies Targeting High-Emitting Trips

Ran Tu[1], Islam Kamel[1], Baher Abdulhai[1], and Marianne Hatzopoulou[1]

Abstract

Traffic emission inventories have been under development for decades, often relying on data from traffic assignment models, ranging from macroscopic models generating average link speeds, to more detailed microscopic models with instantaneous speed profiles. Policy testing within such frameworks has often focused on identifying changes in total emissions, or in emissions aggregated at a zonal or street level. Emissions from specific trips or trajectories are seldom analyzed, although reductions in greenhouse gas (GHG) emissions can be achieved more efficiently when targeting high emitters. In this paper, we propose a different approach to reducing transportation GHG emissions, by catering policies to specific trips based on their emission burden. We focus on the City of Toronto downtown. Using second-by-second speed data for entire trajectories, GHGs (in CO_{2eq}) and nitrogen oxides (NO_x) emissions were estimated. We observe that the destinations attracting the highest trip emissions tend to be in the hospital and financial districts. Trips originating and ending in the downtown area are responsible for a small share of total emissions, although they have high emission intensity. Removing trips with high total emissions and high emission intensity led to significant reductions in CO_{2eq} and NO_x emissions, whereas removing shorter trips, did not have a significant influence on total emissions nor emission intensities.

Transportation has become the second largest source of greenhouse gas (GHG) emissions in 2015 in Canada (*1*), emitting 807.3 kt of nitrogen dioxide (NO_2) in that year, which was the highest among all economic sectors (*2*). In 2013, road transportation became the greatest GHG emitter in the City of Toronto, responsible for 41% of the total GHG emissions (*3*). With increasing travel demand, effective controls are important for the reduction of emissions from road traffic, and accurate emission estimates are an important prerequisite for policy analysis.

Various traffic emission models and approaches have been developed, primarily to improve urban and regional GHG emission inventories. The International Vehicle Emission (IVE) model developed by joint researchers from the International Sustainable System Research Center and the University of California, Riverside, uses vehicle technology distributions, power-based driving factors, vehicle soak distributions, and meteorological factors to estimate emissions. It has been implemented effectively worldwide for emission estimation and local policy analysis (*4*). The Computer Programme to Calculate Emissions from Road Transport (COPERT) (*5*) and the Handbook of Emission Factors for Road Transport (HBEFA) (*6*) developed in Europe use average speed (COPERT) or particular descriptions of the driving cycle (HBEFA) to estimate emissions. These approaches, although relatively easy to use, have been shown to provide inaccurate estimates compared with measured emissions (*7–9*). The on-road vehicle emission factor (EF) model EMFAC developed by the California Environmental Protection Agency (*10*) and Motor Vehicle Emission Simulator (MOVES) developed by the U.S. Environmental Protection Agency (USEPA) (*11*) are two models that use average speed as input. MOVES also has another estimation scheme that derives an operating mode (opmode) distribution directly from a second-by-second drive cycle, then uses the opmode distribution to generate total emissions (*12*). Models such as PHEM (*13*), CMEM (*14*), and VT-MICRO (*15*) calculate EFs from engine power or vehicle operating modes, in which the EF is a function of engine power and instantaneous speed.

Studies on urban scale traffic-related emissions have been conducted in many areas worldwide, with scales down to individual road sections, street corridors, and facilities of an urban landscape (*16–20*). GHG emission models have been

[1]Department of Civil Engineering, University of Toronto, Toronto, ON, Canada

Corresponding Author:
Address correspondence to Marianne Hatzopoulou:
marianne.hatzopoulou@utoronto.ca

applied to different scenarios to analyze the effects of emission reduction policies. Studies have found that managing demand (*21,22*), developing public transit systems (*23*), controlling drivers' behavior (*24–26*), and multimodal traffic networks (*27,28*) generally lead to positive effects on the reduction of traffic-related air pollutants and GHGs.

Although various studies have estimated emissions in different urban areas and under various policy scenarios, they often ignore vehicle trajectories. In this paper, the instantaneous speed profile of each vehicle is generated based on microscopic traffic simulation of the study area. Emissions are estimated at the level of individual trips based on the entire trajectory of the vehicle, from the moment it enters to the moment it leaves the network, with emissions aggregated to trip origins and destinations to identify the most emission-intensive origins or destinations in the study area. Two policies directly working at the level of the individual trip are implemented, in addition to one "conventional" policy, applied generally to the urban area, are examined and compared.

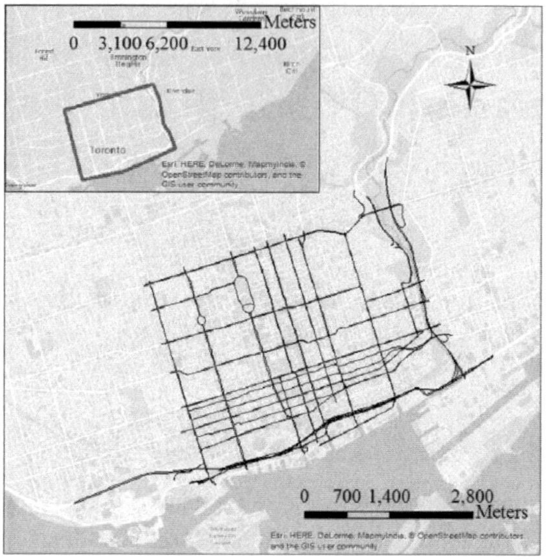

Figure 1. Road network of the study area.

Methods

This paper uses the road network of downtown Toronto, Canada as the study area. Using microscopic simulation with the AIMSUN platform, second-by-second trajectories of each vehicle are generated and used to estimate emissions, both at trip and network level. Three policy scenarios are implemented in the study area and their emissions are compared with the base case.

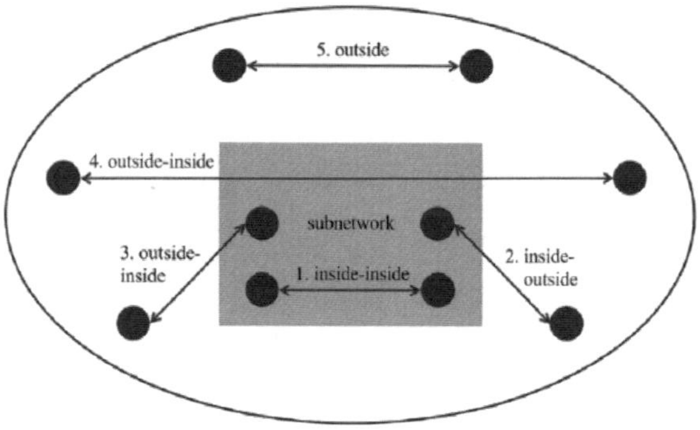

Figure 2. Relationship between trips and subnetwork.

Study Area and Data

The network studied consists of major roads in downtown Toronto, covering an area of 3.31 km from North to South and 4.53 km from West to East. The total road length in both directions is 165.8 km. The boundaries and streets within the network are highlighted in Figure 1.

Trip data are based on the Transportation Tomorrow Survey (TTS), which was used to generate origin–destination (OD) matrices of household trips in the Greater Toronto Area (GTA). As the network in this study does not encompass the entire road network of the metropolitan region, traffic demand of the study area should be retrieved from the GTA OD matrix in AIMSUN by its "OD traversal" function. OD traversal summarizes the total amount of trips that cross the study boundary (in–out, out–in, out–out), or happen inside the boundary (in–in) (*29*), as illustrated in Figure 2. Traffic demand of the study area is generated from 6:00 am to noon.

Traffic Modeling

In this paper, the AIMSUN platform is used for traffic simulation (*30*). The process of microscopic simulation is called "time-based." Time moves according to timesteps and intervening movements, such as varying speed or changing lane, are considered for each vehicle, while the vehicle moves by the distance determined for that timestep. The vehicle instantaneous speed and acceleration of each timestep can be recorded by microscopic simulation (*31*), the full algorithm of microsimulation can be found in Aimsum (*32*).

Emission Modeling

The MOVES model, developed by the USEPA, is used in this paper for the generation of EFs, by determining the vehicle operating mode, which is based on speed, acceleration, and vehicle specific power (VSP) (*33*):

$$P_{vt} = \frac{Av_t + Bv_t^2 + Cv_t^3 + mv_t a_t}{m} \quad (1)$$

where:

P_{vt} = VSP;

v_t = speed at time t (m/s);
a_t = acceleration at time t (m/s^2);
m = mass of the vehicle, usually referred as "weight" (tons);
A, B, and C = track-road coefficients, representing rolling resistance, rotational resistance and aerodynamic drag, in units kW-s/m, kW-s^2/m^2, and kW-s^3/m^3.

Two formats for vehicle activity inputs are available to use in MOVES. First, by providing average speed and road characteristics, a default driving cycle is assigned that best fits the given data, and provides an EF associated with each average link speed. Total emissions are calculated by multiplying EFs with vehicle kilometers traveled (VKT). Second, by providing second-by-second speed profiles, an operating mode (opmode) distribution of the driving cycle is calculated. Combining with meteorology, fuel type, and vehicle type, EFs for each opmode, and total emissions using the EFs and opmode distributions, are generated. In this paper, this mode is applied to the results of microscopic traffic simulation.

To speed up the computation process, a matrix of EFs was derived from MOVES for each set of parameters, and the estimation of total emissions was calculated outside MOVES using the EF matrix. This method has been proven to largely improve the computational speed, compared with the MOVES-GUI method (*34*). The on-road vehicle composition comes from 2016 Statistics and Management Reporting of the Ontario Ministry of Transportation (*35*). The estimation date in MOVES is defined as a weekday of June 2011. The emission results from MOVES in this paper are presented in metric units.

Policy Scenarios

In this paper, three policy scenarios are tested and compared in terms of their emission reduction potential. In the first scenario (S1), trips shorter than 600 m are removed from the OD matrix. This scenario is developed to mimic initiatives that may encourage the shift to active transportation; we did not model the effect of demand here, but removed these trips to assess what their effect on the network would be in terms of speed, congestion, and emissions. In the second scenario (S2), trips with GHG emissions (in CO_{2eq}) higher than 1.75 kg are eliminated; this is expected to mimic the imposition of a tax on high-emitting trips, effects on demand are outside the scope of the analysis. The third scenario (S3) targets trips with high emission intensity, meaning trips that occur under highly congested conditions. In this scenario, trips with CO_{2eq} EFs higher than 500 g/VKT were removed.

The results of microscopic simulation based on the traffic demand of the three scenarios were compared with the base case (S0). The total demand after the implementation of each scenario, was similar across the three scenarios, roughly 86% of the base case demand. A similar reduction in demand helped identify how various trips affect the network-wide emission inventory and where the largest reductions can be achieved.

Results

Base Case Scenario Analysis

According to the results from microscopic simulation, the network has an average delay of 141.23 s/km and an average speed of 31.81 km/h. Figure 3, *a* and *b* show the trends in network flow and average speed, respectively. The peak demand occurs between 7:00 am and 8:30 am, during which the average speed is much lower than other time periods. Figure 3*c*; shows the distribution of instantaneous speeds although the overall network-average speed is over 30 km/h, at most time points, instantaneous speeds are less than 10 km/h.

Using the MOVES EF matrix, the total GHG (in CO_{2eq}) and NO_x emissions produced within the study area were estimated to be 70.1 tons and 28.8 kg, respectively. Trip-averaged emissions for CO_{2eq} were 983.36 g/trip and for NO_x were 0.4 g/trip.

Figure 4 shows the distribution of mean EFs for each trip. The average EF for CO_{2eq} and for NO_x is 358.95 g/VKT and 0.14 g/VKT, respectively. About 86% of the trips have CO_{2eq} EFs lower than 500 g/VKT, and NO_x EFs lower than 0.18 g/VKT. A total of 76.41% of trips have CO_{2eq} EFs between 150 and 400 g/VKT. Within this range, trips with CO_{2eq} EFs between 200 and 300 g/VKT take the highest proportion (41.06%). For NO_x, 66.1% of trips have EFs between 0.1 and 0.16 g/VKT, within which trips with NO_x EFs between 0.12 and 0.14 g/VKT have the highest probability (close to 20%).

To identify the neighborhoods generating and attracting the highest emissions in the study area, trip-level emissions were aggregated to OD zones and the spatial distribution of zonal emissions is presented in Figure 5.

The entire downtown Toronto area, as a destination, attracts 32.4 tons of CO_{2eq} and 11.4 kg of NO_x, which is 46.2% and 39.5% of the total CO_{2eq} and NO_x that the full network generates. This estimate is based on the emissions of all trips ending downtown (starting within or outside). This is more than what downtown Toronto residents generate, including those who take trips within the downtown or travel from downtown to outer areas (18.9% of the total CO_{2eq} and 21.8% of the total NO_x).

Within the downtown area, trips destined to the financial district and hospital zones produce the highest amount of emissions. These zones alone attract 23.6% of total CO_{2eq} emissions and 19.6% of total NO_x emissions. In terms of the total emissions attracted by the downtown, these districts account for 51.0% of total CO_{2eq} emissions attracted and 49.5% of total NO emissions. The amount of emissions generated by these trips is about five times the amount of trips originating from these zones. This analysis begins to pinpoint areas that should become candidates for targeted

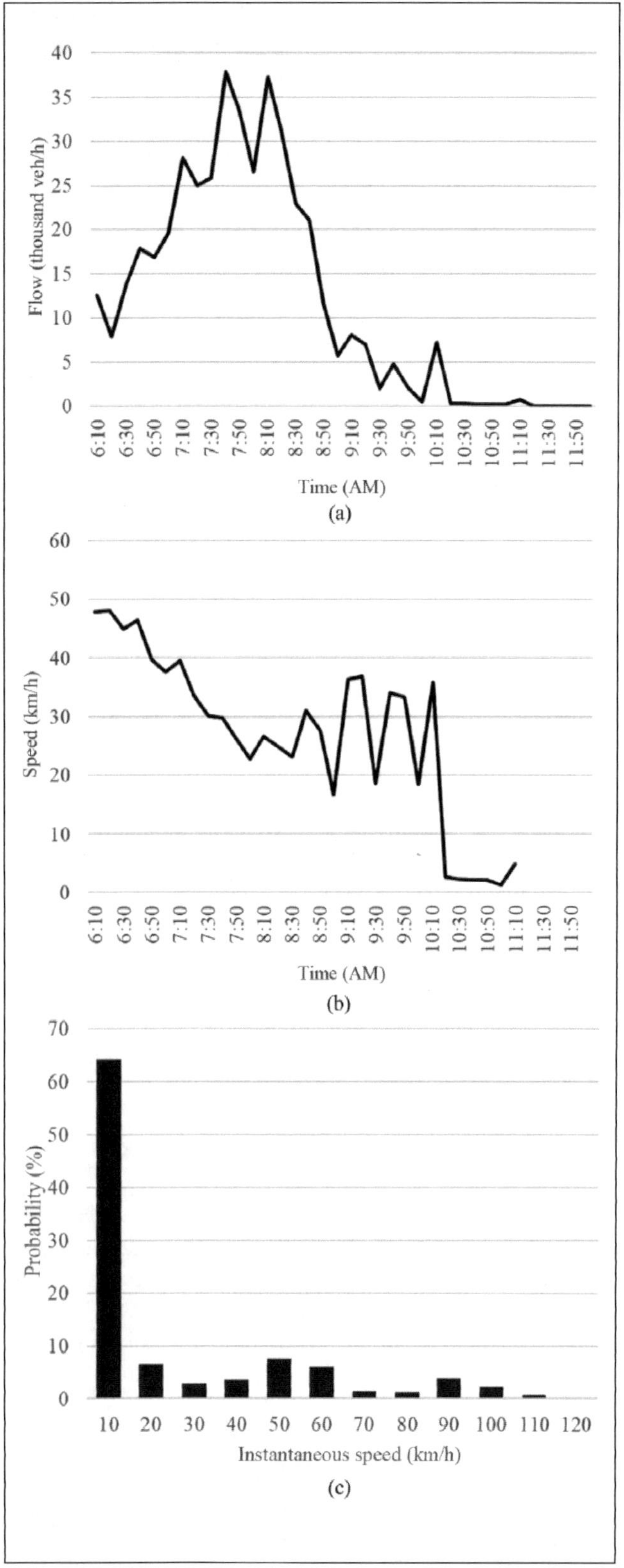

Figure 3. Microscopic simulation results for downtown Toronto road network: (*a*) network flow; (*b*) network average speed; (*c*) distribution of instantaneous speeds (the labels on *x*-axis represent the upper limit of speed intervals).

policies. Information on where the high emitters are destined to, where they originate from, and which neighborhoods contribute to the highest emissions in an area because of the trips they attract, is very valuable. Such information cannot be generated without an approach that tracks emissions by trips and trajectories.

Based on the location of origins and destinations, all trips in our study area were categorized into four groups: traveling inside the downtown area, going from inside to outside, from outside to inside, and from outside to outside (through traffic). Table 1 shows the proportion of each group in terms of count, CO_{2eq} and NO_x emissions, as well as mean CO_{2eq} EFs, which indicate the emission intensity of the trip.

Based on Table 1, we observe that trips starting outside of the downtown area and ending in downtown (out–in) have the highest share of all trips (38.3%), followed by through traffic (out–out, 29.6%). The out–in trips also have the highest proportion of CO_{2eq} emissions (44.1%), whereas through traffic produces the largest amount of NO_x (40.7%). Trips that start and end in the downtown area (in–in), only amount to 7.8% of total trips, and also generate the lowest share of emissions and the second highest mean EF for CO_{2eq}. The out–in trips have the highest average CO_{2eq} EF (421.2 g/VKT).

Scenario Analysis

Three policy scenarios were implemented: first, trips shorter than 600 m were removed from the OD matrix (S1); second, trips with total CO_{2eq} higher than 1.75 kg were removed (S2); third, trips with CO_{2eq} EF higher than 500g/VKT were removed (S3). The results of traffic simulation and emissions were compared with the base case (S0).

Table 2 presents the comparison in terms of the average network traffic conditions. Trips removed from each of the three policies are around 14% of the initial demand in the base case (S0). We observe that the scenario in which trips with CO_{2eq} EFs higher than 500g/VKT are removed (S3), shows the greatest improvement in terms of average traffic conditions: it decreases delay by 60.9 s/km and increases average speed by 4.8 km/h compared with S0. The scenario in which trips with total CO_{2eq} higher than 1.75 kg are removed (S2), also has a great improvement on delay and speed. In contrast, S1 (in which trips shorter than 600 m are eliminated) does not lead to much change to traffic conditions.

Figure 6 summarizes the total CO_{2eq} and NO_x emissions produced in the base case and three scenarios. S2 (in which trips with CO_{2eq} emissions higher than 1.75 kg are eliminated) leads to the highest reduction in trips compared with S0 (base case), and produces the lowest amount of emissions. It reduces CO_{2eq} and NO_x emissions by 41.1% and 29.8%, respectively. S3 (in which trips with CO_{2eq} EFs higher than 500 g/VKT are eliminated) leads to the second highest emission reduction: 32.7% for CO_{2eq} and 19.1% for NO_x. Meanwhile, S1 (in which trips shorter than 600 m are

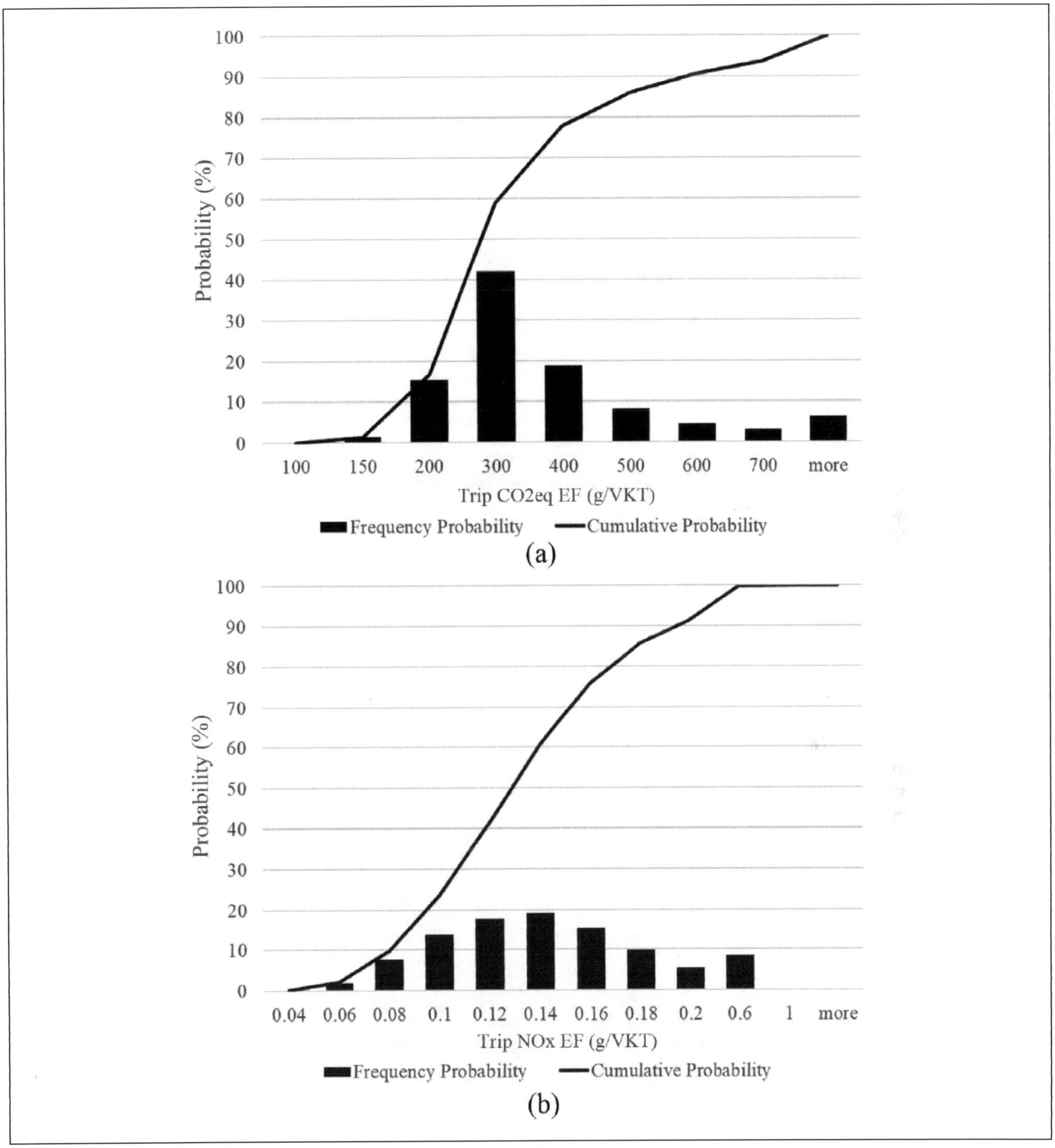

Figure 4. Trip (*a*) CO_{2eq} and (*b*) NO_x EF distributions (the labels on the *x*-axis represent the upper bound of each EF interval).

eliminated) leads to the least emission reduction: 1.6% for CO_{2eq} and 1.8% for NO_x.

Compared with S2 (in which trips with CO_{2eq} emissions higher than 1.75 kg are eliminated) and S3 (in which trips with CO_{2eq} EFs higher than 500 g/VKT are eliminated), which lead to significant improvements in emissions, the influence of S1 (in which trips shorter than 600 m are eliminated) is smaller. It is possible that trips with shorter distance are less efficient, as they do not contribute much to total emissions, so the emission benefits from removing them are not as deep as the other two scenarios. Short-distance trips are more likely to occur along arterials and removing them cannot mitigate the congestion on freeway facilities, which are often oversaturated with traffic. Removing short-distance trips cannot remove the heavy

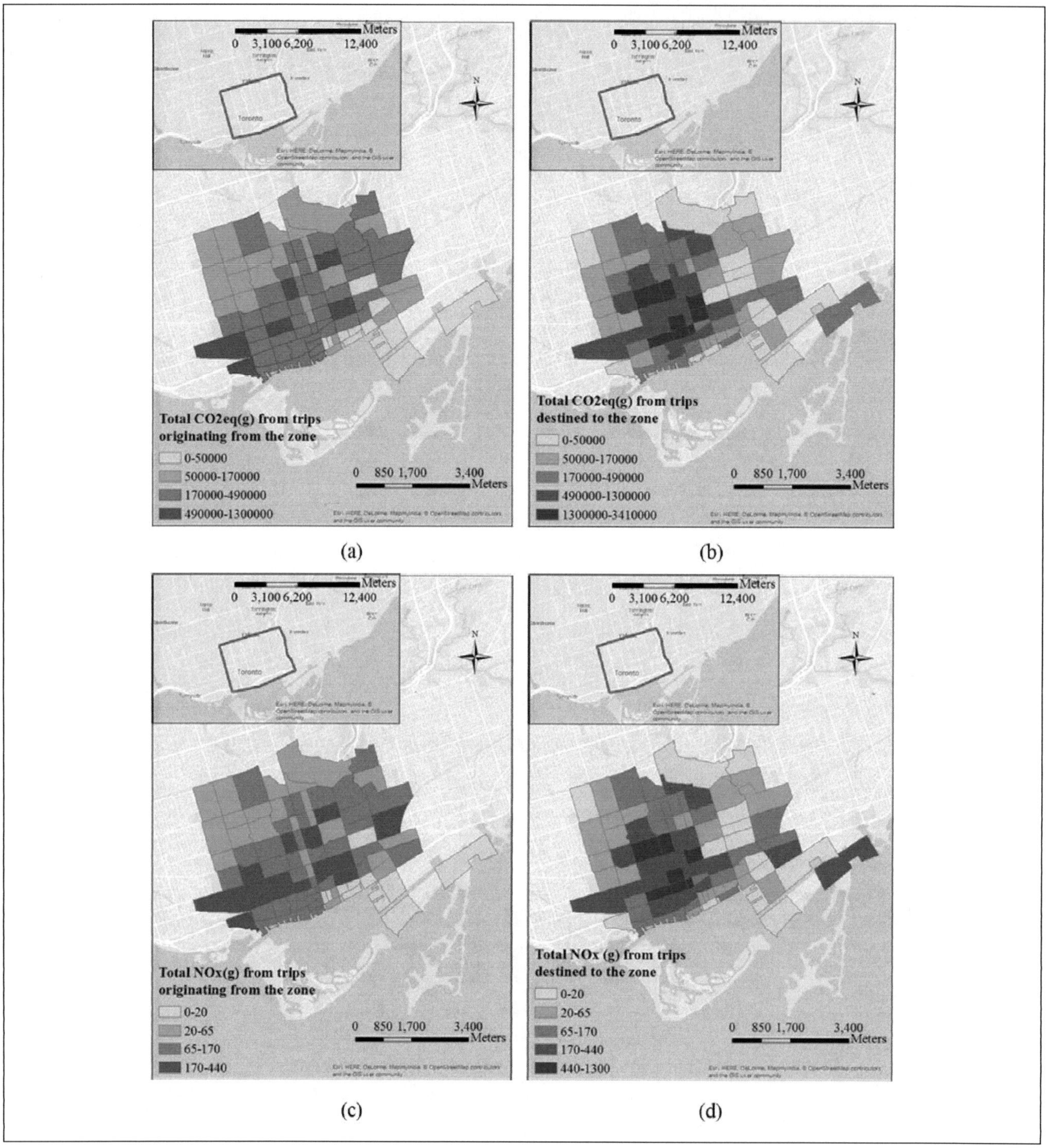

Figure 5. Spatial distribution of emissions generated/attracted by OD zone. GHG emissions aggregated to origin zones (*a*) and destination zones (*b*). NO_x emissions aggregated to origin zones (*c*) and destination zones (*d*).

Table 1. Trip and Emission Proportions for Four Categories of Trips

	% trips	% CO_2eq	% NO_x	Average CO_{2eq} EF (g/VKT)
1. in–in	7.8%	2.1%	2.0%	354.03
2. in–out	24.3%	16.8%	19.8%	276.71
3. out–in	38.3%	44.1%	37.5%	421.20
4. out–out	29.6%	37.0%	40.7%	347.40

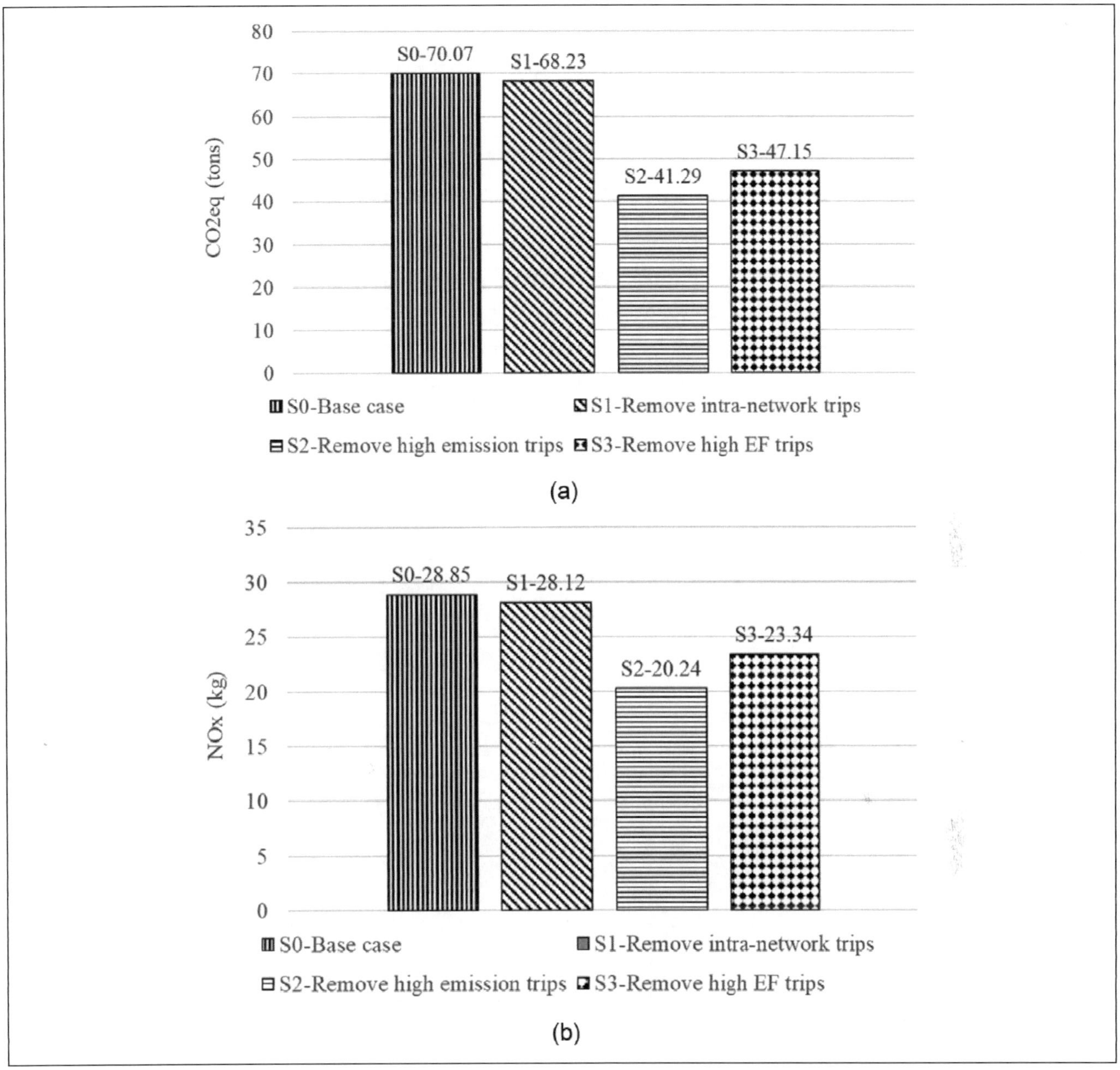

Figure 6. Total emissions generated by base case and policy scenarios: (*a*) network-wide CO_{2eq} emissions; (*b*) network-wide NO_x emissions.

Table 2. Effects of Scenarios on Traffic Conditions Across the Network

	S0	S1	S2	S3
Average speed (km/h)	31.8	32.5	35.3	36.6
Delay (sec/km)	141.2	135.5	95.6	80.3
% trip removed	0%	14%	14%	14%

Note: S0 = base case; S1 = remove short trips; S2 = remove high-emission trips; S3 = remove high EF trips.

pressure from through-network demands, which are more likely to have longer travel distance and produce higher emissions.

Apart from total emissions, EFs are important indicators of the emission intensity and the efficiency of trips. Figure 7 illustrates the distributions of EFs for CO_{2eq} and NO_x in the three scenarios and the base case.

The CO_{2eq} EF is lower than 500 g/VKT for around 95% of trips in S2 (in which trips with CO_{2eq} emissions higher than 1.75 kg are eliminated) and 98% trips in S3 (in which trips with CO_{2eq} EFs higher than 500 g/VKT are eliminated). Trips with EFs within this same range account for around 86% of total trips in S0 (base case) and S1 (in which trips shorter than 600 m are eliminated). Similar to CO_{2eq}, the influence on NO_x EFs in S2 and S3 is much higher than that in S1. A *t*-test result showed that the effect of S1 on both EFs was not

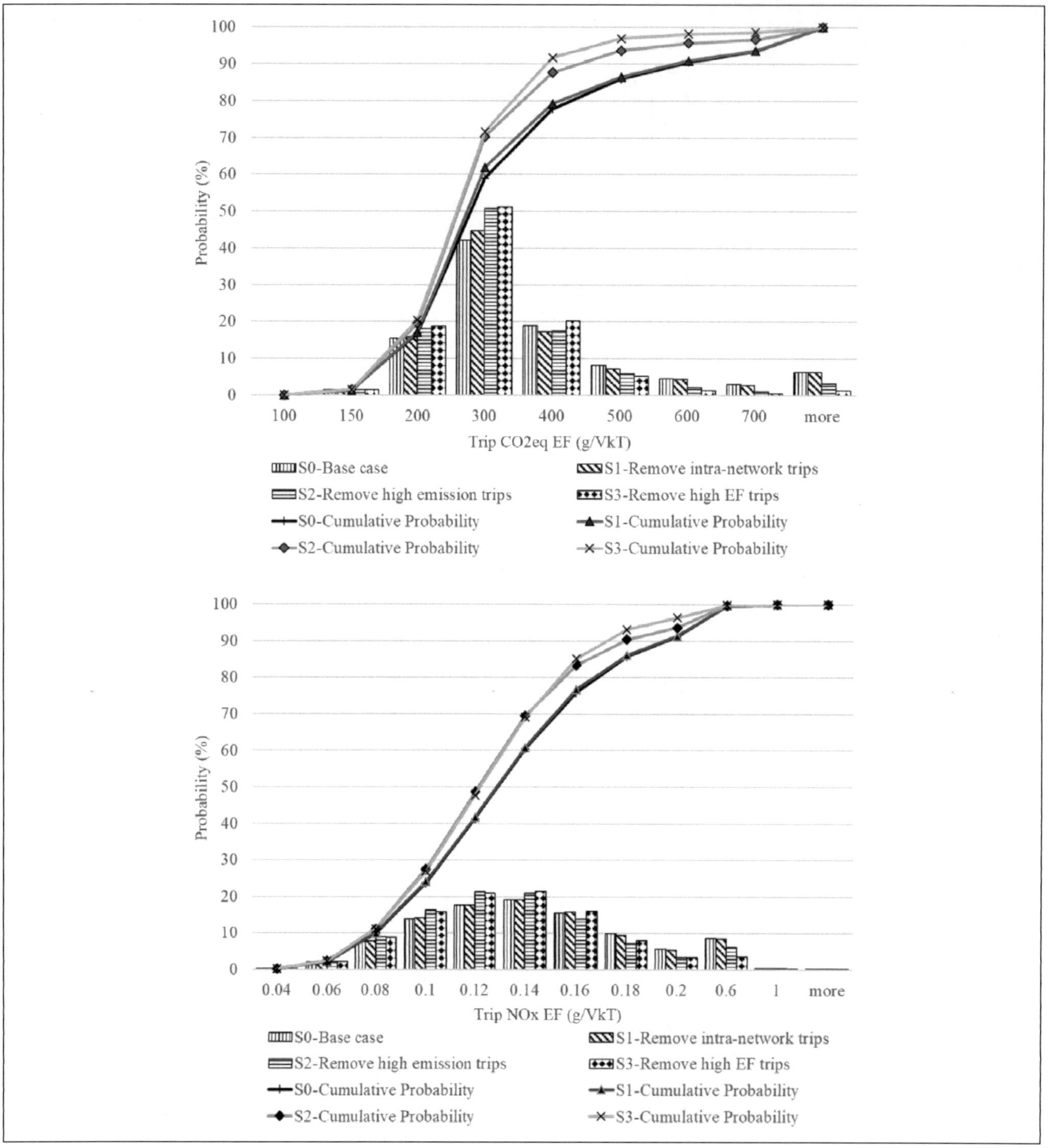

Figure 7. Trip-level EFs across base case and policy scenarios (bars indicate probability of occurrence and lines indicate cumulative probability): (*a*) CO_{2eq} EF distribution; (*b*) NO_x EF distribution.

significant. The comparison shows that even removing a similar number of trips, the policy targeting short trips (S1) is much less effective than policies working on the individual emission footprint of trips (S2 and S3). Trips with shorter distance are more likely to occur along arterials, while remaining trips are still influenced by stop-and-go traffic around signalized intersections where spots of high emission intensity are generally located. In S2 and S3, trips with higher emitters or higher EFs are possible to have longer distances or spend more time in congested areas, so removing these types of trips would increase the efficiency of the network.

Comparing S2 (in which trips with CO_{2eq} emissions higher than 1.75 kg are eliminated) and S3 (in which trips with CO_{2eq} EFs higher than 500 g/VKT are eliminated), traffic congestion becomes less severe under S3 based on higher average speed and shorter delay. From the comparison of emissions, the average EF of S3 is lower than S2, indicating higher efficiency of S3. S2 works on high emitters, and the high emissions of these trips may be due to extremely long travel distance, so that the demand eliminated by S2 does not necessarily affect congestion. Trips with higher EF are more likely to go through congested links, so S3 is another way to mitigate the demand in these areas. Although S3 has higher total emissions than S2, the network efficiency of S3 is higher than S2.

Conclusion

This paper has investigated urban transportation emissions based on a trip and vehicle trajectory analysis. Using microscopic traffic simulation, we were able to keep the information such as instantaneous speed, travel distance, and emissions for individual trips, as well as the information of the origin and destination. Combined with microscopic emission estimation using MOVES, total emissions were estimated at network and individual trip levels.

The results showed that the network total CO_{2eq} and NO_x emissions were 70.1 tons and 28.8 kg, respectively. The trip-average amount of CO_{2eq} and NO_x was 983.36 g/trip and 0.4 g/trip, respectively.

The total emissions attracted by the downtown area were around twice the emissions generated by trips originating from downtown. Inside the downtown area, the hospital district and financial district attracted the majority of emissions.

At the level of individual trips, 76.41% of the trips had CO_{2eq} EFs between 150 and 400 g/VKT, and 66.1% trips with NO_x EFs between 0.1 and 0.16 g/VKT. In addition, trips starting from outside the downtown area and ending in downtown, produced the highest proportion of total CO_{2eq}, whereas through traffic led to the highest NO_x emissions. Intra-downtown trips led to the second highest CO_{2eq} EF with the lowest share of total emissions.

Three policy scenarios were applied. The results indicated that removing short-distance trips (S1) had the least significant influence on both total emissions and EFs. Removing trips with high CO_{2eq} emissions (S2) led to the greatest reduction in total emissions, followed by the scenario that involves removing trips with high EFs (S3). S2 and S3 scenario trips had a statistically significant influence on trip-based EFs.

With the help of microscopic trip-based emission analysis, the paper presents a promising direction to policy analysis. This approach first identifies the types of trips that lead to the highest emissions and develops solutions and policy options that have the potential to target these trips. Conventional policies, which work on identifying general travel demand management measures, cannot ensure large reductions in emissions. Policies restricting spatial access, such as low-emission zones, multi-use lanes, and so forth, view the network as a whole, ignoring the variability of affected trips (*36*). Moreover, policies applied to groups of trips are often based on their spatial or temporal features, whereas their emission characteristics are seldom considered.

In practice, from the perspective of travelers, policies that target specific trips may be aggressive and unfair. This paper does not propose policy scenarios and implementation mechanisms, but it illustrates that an urban emission inventory capable of analyzing trip-level emissions may be a far more efficient tool for policy analysis than standard emission inventories based on segment-level emissions. In future research, an agent-based emission analysis will be investigated by calculating the emission footprint of each traveler rather than each trip.

References

1. EnvironmentandClimateChangeCanada. *CanadianEnvironmental Sustainability Indicators Greenhouse Gas Emissions*. 2017.
2. EnvironmentandClimateChangeCanada. *CanadianEnvironmental Sustainability Indicators Air Pollutant Emissions*. 2016.
3. Chief Corporate Officer. *Toronto's 2013 Greenhouse Gas Inventory*. 2016.
4. Davis, N., J. Lents, M. Osses, N. Nikkila, and M. Barth. Development and Application of an International. *Vehicle Emissions Model Transportation Research Board 81st Annual Meeting*, January, 2005, pp. 1–20.
5. EMISIA. The Popular, Straightforward and Simple to Use Emissions Calculator. http://emisia.com/products/copert.
6. INFRAS. WelcometoHBEFA. http://www.hbefa.net/e/index.html.
7. Borge, R., I. de Miguel, D. de la Paz, J. Lumbreras, J. Pérez, and E. Rodríguez. Comparison of Road Traffic Emission Models in Madrid (Spain). *Atmospheric Environment*, Vol. 62, 2012, pp. 461–471.
8. Colberg, C. A., B. Tona, W. A. Stahel, M. Meier, and J. Staehelin. Comparison of a Road Traffic Emission Model (HBEFA) with Emissions Derived from Measurements in the Gubrist Road Tunnel, Switzerland. *Atmospheric Environment*, Vol. 39, No. 26, 2005, pp. 4703–4714.
9. O'Driscoll, R., H. M. ApSimon, T. Oxley, N. Molden, M. E. J. Stettler, and A. Thiyagarajah. A Portable Emissions Measurement System (PEMS) Study of NO X and Primary NO 2 Emissions from Euro 6 Diesel Passenger Cars and Comparison with COPERT Emission Factors. *Atmospheric Environment*, Vol. 145, 2016, pp. 81–91.
10. California Environmental Protection Agency Air Resources Board. EMFAC Web Database. https://www.arb.ca.gov/emfac/.
11. United States Environmental Protection Agency. MOVES2014a: Latest Version of MOtor Vehicle Emission Simulator (MOVES). https://www.epa.gov/moves/moves2014a-latest-version-motor-vehicle-emission-simulator-moves.
12. Abou-Senna, H., E. Radwan, K. Westerlund, and C. D. Cooper. Using a Traffic Simulation Model (VISSIM) with an Emissions Model (MOVES) to Predict Emissions from Vehicles on a Limited-Access Highway. *Journal of the Air & Waste Management Association*, Vol. 63, No. 7, 2013, pp. 819–831.
13. Hausberger, S., M. Rexeis, M. Zallinger, and R. Luz. Emission Factors from the Model PHEM for the HBEFA Version 3. *Report Nr. I-20/2009 Haus-Em*, Vol. 33, No. 8, 2009, p. 679.

14. Center for Environmental Research & Technology (University of California Riverside). Comprehensive Modal Emission Model (CMEM). http://www.cert.ucr.edu/cmem/.
15. Rakha, H., K. Ahn, and A. Trani. Development of VT-Micro Model for Estimating Hot Stabilized Light Duty Vehicle and Truck Emissions. *Transportation Research Part D: Transport and Environment*, Vol. 9, No. 1, 2004, pp. 49–74.
16. Hao, J., M. Hatzopoulou, and E. Miller. Integrating an Activity-Based Travel Demand Model with Dynamic Traffic Assignment and Emission Models: Implementation in the Greater Toronto, Canada, Area. *Transportation Research Record: Journal of the Transportation Research Board*, 2010. 2176: 1–13.
17. Cen, X., H. K. Lo, and L. Li. A Framework for Estimating Traffic Emissions: The Development of Passenger Car Emission Unit. *Transportation Research Part D: Transport and Environment* Vol. 44, 2016, pp. 78–92. https://doi.org/10.1016/j.trd.2016.02.013.
18. Megenbir, L. E., and M. A. Habib. Estimating Fuel Consumption and Vehicle Emissions Using Multi-Year Travel Survey Data. In Presented at 94th Annual Meeting of the Transportation Research Board, Washington, D.C., 2015.
19. González, C. M., C. D. Gómez, N. Y. Rojas, H. Acevedo, and B. H. Aristizábal. Relative Impact of on-Road Vehicular and Point-Source Industrial Emissions of Air Pollutants in a Medium-Sized Andean City. *Atmospheric Environment*, Vol. 152, 2017, pp. 279–289.
20. Borge, R., I. De Miguel, D. De Paz, J. Lumbreras, J. Pérez, and E. Rodríguez. Comparison of Road Traf Fi c Emission Models in Madrid (Spain). *Atmospheric Environment*, Vol. 62, 2012, pp. 461–471. https://doi.org/10.1016/j.atmosenv.2012.08.073.
21. Shabbir, R., and S. S. Ahmad. Monitoring Urban Transport Air Pollution and Energy Demand in Rawalpindi and Islamabad Using Leap Model. *Energy*, Vol. 35, No. 5, 2010, pp. 2323–2332.
22. Wang, M., Y. Che, K. Yang, M. Wang, L. Xiong, and Y. Huang. A Local-Scale Low-Carbon Plan Based on the STIRPAT Model and the Scenario Method: The Case of Minhang District, Shanghai, China. *Energy Policy*, Vol. 39, No. 11, 2011, pp. 6981–6990.
23. Chavez-Baeza, C., and C. Sheinbaum-Pardo. Sustainable Passenger Road Transport Scenarios to Reduce Fuel Consumption, Air Pollutants and GHG (Greenhouse Gas) Emissions in the Mexico City Metropolitan Area. *Energy*, Vol. 66, 2014, pp. 624–634.
24. Barla, P., M. Gilbert-Gonthier, M. A. L. Castro, and L. Miranda-Moreno. Eco-Driving Training and Fuel Consumption: Impact, Heterogeneity and Sustainability. *Energy Economics*, Vol. 62, 2017, pp. 187–194.
25. Yang, H., H. Rakha, and M. V. Ala. Eco-Cooperative Adaptive Cruise Control at Signalized Intersections Considering Queue Effects. *IEEE Transactions on Intelligent Transportation Systems*, Vol. 18, No. 6, 2017, pp. 1575–1585.
26. Tu, R., J. Du, H. A. Rakha, and H. Yang. System-Wide Impacts of Arterial and Freeway Eco-Cooperative Adaptive Cruise Control. Presented at 96th Annual Meeting of the Transportation Research Board, Washington, D.C., 2017.
27. Sexton, T., R. Deardorf, and S. Helgath. *Drive or Sail: Comparing Cost, Time, and Greenhouse Gas Emissions - Differences in Travel Modes for the Six Busiest Weekday, Commute Routes around the Puget Sound.* Washington State Department of Transportation, 2012.
28. Zhang, W., Y. Qi, Y. Yan, J. Tang, and Y. Wang. A Method of Emission and Traveller Behavior Analysis under Multimodal Traffic Condition. *Transportation Research Part D: Transport and Environment*, Vol. 52, 2017, pp. 139–155.
29. Aimsum. TSS - Transportation Simulation Systems. Estimation of Traversal OD Matrices - Aimsun (Version 8.2) Help.
30. Aimsum. TSS - Transportation Simulation Systems. Aimsun Traffic Modelling Software. https://www.aimsun.com/aimsun/.
31. Aimsum. TSS - Transport Simulation System. Simulation Process - Aimsun (Version 8.2) Help.
32. Aimsum. TSS - Transport Simulation Systems. Aimsun (Version 8.2) Help - Modelling Vehicle Movement.
33. United States Environmental Protection Agency. *Exhaust Emission Rates for Light - Duty On - Road Vehicles in MOVES2014 Final Report.* USEPA, 2015.
34. Liu, H., Y. A. Xu, M. O. Rodgers, A. Akanser, and R. Guensler. *Improved Energy and Emissions Modeling for Project Evaluation (MOVES-Matrix).* Report, UC Davis, 2016. https://ncst.ucdavis.edu/wp-content/uploads/2014/09/10-10-2016-NCST_MOVESMatrix_FinalReport_version-4.pdf
35. Ministry of Transportation Ontario. *Statistics and Management Reporting.* Ministry of Transportation Ontario, 2017.
36. Holman, C., R. Harrison, and X. Querol. Review of the Efficacy of Low Emission Zones to Improve Urban Air Quality in European Cities. *Atmospheric Environment*, Vol. 111, 2015, pp. 161–169.

The Standing Committee on Transportation and Air Quality (ADC20) peer-reviewed this paper (18-01090).

Article

TRR
JOURNAL OF THE TRANSPORTATION RESEARCH BOARD

Transportation Research Record
2018, Vol. 2672(25) 21–34

DOI: 10.1177/0361198118774667
journals.sagepub.com/home/trr
SAGE

The Potential of Metering Roundabouts: Influence in Transportation Externalities

Paulo Fernandes[1], João Teixeira[1], Claudio Guarnaccia[2], Jorge M. Bandeira[1], Eloísa Macedo[1], and Margarida C. Coelho[1]

Abstract

Roundabouts are increasingly being used on busy arterial streets for traffic calming purposes. However, if one roundabout leg is near a distribution hub, for example, parking areas of shopping centers, the entry traffic volumes will be particularly high in peak hours. This paper investigated a partial-metering-based strategy to reduce traffic-related costs in a corridor. Specifically, the resulting traffic performance, energy, environmental, and exposure impacts associated with access roundabouts were studied in an urban commercial area, namely: (a) to characterize corridor operations in terms of link-specific travel time, fuel consumption, carbon dioxide and nitrogen oxides emissions, and noise costs; (b) to propose an optimization model to minimize these outputs; and (c) to demonstrate the model applicability under different traffic demand and directional splits combinations. Traffic, noise, and vehicle dynamics data were collected from a corridor with roundabouts and signalized intersections near a commercial area of Guimarães, Portugal. Microscopic traffic and emission modeling platforms were used to model traffic operations and estimate pollutant emissions, respectively. Traffic noise was estimated with a semi-dynamical model. Link-based cost functions were developed based on the integrated modeling structure. Lastly, a sequential quadratic programming-type approach was applied to find optimal timing settings. The benefit of the partial-metering system, in terms of costs, could be up to 13% with observed traffic volumes. The efficiency of the proposed system increased as entering traffic at the metered approaches increased (~7% less costs). The findings enable quantification of metering benefits near shopping areas.

Previous studies have demonstrated the commercial, healthy environment, and safety benefits of replacing signalized corridors with functionally interdependent roundabouts near commercial locations (*1*, *2*). Consequently, some authorities have adopted roundabouts in series along corridors (*3*, *4*), some of these located near business and commercial areas.

Nevertheless, some evidence suggests roundabouts may reach saturation under moderate traffic volumes (*5*). Unbalanced flows among legs may not be a problem when the overall demand level is low. However, if one leg of a roundabout provides access to parking areas of shopping mall, the entry volumes of the corresponding approach will be high. This fact may increase delay on the next approaches, resulting in unnecessarily long queues and congestion (*6*).

Among possible solutions, such as changing roundabout design, implementing other forms of intersection, or installing metering signals, the latter strategy is the most cost-effective measure (*5*). Metering signals regulate flow into the circulating area of roundabouts from one approach and thereby create larger gaps for downstream entries, thus alleviating vehicle delays on roundabout legs (*7*). Full-time and partial-time metering strategies can be adopted. Roundabouts with full leg-by-leg control are not suitable for these cases as approaches must split in few seconds after signals are implemented (*8*). In turn, partial metering can control a specific roundabout leg (*7*) during heavy demand periods and optimize delay for other legs.

Design guidelines available for transportation planners to implement metering signal systems at roundabouts are lacking. Applications of metered roundabouts have been mostly conducted in Australia, the United Kingdom, and in the United States (*7*).

Typically, traffic lights are installed on roundabout approaches, but a second stop line to control left-turning traffic can be adopted (*9*). In this context, Fahmy

[1]Department Mechanical Engineering/Centre for Mechanical Technology and Automation (TEMA), University of Aveiro, Aveiro, Portugal
[2]Department of Civil Engineering, University of Salerno, Fisciano (SA), Italy

Corresponding Author:
Address correspondence to Paulo Fernandes: paulo.fernandes@ua.pt

introduced an adaptive traffic signaling method based on fuzzy logic for roundabouts with four legs, which recorded improvements in both waiting and moving times (*10*). Ma et al. also proposed an optimization model to improve signal timings and design lane markings for signalized roundabout legs and circulating lanes (*11*). It was found that the optimal plan reduced construction costs and vehicle delay.

Akçelik studied partial-metering roundabouts with one, two, and three circulating lanes using the aaSIDRA model (*8*). The findings indicated that short cycle times (lower than those used in practice) improved delay, operation costs, fuel consumption, and carbon dioxide (CO_2) emissions.

Hummer et al. developed a macroscopic model based on the *Highway Capacity Manual* (HCM) formulas and validated it using the VISSIM traffic model (*7*). Vehicle delays were computed for one cycle length under different signal locations in both single- and two-lane roundabouts. The proposed model was not subjected to field testing, and did not include an emissions or noise criterion.

In a recent study on traffic operations and capacity at roundabouts by Marilo et al. (*5*), the impacts of partial-metering strategies were investigated in two single-lane roundabouts in Spain. The authors suggested a reduction in average delay up to 60% depending on the combination of controlling traffic volumes and conflicting traffic flow.

Literature specifically about the use of partial metering is scarce and mostly focused on achieving specific performance goals (*7*). Selection and design of a partial strategy may involve the balancing of other competing objectives, such as environmental, energetic, and noise-related aspects.

With these concerns in mind, this study introduces a signal metering-based strategy with the main aim of reducing corridor-specific costs. These costs are represented by travel time, fuel consumption, global and local pollutant emissions, and traffic noise. The novelty of this paper is that it not only optimizes traffic performance measures but also environmental, energy, and noise-related criteria on a link basis, which is crucial for sustainable road transportation infrastructure.

This paper tests a signal partial-metering strategy at a real-world corridor near a shopping mall that often experiences congestion problems arising from heavy flow approaches. The objectives of this paper are threefold:

1) To evaluate corridor operations in terms of link-specific travel time, fuel use, CO_2, nitrogen oxides (NO_X), and noise costs;
2) To propose an optimization model to minimize above outputs;
3) To demonstrate model applicability under different traffic demand and directional split scenarios.

Methodology

The fundamental idea of the research methodology was to develop a modeling framework to optimize link costs with partial metering (Figure 1) involved the following tasks: (*i*) to collect traffic, noise, and vehicle dynamic data in one real-world corridor; (*ii*) to calibrate and validate traffic model; (*iii*) to define a methodology to estimate costs; (*iv*) to implement and optimize partial-metering signal according to link-specific costs; (*v*) to compare optimal metered and unmetered solutions under variations in traffic demand and directional split distributions.

Field Study

Field measurements were conducted on a corridor near Guimarães (Portugal), which comprises a single-lane roundabout (RBT1), a signalized intersection (I1), and three two-lane roundabouts (RBT2-RBT4), as shown in Figure 2.

The corridor is a stretch along N206 national road (~2.2 km length) and is located near major industrial areas. It has one lane between RBT1 and RBT2 and two lanes on the other arterials, and posted speed limits range from 40 (roundabout approaches) to 70 km/h. All roundabouts are suburban with small pedestrian impedance. I1 has a fixed cycle with the same setup during the day (overall cycle time is 83 s).

Depending on the day and demand period, the main stream is between RBT1 westbound and RBT4 southbound. RBT2 and RBT3 have unbalanced traffic flow among approaches and their eastbound legs are distribution hubs. The shopping center nearby has 1,960 available parking lots, and these roundabouts record high traffic volumes in some periods (especially on weekends and lunch/dinner periods).

One set of traffic, vehicle dynamic, and acoustic data was collected during a 12-h period (9:00 AM–9:00 PM) on a Sunday (which is the day with highest number of trips to the shopping mall) in June 2017 under dry weather conditions.

Cameras were installed at each intersection to gather intersection-specific direction split patterns. GPS devices were mounted on one passenger car to record vehicle dynamic data (second-by-second speed and acceleration-deceleration). These data were mostly collected during morning (9:00–11:00 AM) and afternoon (5:00–7:00 PM) peak periods. About 100 km of data coverage (more than 40 GPS runs) for each through movement (RBT1–RBT4 and RBT4–RBT1) were acquired for this research, according to practice suggested in Fries et al. (*12*). Four

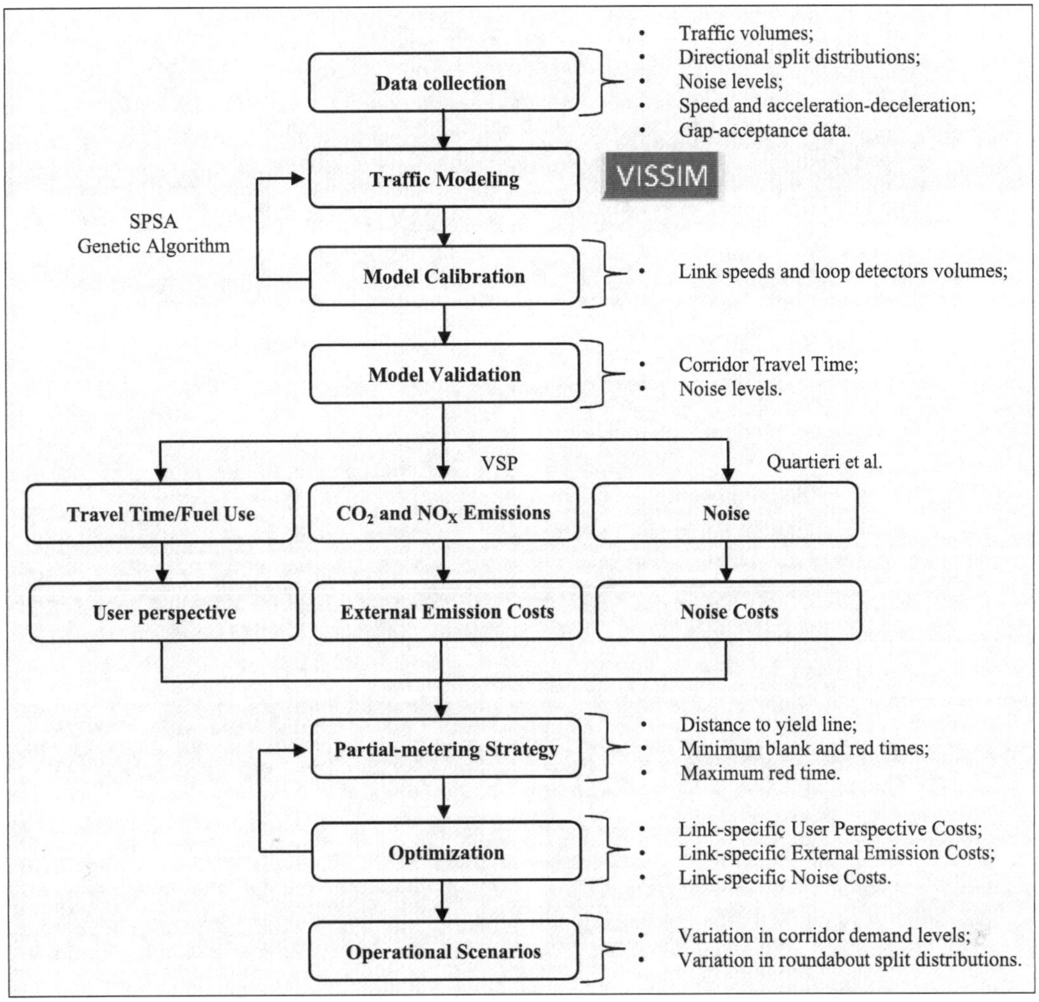

Figure 1. Overview of the research methodology (SPSA: simultaneous perturbation stochastic approximation; VSP: vehicle-specific power).

different drivers (three males and one female, ages 27 to 35) performed these routes to ensure driving variability.

Noise data were collected using an integrating sound level meter RION-NL52 installed at locations near RBT2–RBT3, and followed the ISO 11819-1:1997 standard. Tests were conducted with wind speeds lower than 4 km/h without the effects of other external sources, such as reflection and traffic from minor roundabout legs. The microphone was in the acoustic field at 1.2 m from the ground and at 15 m from the main road axis. The sound pressure levels were recorded every 1 s.

It is worthwhile mentioning that because of limitations in terms of equipment availability to obtain air pollution exposure data, shopper numbers, different traffic volumes, or variable weather conditions, only the above-mentioned data were collected.

Traffic, Emissions, and Noise Modeling

Corridor Coding. The studied corridor was implemented in the VISSIM9 package tool (*13*). Link/connectors coding was made following good practices for signalized intersections and roundabouts (*14*). The minimum dimension of link has been defined so that vehicles stayed at least 1 s (time resolution of the emission methodology) in each link. Thus, the link-specific vehicle dynamic and traffic values will not result in undervalued noise and emission-related costs (i.e., link without traffic information). Priority rules were introduced to reflect local driving habits on the roundabout approaches (*15*).

All simulation experiments assumed a 15-min "warm-up" prior analysis period to load the study domain (vehicles do not spend more than 10 min crossing the entire corridor).

Figure 2. Aerial view of the studied corridor with suggested metering legs, equipment location, and intersections identification [*Source*: Bing maps].

Emissions assessment. Emission estimates were based on vehicle-specific power (VSP), a methodology providing instantaneous power per unit mass of vehicle taking into account aerodynamic drag, speed, acceleration, road grade, and rolling distance effects (*16*). VSP values are categorized into 14 modes, and an emission factor for each mode was used to estimate the footprints of CO_2 (impact on global warming), and NO_X (precursor to troposphere ambient ozone and with demonstrated effects in human health) emissions (*16*).

The team fitted, as much as possible, the emission rates to the Portuguese car fleet, using five different vehicle types with the following composition (*17*): 39% (1.4L 33% + 1.8L 5.95% + 2.2L 0.05%) light duty gasoline vehicles, 40% light duty diesel vehicles (1.9L), and 21% light commercial diesel vehicles (2.5L). Their emission factors can be found elsewhere (*18*, *19*).

Although significant differences may occur in total absolute emissions, the relative differences in emissions associated with above set of vehicles reflected vehicle dynamics patterns.

Noise Assessment. Quartieri et al.'s (*20*) methodology was used to calculate noise levels produced by road traffic. This includes speed data information which notably increases the precision of traffic noise estimates. Thus, source power level ($L_{w,i}$) was first analyzed for all traffic flow in each link, and then overall corridor noise levels at a fixed distance were computed. Equation 1 gives the $L_{w,i}$ results for car passenger vehicles:

$$\begin{cases} 82, \text{if } v < 11.5 \text{km.h}^{-1} \\ \alpha + \beta \log v, \text{if } v > 11.5 \text{km.h}^{-1} \end{cases}, \quad (1)$$

where

α = 53.6 ± 0.3 dBA;

β = 26.8 ± 0.2 dBA (*20*).

Once the average link-speed data are obtained, the hourly equivalent noise level can be calculated using Equation 2 (*21*):

$$L_{eq,i} = 10\log N + \alpha + \beta \log v - 20 \log d - 47.563, \quad (2)$$

where:
$L_{eq,i}$– Link-specific equivalent noise level (dBA);
N– Link-specific hourly traffic volume (vehicles per hour – vph);
v– Link-specific average speed (km.h^{-1});
d– Distance between the road axis and the receiver (m) = 7.5 (*20*).

The methodology was validated by comparing measured and estimated noise data in different corridor locations.

Model Calibration and Validation. The traffic model was calibrated and validated using different data sets (there was a random selection of 70% for calibration and 30% for validation). Three steps were performed, namely:

- Adjust driver behavior parameters [average standstill distance, additive and multiple parts of safety distance, time before diffusion, front and rear gaps, safety distance factor, and simulation resolution (*13*)] to assess their impacts on speeds by link (*22*), and traffic flows for each intersection entry and exit leg (loop detector);
- Use the SPSA Genetic Algorithm to optimize the adjusted parameters. Calibration stopped after root mean squared normalized error (RMSNE) was lower than 15% (*23*) at each point. RMSNE was computed using Equation 3:

$$RMSNE_i = \sqrt{\frac{1}{N}\sum_{j=1}^{N}\left(\frac{SV_j - OV_j}{OV_j}\right)^2}, \quad (3)$$

where:
N– Number of loop detectors in the coded network;
SV_j– Simulated traffic volume in the loop detector j (km.h^{-1});
OV_j– Observed traffic volume in the loop detector j (km.h^{-1}).

- Validation compared corridor-specific simulated and observed travel time using the optimal calibration parameters with 10 random seed runs (*22*).

Development of Link-based Costs

External Emission Costs. The external emission cost approach focused on the quantification of the unequivocal impacts that emissions have on human health, environment, and economic activity.

The estimate damage NO_X and CO_2 costs per ton from transport is around €1,957/ton and €90/ton, respectively, for Portugal (*24*). Since exposure assessment is related to the population being exposed to air pollutant emissions, the local (parish level) population density is used to adjust average national costs suggested in the literature for NOx. Specifically, it corresponds to the ratio between local (508.4 inhabitants.km^{-2}) and national (114.5 inhabitants.km^{-2}) (*25*) population densities. For the purpose of analysis, all vehicle types are assumed to have a same dynamic pattern. Thus, the second-by-second cost estimation also includes the share of vehicle type. Finally, all costs are summed and further assigned to a coded link. The integrated emission costs of a representative vehicle for each VSP mode was given by Equation 4:

$$IEC_i = \sum_{j=1}^{5}\left(c_1 \mu v_j \times ef_{NO_X,j,i} + c_2 v_j \times ef_{CO_2,j,i}\right), \quad (4)$$

where:
IEC_i– Integrated emission costs for a representative vehicle and VSP mode (i = 1,…,14) (€.s^{-1});
c_1 = 1,957 – National damage cost of NO_X (€/ton);
c_2 = 90 – National damage cost of CO_2 (€/ton);
μ = 4.44 – Ratio between local and national population density;
v_j– Share of the vehicle type j in the vehicle park fleet;
$ef_{j,i}$– Emission factor for vehicle type j for each VSP mode i (g.s^{-1}).

It follows:

$$IEC_i = \sum_{j=1}^{5} v_j\left(8,689 ef_{NO_X,j,i} + 90 ef_{CO_2,j,i}\right), \quad (5)$$

Total external link-specific costs per kilometer (*IEC*) are obtained by summing IEC_i for time spent in each VSP mode.

Noise Costs. Scarce information is available about noise-related costs from road traffic. Some methodologies do not account for speed effects, thus resulting in varied results in noise costs estimates. A naive approach was used in this paper to estimate noise costs (*26*). Average link noise cost was computed by multiplying the number of individuals in a noise range with the cost of noise in €/dBA per exposed person and per hour for a range of noise divided by traffic in kilometers traveled (Equation 6):

$$NC_i = \frac{Costs(L_i)}{a \times b \times T} \times pop, \quad (6)$$

where:
NC_i– Average link noise cost (€/dBA.veh.km)
Costs (L_i) – Cost of a given noise level L_i (€/dBA per person and per year);
a = 365 – Number of days;
b = 24 – Number of hours;
pop– Number of person exposed to the noise level L_i;
T– Traffic in vehicle kilometers (veh.km)

A value of 2,128 inhabitants (*27*) was used as reference of *pop* (local population); monetary values to express *Costs* (L_j) are presented elsewhere (*26*). It follows:

$$NC_i = \frac{266}{1,095}\frac{Costs(L_i)}{T}, \tag{7}$$

Lastly, total links costs (*NC*) are obtained by summing NC_i by each corridor link. It must be stressed that this methodology assumes that only the local population is exposed to noise and it ignores people who are arriving at or leaving the shopping area. There are technologies such as radiofrequency, remote sensing, and activity-space analysis to quantify the presence of vulnerable users in the study area; however, this is beyond the scope of the paper.

User Perspective. With respect to the drivers' perspective, the impacts on travel time and fuel costs were estimated by using different criteria and sources to examine the chosen parameters. A value of 3/4 of the ratio between local average wage and monthly hours was suitable to provide an approximation of the average value of travel time (VTT) for recreational trips (*28*). This study considered a €713 local average wage (*29*) and approximately 154 h monthly labor hours of (*30*), ending up with a VTT of 0.0009677€.s^{-1}. The procedure for computing fuel-related costs is described by Equation 8 and includes fuel price, fuel consumption factor for each vehicle type associated with a VSP bin.

$$IDC_i = 0.0009677 + \sum_{j=1}^{5} v_j \times FC_{j,i} \times P_j, \tag{8}$$

where:
IDC_{VSPi}– Integrated Driver Costs by VSP mode (i = 1,...,14) (€.s^{-1});
v_j– Share of the vehicle type j in the vehicle park fleet;
$FC_{j,i}$– Fuel consumption factor for vehicle type j (L.s^{-1});
P_j– Price of Fuel (Diesel or Gasoline) for vehicle type j (€.L^{-1}).

Total link-specific costs as user perspective per kilometer (*IDC*) are computed by summing IDC_i for time spent in each VSP mode.

Operational Scenarios

To measure the merits of the partial-metering strategy, two main demand scenarios for both baseline (unmetered) and metered cases were defined:

1) Different traffic volumes at the RBT2 and RBT3 East entries, assuming no changes in the directional splits at all intersections and traffic volumes on the other RBT2-RBT3 legs;
2) Different left-turning rates at the RBT2 and RBT3 East entries, assuming no changes in the total entry flow and directional splits distributions on the other intersections.

The effects of both the uniform traffic growth and directional split distributions were evaluated at three levels each: traffic growth consisting of demand factors of 100% (observed), 150%, and 200%; and left-turning rates of 60%, 70%, and 80%. It is worth noting that the highest traffic growth scenario had a lower demand than the available shopping parking capacity.

Metered Strategy and Optimization

Partial-Metering System. An indirect (not controlled circulating traffic with priority) and part-control (one roundabout approach was under control) metering system was used. RBT2 and RBT3 East approaches were the candidates for the implementation of this system.

Figure 3 shows the scheme of signal setting for both roundabouts. The signal is set on the East approaches, also named the "metered approach." Conceptually, if signal is red, approaching vehicles will stop before the signal head. If it is blank, vehicles will drive normally and will enter the roundabout when an appropriate gap emerges (they must give way to circulating vehicles at any time). During the blank signal, the light will be off so that it will not mislead the entering traffic. The minimum blank and red times should be sufficient to ensure a trustworthy system for drivers when leaving the shopping mall, while the maximum red time must avoid excessive delays on the controlling approaches.

Once the partial-metering system was implemented in VISSIM (*13*), the following design parameters and corresponding reference values (*5*) were defined: (*a*) distance to yield line: 14–24 m; (*b*) minimum blank time setting: 20–50 s; (*c*) minimum red time setting: 10–20 s; and (*d*) maximum red time setting: 30–80 s.

Because the traffic flow at the candidate approaches was stationary over short periods (1 h), fixed-time signal timing was used. To simplify the optimization, the distance to yield line and maximum red time setting were set at 15 m and 30 s, respectively, for all operational

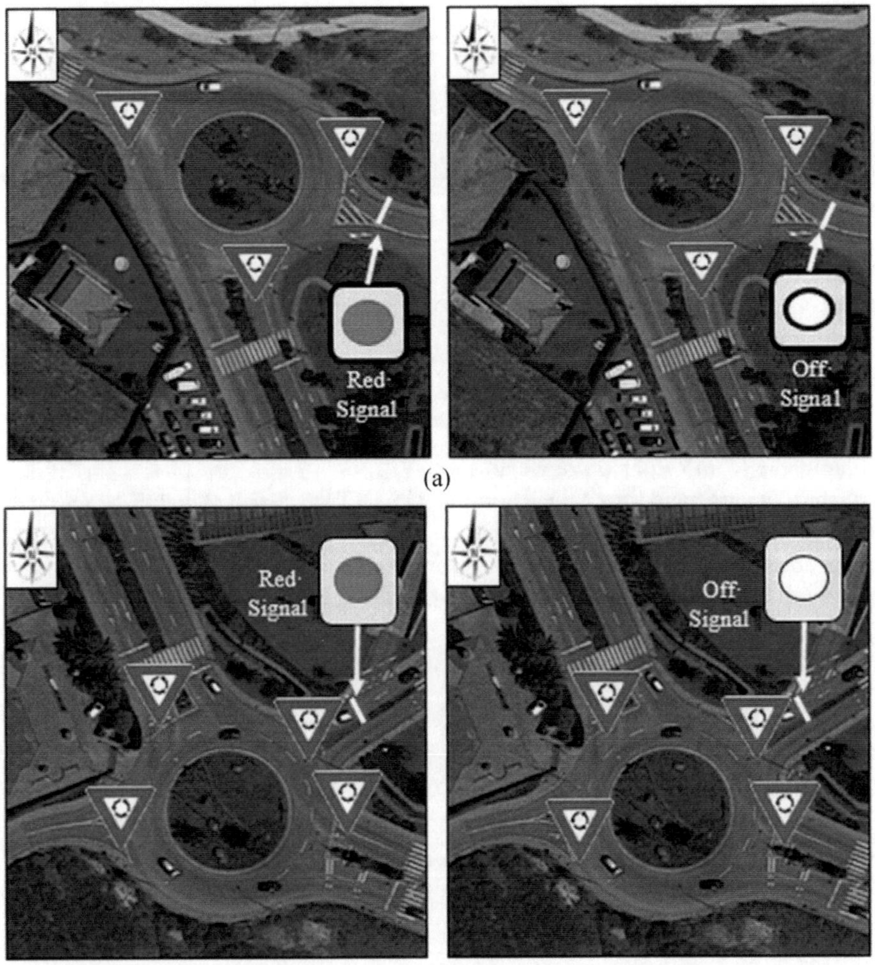

Figure 3. Scheme of roundabout with metered approach: (*a*) RBT2; (*b*) RBT3 [*Source*: Google maps].

scenarios (*5*). Also, the time settings were assumed to be equal in both metered approaches.

Optimization Formulation. The implementation of the partial-metered signals results in a trade-off. On one hand, the control delay on metered approaches is higher than that observed in baseline conditions. On the other hand, some links may record fewer costs because vehicles in the adjacent approaches may have more gaps in the circulating stream.

The main goal of the proposed multiobjective model was to minimize the overall costs based on Equations 5, 7 and 8, considering times blank (t_{blank}) and red (t_{red}) as decision variables bounded as follows:

$$20 \leq t_{blank} \leq 50 \tag{9}$$

$$10 \leq t_{red} \leq 30 \tag{10}$$

In addition, the candidate metered approaches have to yield medium-low volume-to-capacity ratios (v/c <0.7) to maintain tolerable operating conditions (*5*) when partial-metering signals are introduced. The estimation of capacity for both RBT2 and RBT3 East approaches was based on the HCM procedure (*31*). Combining the objective functions and main constraints, the optimization problem can be formulated as follows (Equation 11):

$$\min f(IEC, NC, IDC) = (f(IEC(t_{blank}, t_{red})), f(NC(t_{blank}, t_{red})), f(IDC(t_{blank}, t_{red})))$$

$$\text{subject to} \begin{array}{c} 20 \leq t_{blank} \leq 50 \\ 10 \leq t_{red} \leq 30 \\ v/c_{RBT2} < 0.7 \\ v/c_{RBT3} < 0.7 \end{array}, \tag{11}$$

Objective functions were constructed using multiple linear regressions, whose coefficients were obtained taking into account several simulations in 5-s increments for the decision variables, for each scenario. A total of 35 combinations of control metering signals parameters were performed to obtain optimal settings (total of 2,100 simulations in VISSIM).

Optimization Algorithm. The above partial-metered signal multiobjective optimization can search the optimal timing scheme, expressed as t_{blank} and t_{red}. Equation 11 is a bound constrained multiobjective optimization problem that was solved by using an SQP-type approach (*32*). The algorithm performs sequential quadratic programming-type iterations to build an approximation to the Pareto front. Numerical results in the form of performance and data profiles show that it outperforms the non-dominated sorting genetic algorithm for a considerable number of benchmark problems (*32*).

The main difference to other algorithms is that this new algorithm updates a finite set of points at each iteration, instead of a single one. The procedure is divided into three stages: (*i*) initialization; (*ii*) spread; and (*iii*) optimality-refining (*32*). In the initialization stage, initial guesses for Pareto points are provided. Then, in the spread stage, a set of new points, enriched with non-dominated points, is computed such that the set of their images is spread along the Pareto front. Finally, a refining process is applied to construct a sequence of converging points. During the procedure, derivatives of objective functions are assumed to be available, quadratic approximations to the objectives are considered, and various auxiliary optimization subproblems have to be solved.

The team adapted the multiobjective sequential quadratic programming solver, which is a publicly available MATLAB implementation of the algorithm (*32*).

Results and Discussion

Field Measurements

The analysis of corridor-specific demand patterns showed that the period with highest traffic volumes occurred between 5:30 and 6:30 PM. Thus, this period was selected for the implementation of the partial-metering strategy.

Table 1 lists observed values of approach, the level-of-service (LOS) criteria (based on average delay for all lanes), average and stopping delay per vehicle, and queue length (*31*). The number of vehicles entering each intersection ranged from 1,356 to 2,298 vph for I1 and RBT4, respectively, and 1% of traffic volume was composed of heavy-duty vehicles (HDV). The findings showed that roundabout approaches mostly operated with LOS A, that is, control delays lower than 10 s (*31*). As suspected,

Table 1. Traffic Performance Results Between 5:30 and 6:30 PM

	North approach					South approach					West approach					East approach				
ID	Entry flow [vph]	LOS	Total delay [s]	Stop delay [s]	Queue length [m]	Entry flow [vph]	LOS	Total delay [s]	Stop delay [s]	Queue length [m]	Entry flow [vph]	LOS	Total delay [s]	Stop delay [s]	Queue length [m]	Entry flow [vph]	LOS	Total delay [s]	Stop delay [s]	Queue length [m]
RBT1	416	A	4.7	1.1	61.1	281	A	3.2	0.9	31.8	476	A	3.8	0.9	54.4	466	A	3.6	0.5	62.4
I1	82	D	39.9	31.1	34.5	13	C	20.1	11.1	14.5	640	C	27.3	9.8	90.7	519	D	35.9	15.3	87.0
RBT2	742	B	13.0	4.9	62.4	544	A	2.7	0.4	57.8	N/A					267	A	1.3	0.1	30.6
RBT3	795	B	10.0	3.8	40.9	839	A	2.4	0.9	45.5	25	A	2.6	0.9	14.8	339	B	11.7	4.6	43.7
RBT4	1013	A	3.1	1.0	36.0	963	A	2.5	0.3	22.5	N/A					322	B	11.3	6.2	23.7

Note: N/A = Not Applicable; Values of total and stop delay are per vehicle.

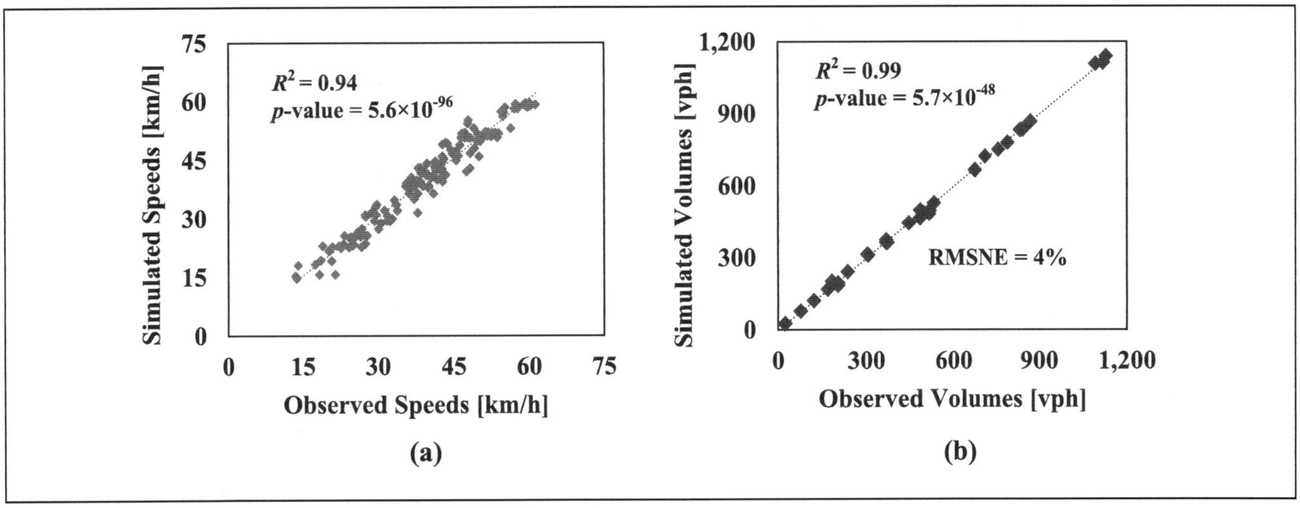

Figure 4. Calibration of traffic model: (*a*) speeds; (*b*) traffic volumes.
Note: p-value is *F*-test (ANOVA) performed in R^2 coefficients.
Calibrated Model Parameters (1-hour resolution): Average standstill distance: 0.5 m; Additive part of safety distance: 0.95; Multiple part of safety distance: 1.15; Time before diffusion: 120 s; Front Gap: 0.35 s; Rear Gap: 0.35 s; Safety Distance Factor: 1.15; Simulation Resolution: 10 time steps per simulation seconds.

RBT2 and RBT3 East approaches induced slight delays on the North approaches (LOS B). For these entry legs, the stopping delay accounted for approximately 40% of the approach average delay. This happened because approximately 50% of RBT2 and RBT3 traffic from East approaches went left. I1 West approach operated with an unstable flow (LOS D), whereas I1 East approach had flow restrictions concerning maneuverability and speed (LOS C) (*31*). Yet, the resulting queue lengths were lower than the distance to upstream RBT1 (˜750 m) and RBT2 (˜800 m).

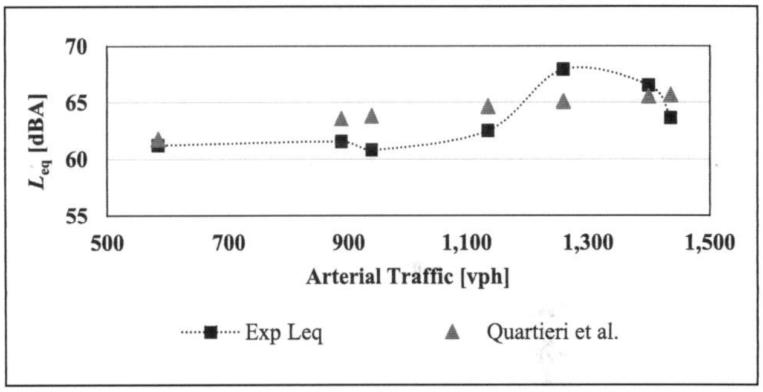

Figure 5. Validation of traffic noise methodology.

Calibration and Validation

Traffic Model. Figure 4 shows observed and estimated vehicle speeds (147 links) and traffic volumes (34 loop detectors) after calibration of the traffic model. The results confirmed good fit between simulated and observed data using a linear regression. The predicted R^2 was 0.94 and 0.98 for simulated speeds and traffic volumes, respectively. Also, the calibration target was reached since RMSNE was only 4% (*23*).

The comparison of observed and simulated travel time was performed using 30 floating car runs (*12*). The average travel time differences were, respectively, 7% (p-value = 0.29 >0.05, and thus, not statistically significant) and 6% (p-value = 0.07, also not statistically significant) in the directions RBT1–RBT4 (observed – 209s; simulated – 233s) and RBT4–RBT1 (observed – 208s; simulated – 220s). The calibrated parameters at the bottom of Figure 4 were then applied to the subsequent operational scenarios.

Noise Model. This section analyzed the capability of the noise methodology to estimate site-specific noise. Seven data sets of 15 min (L_{eq} and respective arterial traffic in front of sound meter) were selected in different corridor locations to ensure variability in the comparison (Figure 5). The estimated noise approach from Quartieri et al. (*20*) fit the field data (differences ranged from 1% to 5%). The highest differences between estimated and observed noise (~3 dBA) may be as a result of the presence of HDV that were not included in the L_{eq} formula. The decrease in experimental noise in high-volume values (1,300 and 1,500 vph) was explained by traffic congestion (lower speeds).

Comparison between Baseline and Optimal Metered Conditions

The main results of the multiobjective optimization of the partial-metered system with observed traffic demand

(100%) are presented in this section. The following optimal signal timing settings were obtained for both signals at RBT2 and RBT3 East approaches:

- $t_{blank} = 26s$; $t_{red} = 10s$ (v/c < 0.7)

Under the optimal model, overall *IEC, NC*, and *IDC* costs decreased by 6%, 11%, and 2%, respectively, compared with the baseline (unmetered). Also, independent two-sample *t*-test confirmed that the differences in average *IEC* and *NC* between baseline and metered system were statistically significant at the 95% confidence level. The implementation of the metered solution allowed *IEC* and *IDC* costs to be reduced by more than 10% on links upstream RBT2 and RBT3 (North direction). In contrast, overall costs at metered approaches increased by 50%.

Figure 6, *a–c* depicts the hotspot costs location on the shopping mall accesses with the baseline scenario. Analysis of the results showed that links with highest *IEC* were found at the upstream, downstream, and circulating areas of RBT2, RBT3, and RBT4. This was 60% more than the average *IEC* corridor value (13 €.km^{-1}). *NC* had a similar distribution near the shopping mall (Figure 6*b*) but high values (>10€/dBA.veh.km) were observed in the RBT3 West approach. This happened because these links had short length (< 10 m) and low traffic volumes. The findings from *IDC* showed an identical trend, as *IEC* did (Figure 6*c*). Fuel consumption and travel time-related costs along the downstream and upstream were 70% higher than the average corridor value.

When looking at link-specific range values with the partial-metered strategy (Figure 6, *d–f*), both RBT2 and RBT3 North approaches were notably improved. Specifically, *IEC* and *IDC* had yellow or green colors, whereas in the unmetered case these were orange or red (highest range of values). Another benefit was observed at the mid-block section between RBT2 and RBT3. This point was explained by low stop-and-go situations in North approach of RBT3 (vehicles were not retained by exiting shopping parking traffic), which in turn benefited RBT2 operations.

Sensitivity Analysis

To quantify the benefit at different operational scenarios, a comparison between the partial-metered system and baseline was conducted. The optimization model best solutions by each scenario were the following:

- 150%: $t_{blank} = 20s$; $t_{red} = 10s$;
- 200%: $t_{blank} = 46s$; $t_{red} = 28s$;
- 60%_Left: $t_{blank} = 50s$; $t_{red} = 10s$;
- 70%_Left: $t_{blank} = 50s$; $t_{red} = 10s$;
- 80%_Left: $t_{blank} = 45s$; $t_{red} = 26s$.

These values are in accordance with previous studies in the partial-metered strategy conducted at roundabouts (*5*). Almost all scenarios yielded optimal signal timing settings with long t_{blank} and short t_{red}. For the 200% traffic growth scenario, however, an optimal t_{red} of 28 s was suggested. This occurred because East approach traffic reached a demand so that the adjacent legs (RBT2 and RBT3 North approaches) were not able to discharge their queues with unmetered conditions. Even though metered approaches have performed poorly with long red times, it seems this setting brings benefits to the overall corridor.

Figure 7, *a–c* shows *IEC, NC*, and *IDC* costs for both unmetered and metered systems for each testing scenario. Some conclusions were:

- Metered signals improved corridor operations regardless of traffic growth scenarios (100%, 150%, and 200%). For the 150% growth scenario, benefits could be up to 10% for all costs;
- The partial-metered system became more effective in reducing costs when entering demands at RBT2 and RBT3 East approaches increased. It had reductions in average external costs (*IEC* and *NC*) of about 6%, and user perspective costs decreased by more than 7%. For *IEC* and *IDC*, the average costs were statistically significant (*p*-value <0.05);
- The differences in both *IDC* and *IEC* costs between solutions were found to be small (<2%) and not statistically significant (*p*-value >0.15) with different left-turning rates at the RBT2 and RBT3;
- The partial-metered system offered benefit in reducing *NC* under very high left-turning rates. It yielded 6% lower noise-related costs compared with those obtained without metering.

Conclusions

This paper explored the benefits of implementing a partial-metering strategy on roundabout legs with access to shopping mall parking. The system was designed to minimize link-specific external damage, noise, and user damage costs. The study also examined the system applicability under several combinations of traffic demand and left-turning rates.

A corridor with roundabouts and a signalized intersection in Portugal was characterized. Site-specific operations were calibrated and validated in the VISSIM model. Then, different combinations of blank and red

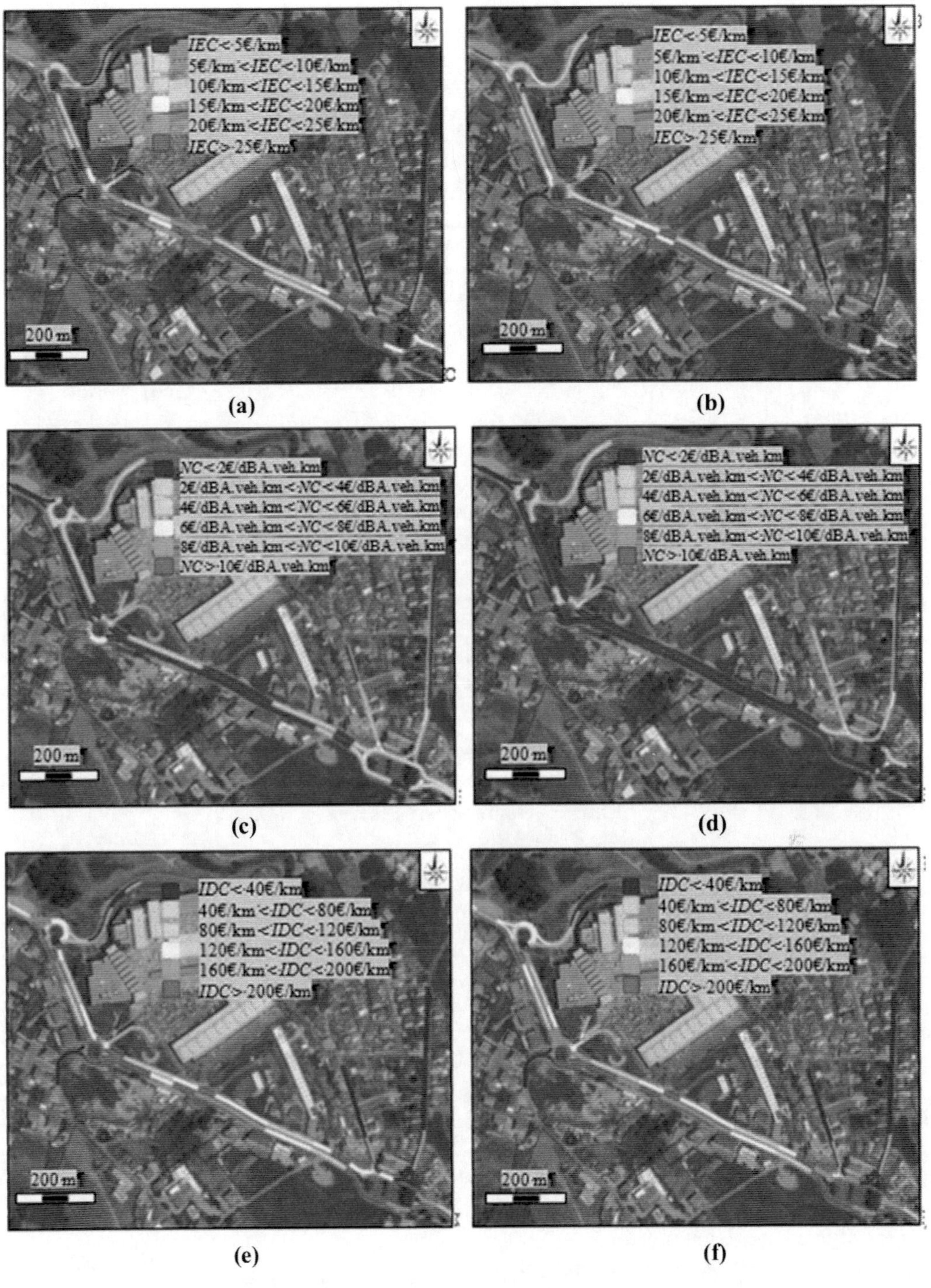

Figure 6. Overview link costs near shopping mall: (*a*) *IEC* – Metered; (*b*) *NC* – Baseline; (*c*) *NC* – Baseline; (*d*) *NC* – Metered; (e) *IDC* – Baseline; (*f*) *IDC* – Metered (*Source*: Bing Maps).

times were tested and their impacts on overall costs quantified. As a solution for the proposed problem, an SQP-type approach was used to search for optimal timing settings.

It was concluded that, under existing traffic demand, overall costs decreased up to 11% compared with the unmetered condition. The adoption of short red times (10 s) and long blank times (50s) had a better impact on

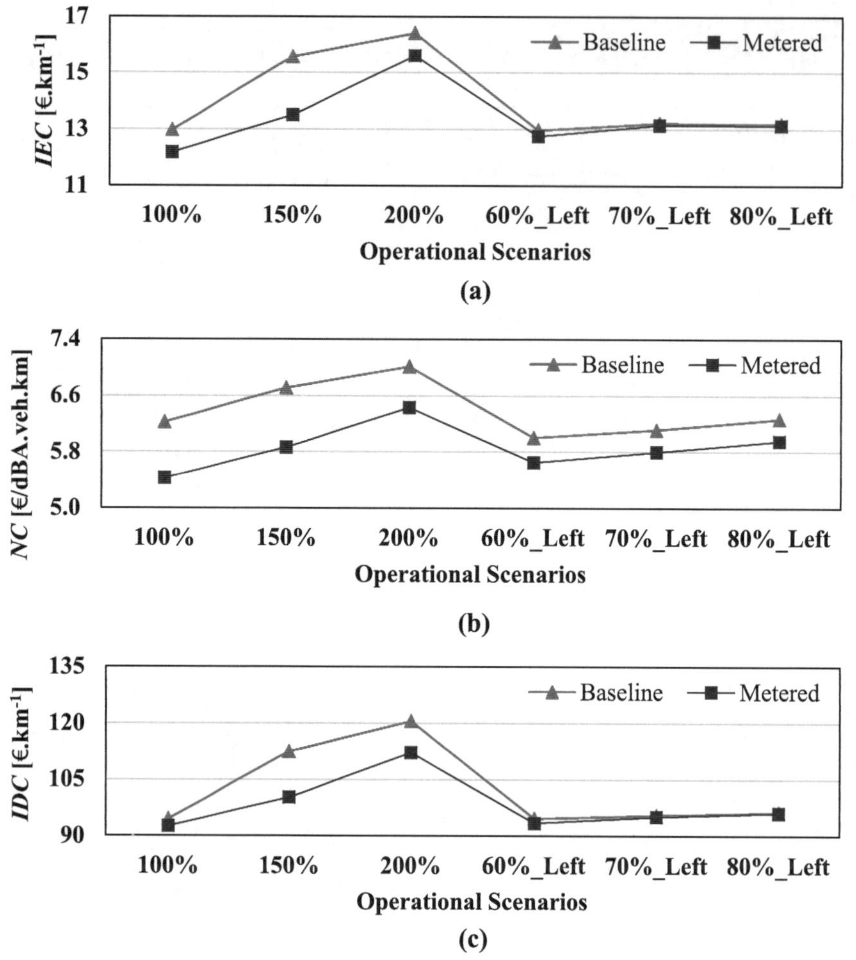

Figure 7. Link costs under different operational scenarios: (*a*) *IEC*; (*b*) *NC*; (*c*) *IDC*.

the controlling approach in almost scenarios. The findings suggested that the partial-metered system was effective when entering demands at the metered approaches increased. In such cases, external and internal damage costs reduced approximately 6% and 7%, respectively. With the exception of noise-related costs, the difference in average costs were statistically significant compared with the present conditions, thus demonstrating the benefits of metering the studied approaches.

This study contributes to determining the need of a metered-based strategy at roundabouts near a shopping mall, and its expected benefits in improving a wide range of parameters besides traditional traffic performance measures. This includes environmental, energetic, or traffic noise criteria, which are essential for a better economy in future road transportation. The proposed system can be used straightforwardly by practitioners, and can be adapted by the research community to include other transportation-related externalities.

Although the utility of the metering system on the candidate corridor has been demonstrated, there are some limitations that must be outlined: (*i*) findings were based on a simulation environmental with identical gap acceptance and car-following behaviors; (*ii*) the partial system operated with fixed-controlled settings; (*iii*) signal optimization only accounted for timing; (*iv*) pedestrian or cyclist impedance effects were discarded.

Acknowledgments

The authors acknowledge the projects: PTDC/EMS-TRA/0383/2014, funded within the project 9471-Reinforcement of RIDTI and funded by FEDER funds; Strategic Project UID-EMS-00481-2013-FCT and CENTRO-01-0145-FEDER-022083; CISMOB Project (PGI01611, funded by Interreg Europe Programme); MobiWise project: From mobile sensing to mobility advising (P2020 SAICTPAC/0011/2015), co-financed by COMPETE 2020, Portugal 2020 – Operational Program for Competitiveness and Internationalization (POCI),

European Union's ERDF (European Regional Development Fund), and the FCT. Jorge Bandeira acknowledges the support of FCT Postdoctoral Fellowship SFRH/BPD/100703/2014. Finally, the cooperation of Toyota Caetano Auto is appreciated which allowed the use of vehicles for data collection.

Author Contributions

The authors confirm contribution to the paper as follows. Study conception and design: Paulo Fernandes, Claudio Guarnaccia; data collection: Paulo Fernandes, João Teixeira, Eloísa Macedo; analysis and interpretation of results: Paulo Fernandes, João Teixeira, Claudio Guarnaccia, Jorge Bandeira, Eloísa Macedo, Margarida C. Coelho; draft manuscript preparation: Paulo Fernandes, João Teixeira, Claudio Guarnaccia, Jorge Bandeira, Eloísa Macedo, Margarida C. Coelho.
All authors reviewed the results and approved the final version of the manuscript.

References

1. Ariniello, A. *Are Roundabouts Good for Business?* Transportation Consultants, Inc., Denver, CO. www.cityofgolden.net/media/roundaboutpaper.pdf. Accessed May 18, 2017.
2. Russell, E., E. Landman, and R. Godavarthy. *A Study of the Impact of Roundabouts on Traffic Flows and Business.* Report No. K-TRAN: KSU-09-10. Kansas State University Transportation Center, Kansas, 2012.
3. Bugg, Z., B. Schroeder, P. Jenior, M. Brewer, and L. Rodegerdts. A Methodology to Compute Roundabout Corridor Travel Time. Presented at 94th Annual Meeting of the Transportation Research Board, Washington, D.C., 2015.
4. Isebrands, H., S. Hallmark, E. Fitzsimmons, and J. Stroda. *Toolbox to Evaluate the Impacts of Roundabouts on a Corridor or Roadway Network.* Report No. K-TRAN: KSU-09-10. Kansas State University Transportation Center, Kansas, 2008.
5. Marilo, G., G. Alfredo, and A. Moreno. Benefits of Metering Signals at Roundabouts with Unbalanced Flow. *Transportation Research Record: Journal of the Transportation Research Board*, 2015. 2585: 20–28.
6. Valdez, M., R. Cheu, and C. Duran. Operations of Modern Roundabout with Unbalanced Approach Volumes. *Transportation Research Record: Journal of the Transportation Research Board*, 2011. 2265: 234–243.
7. Hummer, J., J. Milazzo, B. Schroeder, and K. Salamati. Potential for Metering to Help Roundabouts Manage Peak Period Demands in the United States. *Transportation Research Record: Journal of the Transportation Research Board*, 2014. 2402: 56–66.
8. Akçelik, R. Roundabout Metering Signals: Capacity, Performance and Timing. *Procedia - Social and Behavioral Sciences*, Vol. 16, 2011, pp. 686–696.
9. Xiaoguang, Y., L. Xiugang, and X. Kun. A new traffic-signal control for modern roundabouts: method and application. *IEEE Transactions on Intelligent Transportation Systems*, Vol. 5, No. 4, 2004, pp. 282–287.
10. Fahmy, M. An Adaptive Traffic Signaling For Roundabout With Four Approach Intersections Based On Fuzzy Logic. *Journal of Computing and Information Technology*, Vol. 15, No. 1, 2007, pp. 33–45.
11. Ma, W., Y. Liu, L. Head, and X. Yang. Integrated optimization of lane markings and timings for signalized roundabouts. *Transportation Research Part C: Emerging Technologies*, Vol. 36, 2013, pp. 307–323.
12. Fries, R., Y. Qi, and S. Leight. How many Times Should I Run the Model? Performance Measure. Specific Findings from VISSIMmodels in Missouri. Presented at 96th Annual Meeting of the Transportation Research Board, Washington, D.C., 2017.
13. PTV, AG. *PTV VISSIM 9 User Manual.* Planung Transport Verkehr AG, Karlsruhe, Germany, 2016.
14. Fontes, T., S. R. Pereira, P. Fernandes, J. M. Bandeira, and M. C. Coelho. How to Combine Different Microsimulation Tools to Assess the Environmental Impacts of Road Traffic? Lessons and Directions. *Transportation Research Part D: Transport and Environment*, Vol. 34, 2015, pp. 293–306.
15. Vasconcelos, A., A. M. Seco, and A. B. Silva. Comparison of Procedures to Estimate Critical Headways at Roundabouts. *Promet –Traffic&Transportation*, Vol. 25, No. 1, 2013, pp. 43–53.
16. US EPA. *Methodology for Developing Modal Emission Rates for EPA's Multi-scale Motor Vehicle & Equipment Emission System.* Report No. EPA420, North Carolina State University for US Environmental Protection Agency, Ann Arbor, MI, 2002.
17. EMISIA. *COPERT Countries Data.* http://emisia.com/products/copert-data. Accessed May 16, 2017.
18. Anya, A. R., N. M. Rouphail, H. C. Frey, and B. Liu. Method and Case Study for Quantifying Local Emissions Impacts of Transportation Improvement Project Involving Road Realignment and Conversion to Multilane Roundabout. Presented at 92nd Annual Meeting of the Transportation Research Board, Washington, D.C., 2013.
19. Coelho, M. C., H. C. Frey, N. M. Rouphail, H. Zhai, and L. Pelkmans. Assessing Methods for Comparing Emissions from Gasoline and Diesel Light-duty Vehicles Based on Microscale Measurements. *Transportation Research Part D: Transport and Environment*, Vol. 14, No. 2, 2009, pp. 91–99.
20. Quartieri, J., G. Iannone, and C. Guarnacci. On the Improvement of Statistical Traffic Noise Prediction Tools. *Proc., 11th WSEAS International Conference on Acoustics & Music: Theory & Applications*, Iasi, Romania, June 13–15, 2010. pp. 201–207. World Scientific and Engineering Academy and Society (WSEAS), Stevens Point, WI.
21. Guarnaccia, C. Advanced Tools for Traffic Noise Modelling and Prediction. *WSEAS Transactions on Systems*, Vol. 12, 2013, pp. 121–130.
22. Winnie, D., B. H. Christine, and P. Serge. *Traffic Simulation and Data: Validation Methods and Applications.* CRC Press, Taylor & Francis Group, Boca Raton, FL, 2014.
23. Florida Department of Transportation. *Traffic Analysis Handbook: Reference for Planning and Operations.* Florida Department of Transportation, Tallahassee, FL, 2014.

24. Artem, K., D. Nicola, B. Johannes, H. Michael, M. Henning, G. Gena, V. Adarsh, and C. Victoria. *Update of the Handbook on External Costs of Transport*. Final Report MOVE/D3/2011/571. DG Mobility and Transport for the European Commission, London, 2014.
25. Statistics of Portugal. *Population's Density by Place of Residence*. https://ine.pt. Accessed July 12, 2017. Instituto Nacional de Estatística (Statistics of Portugal), Portuguese Public Institute, Lisbon, Portugal, https://ine.pt. Accessed July 12, 2017.
26. Le Maître, H. *Noise Costs for Road Traffic*. Presented at 94th Annual Meeting of the Transportation Research Board, Washington, D.C., 2015.
27. Statistics of Portugal. *Resident Persons (No.) in Family Nuclei by Place of Residence*. https://ine.pt. Accessed July 12, 2017. Instituto Nacional de Estatística (Statistics of Portugal), Portuguese Public Institute, Lisbon, Portugal, https://ine.pt. Accessed July 12, 2017.
28. Fezzi, C., I. J. Bateman, and S. Ferrini. Using Revealed Preferences to Estimate the Value of Travel Time to Recreation Sites. *Journal of Environmental Economics and Management*, Vol. 67, No. 1, 2014, pp. 58–70.
29. PORDATA. *Average Monthly Basic Remuneration of Employees: Total and by Sex*. Francisco Manuel dos Santos, Foundation, Lisbon, Portugal. http://www.pordata.pt/en/Municipalities/Average + monthly + basic + remuneration
+ of + employees + total + and + by + sex-235. Accessed July 12, 2017.
30. EUROSTAT. *Wages and Labour Costs*. http://ec.europa.eu/eurostat/statistics-explained/index.php/Wages_and_labour_costs. Publications Office of the European Union, Luxembourg, Luxembourg, Accessed July 12, 2017.
31. *Highway Capacity Manual 2010*. Transportation Research Board of the National Academies, Washington, D.C., 2010.
32. Fliege, J., and I. F. Vaz. A Method for Constrained Multiobjective Optimization Based on SQP Techniques. *SIAM Journal on Optimization*, Vol. 26, No. 4, 2016, pp. 2091–2119.

The Standing Committee on Transportation and Air Quality (ADC20) peer-reviewed this paper (18-00301).

TRR
JOURNAL OF THE TRANSPORTATION RESEARCH BOARD

Article

Transportation Research Record
2018, Vol. 2672(25) 35–45

DOI: 10.1177/0361198118791614
journals.sagepub.com/home/trr
SAGE

Digital Signal Processing Approach to Identify the Boundary of Vehicle Idle Status during Operation

Qing Li[1], Fengxiang Qiao[1], Lei Yu[2,3,4], Shuyan Chen[5], and Tiezhu Li[5]

Abstract
The MOVES is a tool to estimate on- and off-road emissions, in which 23 operating mode identification bins were defined based on vehicles' specific power, speed, and acceleration. Bin 1 indicates an idling mode with the speed within 1.0 mph. However, the speed boundary in an earlier model of MOBILE 6.2 was 2.5 mph. Neither the change in the idling definition of the two models nor the speed boundary were investigated and discussed. This study proposed a method to theoretically redefine the idle boundary by characterizing vehicle emission rates. Vehicle speeds close to 0 mph were carefully studied based on 10,000-mile on-board emission tests in the state of Texas. A portable emission measurement system was used to detect real-time emissions from a 12-year-old gasoline light-duty vehicle, while the vehicle's activity information was collected from an On-Board Diagnostic (OBD) II port. Power spectral density analysis was conducted on the collected emission and fuel consumption rates to identify a cut-off point that separates the frequency period with higher and lower energy. A Chebeshev I filter was designed to remove the high-frequency component to visualize the variables of emissions and fuel consumption on the vehicle's moving trend lines. Based on observation and analysis results, 2.26 mph was identified as a boundary for an idle mode at an acceptance level of 95% significant change. It is recommended that the proposed method be applied to the emissions of more different types of vehicles with a wide range of mileages to validate the newly defined idle boundary.

Quantifying traffic-related air quality is essential for decision-makers to improve the living environment, mobility, and even public health (*1*, *2*). Vehicle emissions are nonpoint or mobile sources, making it difficult to monitor and control the daily discharge by 263.6 million vehicles registered in the United States in 2015 (*3*). The estimation of the nonpoint emissions from the enormous number of motor vehicles is a complex process (*4*–*6*). The intelligent transportation system (ITS) application that was initially designed to improve safety and mobility in various traffic conflict areas, such as work zones and intersections, can also significantly reduces exhaust emissions for an equivalent amount of mileage mostly by adjusting driving speeds (*7*–*10*). The change in emission patterns for the adjusted driving speeds could further increase the complexity of emission qualification.

In practice, vehicle emissions can be directly obtained by a real-detected emission device, such as a portable emission measurement system (PEMS) and optical remote sensing (ORS). The accuracy of the PEMS emission measurement satisfies both the relative accuracy and the stringent statistical test requirements, including F-test, t-test, and r-correlation (*11*). Several microscopic emission studies have been conducted using PEMS. For instance, Frey et al. (2005) used a PEMS to measure the emissions from heavy duty vehicles fueled with B20 biodiesel and petroleum diesel (*4*). The PEMS was also used to explore the variables for emissions by measuring the emissions from light-duty vehicle passenger cars and trucks under different conditions, in terms of roadway facility and operational modes. The research findings were significantly different from the estimation by MOBILE 6.2 (*12*), though the MOBILE 6.2 was upgraded to motor vehicle emission simulator (MOVES, *13*) in 2010. Moreover, Kousoulidou et al. (2012) used

[1]Innovative Transportation Research Institute, Texas Southern University, Houston, TX
[2]Xuchang University, Henan, China
[3]Beijing Jiaotong University, Beijing, China
[4]Texas Southern University, Houston, TX
[5]College of Transportation, Southeast University, Nanjing, Jiangsu, China

Corresponding Author:
Address correspondence to Fengxiang Qiao: fengxiang.qiao@tsu.edu

the PEMS-measured emissions to develop and validate the emission factors for passenger vehicles (*14*). Recently, the emission factor estimation was improved by introducing real-time pavement roughness to the PEMS-measured emissions (*15*). For regional emission measurement, remote sensing was often used. For instance, Sjödin and Jerksjö (2007) used a remote sensing (FEAT) emission measurement to evaluate three major European road transport emission models (*16*).

The typical difference between MOVES and MOBILE 6.2 is the definition of an idle boundary. By original definition, vehicle idling occurs when a vehicle is left running unnecessarily while stopped. This status is claimed when vehicle speed is within a certain small range: small or equal to 1.0 mph and 2.5 mph for MOVES and MOBILE 6.2, respectively (*15*). Beyond such speed ranges, a significant change in vehicle emissions may be observed. However, there is only a limited amount of literature focusing on the different boundaries for an idle status. It seems that these definitions of an idle boundary are arbitrary. However, different idle boundaries can result in enormous differences in the exhaust emission estimation, the accuracy of which is unconvincing.

Meanwhile, idling emissions have drawn significant attention from a considerable number of emission research teams looking at fuel consumption efficiency and exhaust emissions. Previously, Qiao et al. (2006) chareacterized truck idling emissions for real-world conditions and Gaines and Brodrick (2009) compared the difference in fuel-cycle vehicle emissions between long-duration idle from Class 8 trucks and short-duration idle achieved by various idling reduction technologies (*18*, *19*). Later, another study by Gaines and Burnham compared the idling emissions from new trucks and from the trucks installed with idling reduction devices (*20*). A different definition of the idle boundary in these studies could contribute significantly to the error in emission estimation.

In addition, exhaust emissions vary according to the type of vehicle and the vehicle's age or mileage which affects the deterioration of the engine controls, the catalyst, and potentially the particle filter (*21*). Whether a unique idle boundary can be applied to all vehicle exhaust emissions needs to be validated. Further, an idle status is associated with nonrecurring congestion for crashes or with stopping under improper signal timing at intersections (*22*). It is thus important to develop a better understanding of idling.

This paper is intended to propose a method to theoretically redefine a speed boundary for an idle mode, in terms of exhaust emission features. The specific change in exhaust emissions and consumption rates at a very low speed range was explored, while a digital signal processing (DSP) technique was selected to characterize the significant change in response to the lower frequency of the very low speed range.

Test and Analytical Methodology

Test Procedure and Equipment

A series of on-road emission tests were conducted in five Texas metropolitan areas: Houston, El Paso, Austin, San Antonio, and College Station, which are illustrated in Figure 1. A dedicated vehicle was employed to drive along the test route covering a total of 10,000 miles of roadways. As vehicle emissions are subject to a vehicle's age or mileage due to the deterioration of the engine controls, the catalyst, and potentially paricle filter, an old vehicle was selected (*21*). The specification of which are listed in Table 1.

Table 1 shows that the tested vehicle was 12 years old with a typical feature of a light-duty gasoline passenger car with automatic transmission in the form of a 2.5 liter displacement in a four-cylinder configuration. A PEMS was installed on the back seat of the vehicle for detecting real-time second-by-second emission rates, including carbon dioxide (CO_2), carbon monoxide (CO), nitrogen dioxides (NO_x), and hydrocarbons (HC), together with fuel consumption (FC). Figure 2 illustrates the placement of the PEMS in the tested vehicle.

Meanwhile, the tested vehicle's real-time operational information, such as speed, acceleration rate, and revolutions per minute (rpm), was collected by an On-board diagnostic (OBD) II at a sampling rate of 1 Hz.

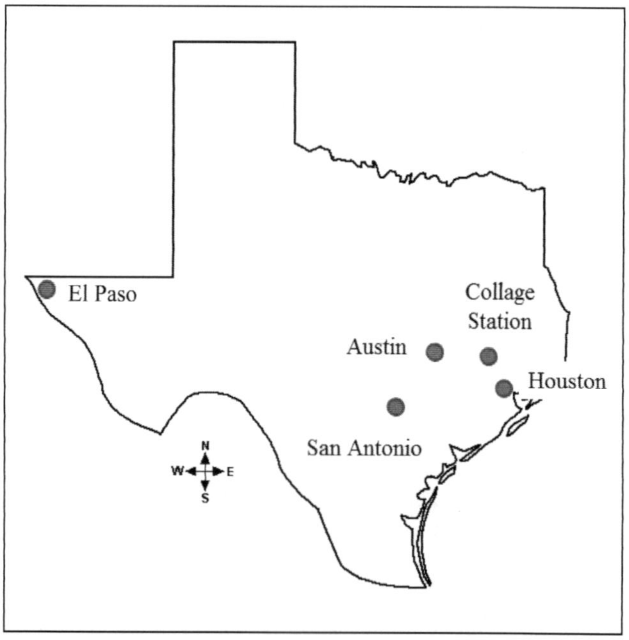

Figure 1. The five test areas in Texas.

Table 1. Tested Vehicle Specification

Vehicle information		Engine information	
Year	2004	Displacement (L)	2.5
Make	Subaru	Cylinder configuration	4
Model	Forester	HP@RPM	165 @ 5699
Vehicle weight	3100 lb	Torque@RPM	225 Nm @ 4000
Test weight	3500 lb	Fuel delivery	Gas

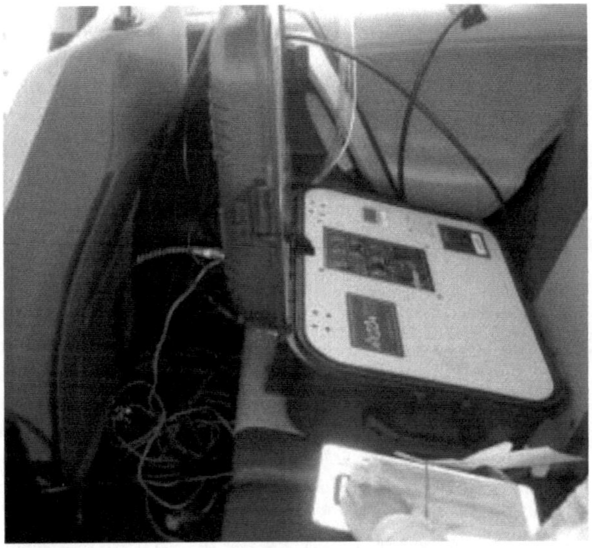

Figure 2. Test equipment PEMS, which is placed inside the test vehicle with cables connected to the tailpipe of the vehicle.

The tested vehicle's emission rates and activity data from PEMS, OBD, and other devices should first be checked through a quality control process, to ensure data consistency, completeness, accuracy, and precision, as well as the removal of outliers, and the filling of missing data. Readers may refer to references such as (*23*) for detailed data quality control methods.

Fundamental Mechanism of the Proposed Methodology

There is discretionary and non-discretionary idling. Discretionary idling takes place for an intentional stop with the engine running, which drivers often conduct at the first time of use of a vehicle in a day. Particularly, on a winter morning, drivers are inclined to warm up their vehicle by leaving it idling before departing on a trip. Nevertheless, drivers are more likely to carry out the non-discretionary idling unintentionally, which is mostly due to the resitrictions of traffic signals, signs and congestion (*24*). During idling, a vehicle either physically stands still or creeps along very slowly (*25*).

An automatic transmission vehicle is equipped with a torque converter, which is a type of fluid coupling that allows the engine to spin independently of the transmission (*26*). Under an idle mode, the automatic transmission vehicle is in gear with the engine running, which means there is still torque passed through the torque converter, but only a very small amount. The small amount of torque input keeps the vehicle moving forward slowly. To hold the vehicle still or creeping, a light or intermittent pressure on the breake pedal is always necessary. However, fuel continues to be burnt to create power, which is not efficiently applied to the purpose of motion. Instead, the power is released into the atmosphere, in the form of heat, vibration, and exhaust emissions. Therefore, the efficiency of the fuel consumption is lowest during an idle status. Further, the emission pattern under the non-discretionary idle mode becomes more complex than other operational modes, such as coasting. It is hard to physically define a boundary to separate an idle and non-idle mode based on such a complex emission pattern.

Probably because of this complexity, to the best of our knowledge, there is only a limited amount of literature that focuses on defining the boundary of idling. Thus, we are trying to apply a kind of data driven approach to identify the idle boundary based on the measured vehicle activity information (e.g., speed) and emissions. The digital signal processing (DSP) technique is one such method which works by clarifying or standardizing the levels or states of a digital signal (i.e., tested data sets in this paper) with enormous noise (i.e., the errors and frequent variations), and has superior control of accuracy.

The DSP has been widely used in processing complex signals, such as speech and geophysical message analysis and processing (*27*). Therefore, the DSP technique should be able to locate an idle boundary based on the change in fuel consumption and exhaust emissions in relation to the corresponding driving speeds.

As measured emissions and other associated information were recorded second by second, there could be significant variation among all measured data. The DSP is able to characterize the changes in emissions within a lower speed range by decomposing the trend out of the highly varying component. This process can be implemented through a technique called filtering, which is a typical process in DSP (*28*).

The variation of data sets can be measured and characterized by its power spectra of frequency, where the lower-frequency part is for less change, and the higher-frequency part is for more frequent change. The separation point is called the cut-off frequency. Once the higher-frequency components are filtered out, the change in the emissions from zero to lower speed performed by the remaining lower-frequency part would be clearly observed. The range of the ideal stage would be easily defined from the change.

Power Spectral Density Function

By DSP, all data sets can be treated as signals, which can be further decomposed into a series of sine functions, each of which has a unique frequency. A measured data set in engineering, such as the emission data set collected from field tests, should normally contain many frequencies. However, the magnitude of each frequency is different. Such a difference is normally measured by the energy for each frequency component. The statistical average of a certain signal or sort of signal (including noise) as analyzed in terms of its frequency content, is called its spectrum. The energy E of a signal $x(t)$ is defined as in Equation 1:

$$E = \int_{-\infty}^{+\infty} x(t)^2 dt \tag{1}$$

The energy spectral density is most suitable for transients (i.e., pulse-like signals) having a finite total energy. Parseval's theorem (*29*) provides an alternate expression for the energy of signal in terms of its Fourier transform (*30*) in:

$$\hat{x}(f) = \int_{-\infty}^{+\infty} e^{-2\pi i f t} x(t) dt \tag{2}$$

$$\int_{-\infty}^{+\infty} x(t)^2 = \int_{-\infty}^{+\infty} \hat{x}(f)^2 df \tag{3}$$

The variable f is the frequency in Hz. The energy spectral density $S_{xx}(f)$ of the signal $x(t)$ is defined as in:

$$S_{xx}(f) = \hat{x}(f)^2 \tag{4}$$

For continuous signals, the power spectral density (PSD) is used to describe how power of a signal or time series is distributed over frequency. The average power P of a signal $x(t)$ over time T is calculated by:

$$P = \lim_{T \to \infty} \frac{1}{2T} \int_{-T}^{+T} x(t)^2 dt \tag{5}$$

The power of the signal in a given frequency band $[f_1, f_2]$ can be calculated using:

$$P_{bandlimitd} = 2 \int_{f_1}^{f_2} S_{xx}(2\pi f) df \tag{6}$$

Welch (*31*) proposed an improved estimator of the PSD, which divides the time series data into (possibly overlapping) segments, computing a modified periodogram of each segment, and then averaging the PSD estimates. This is known as Welch's PSD estimate. By default, the data are divided into eight segments with 50% overlap between them. A Hamming window (*32*) is used to compute the modified periodogram of each segment. The related function for Welch PSD in computer language MATLAB is $[p_{xx}, f_x]$ = pwelch (y, [ ], [ ], [ ], f_s).

One of the good features of the PSD diagram is that it would be easy to see the dominant frequency(ies) that occupy higher powers. If the desired frequency is in the lower-frequency range, then a low-pass filter is needed to filter out the high-frequency part, while keeping the information needed under lower frequency.

Filter Design

Chebyshev filters are often used to separate different bands (e.g., lower band versus higher band) of frequencies. The primary attribute of Chebyshev filters is their speed, typically more than an order of magnitude faster than the windowed-sinc (*33*). There are four parameters needed to design a Chebyshev filter: (1) a high- or low-pass response, (2) cut-off frequency, (3) the percent ripple in the passband, and (4) the number of poles.

Figure 3 is the frequency response of three low-pass Chebyshev filters with passband ripples of 0%, 0.5%, and 20%. As the ripple increases (bad), the roll-off becomes sharper (good). The Chebyshev response is an optimal trade-off between these two parameters. When the ripple is set to 0%, the filter is called a maximally flat or Butterworth filter (*34*). Practically, a ripple of 0.5% matches the typical precision and accuracy in engineering.

The gain (or amplitude) response as a function of angular frequency ω of the nth order low-pass filter is equal to the absolute value of the transfer function $H_n(j\omega)$ in:

$$G_n(\omega) = |H_n(j\omega)| = \frac{1}{\sqrt{1 + \varepsilon^2 T_n^2\left(\frac{\omega}{\omega_0}\right)}} \tag{7}$$

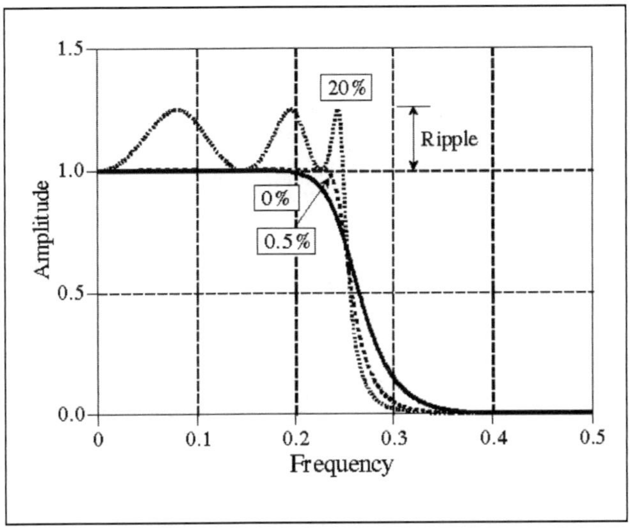

Figure 3. Features of the Chebyshev filter with three different ripples (source: *33*).

where

ε = a ripple factor

ω_0 = the cut-off frequency

T_n = a Chebyshev polynomial of the nth order

Vehicle Emission Normalization

The outputs from a PEMS include the emission rates of CO_2, CO, NO_x, and HC, which are on different scales. A normalization process is adopted to obtain the same scale for a fair comparison among them. A normalized emission rate (NER) is proposed to provide a comprehensive emission level, which is an average of all the four normalized emission rates x_i' for every second, as shown in:

$$NER = \frac{1}{4}\sum_{i=1}^{4} x_i' \quad (8)$$

The normalized emission rates x_i' for each emission component i are calculated using:

$$x_i' = \frac{x_i - x_{i,min}}{x_{i,max} - x_{i,min}} \quad (9)$$

where:

i = the emission component, there are four major components, namely CO_2, CO, NO_x, and HC

$x_{i,max}$ / $x_{i,mini}$ = the maximum / minimum emissions rate of each component

Operational Process

The proposed method is a seven-step procedure:

Step 1: Control data quality, and sort emission data based on the associated speed from 0 to maximum

In each second, a PEMS delivers the information on driving speed, four emission rates (e.g. CO_2, NO_x, HC and CO), and FC rate. The quality of all collected emission and vehicle activity data from the PEMS and other devices is controlled before further analyses. Together with the calculated NER, all outputs are sorted based on driving speed from 0 to maximum, while the lower speed range is of the special interests in this study.

Step 2: Identify the cut-off point frequency through power spectra density (PSD) function

All emission and FC rates are subtracted by their individual mean value to avoid a significant value of "zero frequency" component appearing in the PSD later on. The remain components are treated as signals to be further decomposed into a series of sine functions. Each of the sine function has its unique frequency. The frequency with a significant change at low speed will be identified as the cut-off point.

Step 3: Design a suitable filter

A cross-check can be conducted to determine the type of fitler and its order of a low-pass frequency. The type of filter could be butterworth, Chebyshev types 1 and 2, and elliptic. A suitable fitler is identified by the highest amplitude and the lowest frequency for all the emission rates, FC and NER.

Step 4: Filter-sort emission data

In this step, the designed fitler from step 3 is applied to filter the noise from the sorted and mean subtracted emission and FC data from step 2. The remain is the trend of the data.

Step 5: Make sure the high-frequency portions are fully filtered by comparing the PSD functions before and after filtering

This step is to exam if a higher-frequency component that is higher than the cut-off point, is fully removed by comparing the PSD of the emission and FC data in the same figure.

Step 6: Locate the lower speed region of the filtered data and determine the acceptance level

The low frequency component from step 5 can be plotted to illustrate the emission and FC data changes along with the lower driving speed, in which the lower speed region is located. Meanwhile, an acceptance level of significant change, such as 95%, is determined.

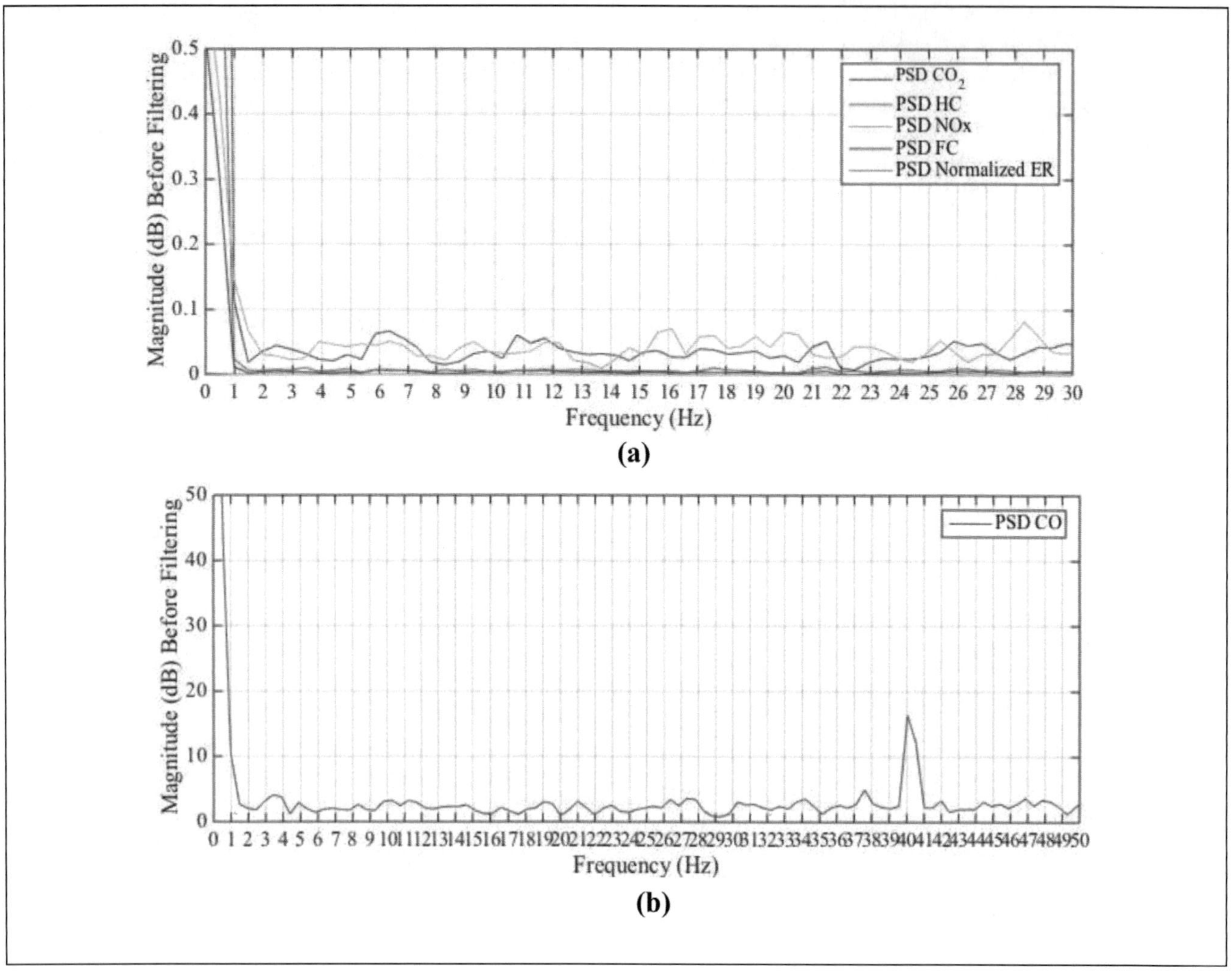

Figure 4. Power spectral density (PSD) estimates of emission rates (CO, CO_2, NO_x, HC, FC, and normalized emission rates).

Step 7: Identify the boundary that can be used to separate idle and non-idle regions

The boundary is identified by the driving speed in response to the determined acceptance level in the plotted figure in step 6.

Results and Discussion

A seven-step procedure was applied to the PEMS measured emission data together with the driving speed. For an idle process, the emission rate is more suitable than the emission factor as a measurement (*35*). As required by Step 1, all emission data, including CO_2, CO, NO_x, and HC, as well as FC, were sorted based on the corresponding speed from the lowest (zero) to the maximum. A "mean subtraction" process was conducted to all emission and FC data to avoid a significant value of "zero frequency" component appearing in the PSD later on. "Zero frequency" and high freuqncy are excluded in this study.

Following Step 2, the PSD of sorted emission rates for all emission and FC data were calculated and plotted in Figure 4, where the PSD for CO is plotted separately (in Figure 4*b*) because its energy is significantly higher than the others.

Figure 4, *a* and *b* clearly indicate that the energy of signals with frequencies larger than 2 Hz is significantly reduced, while most of the energies were within 2 Hz. This implies that the cut-off frequency could be naturally selected as 2 Hz.

In addition, starting with the 1st order of a filter, a cross-check was conducted to seek an optimal order of a low-pass frequency filter. At the 5th order of Chebyshev Type 1, the low-pass frequency response can provide the highest amplitude and the lowest frequency for all the measured emission rates, fuel consumption rates, and the total normalized emission rates. Therefore, a 5th order

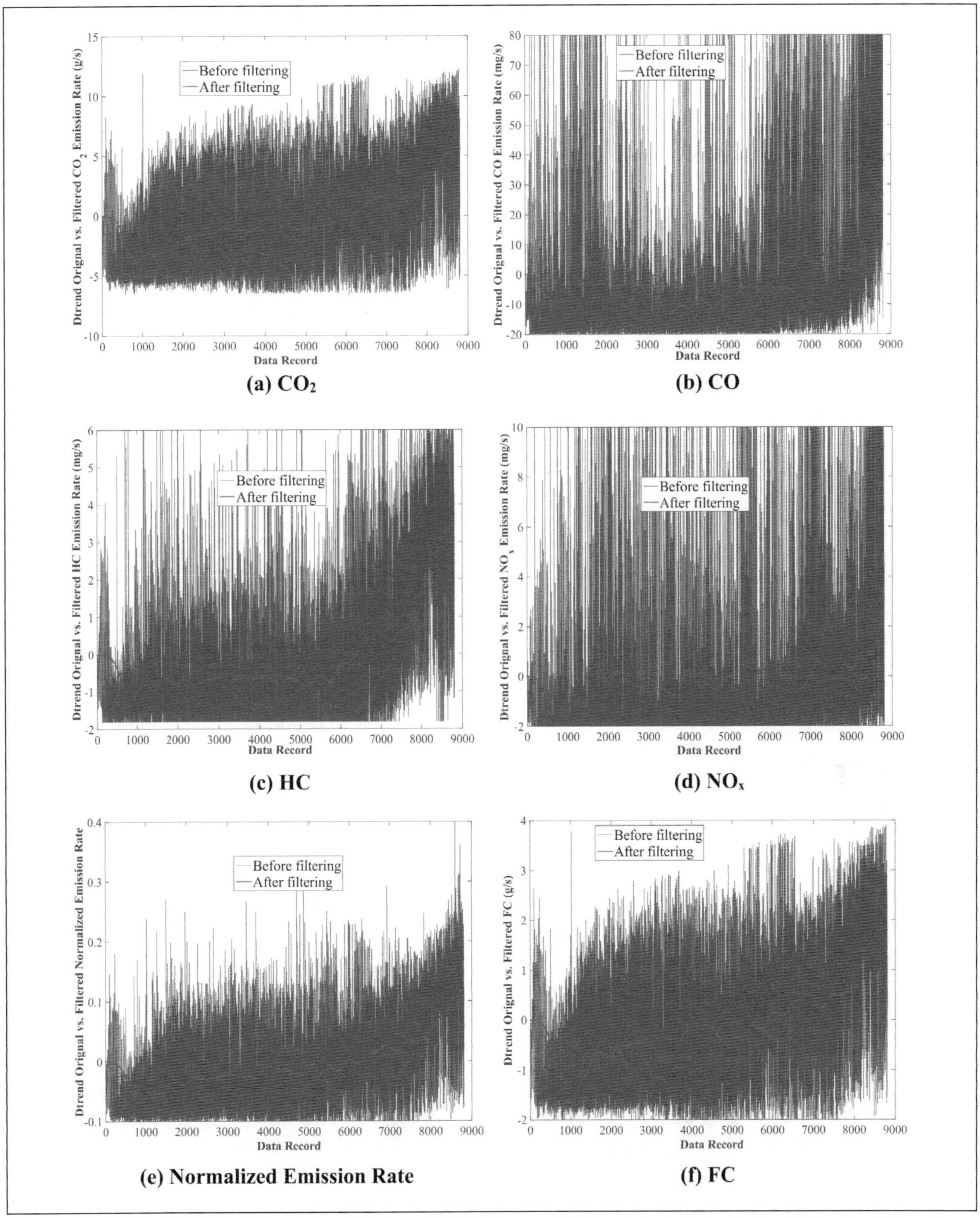

Figure 5. Original (blue) and filtered (red) emission rates.

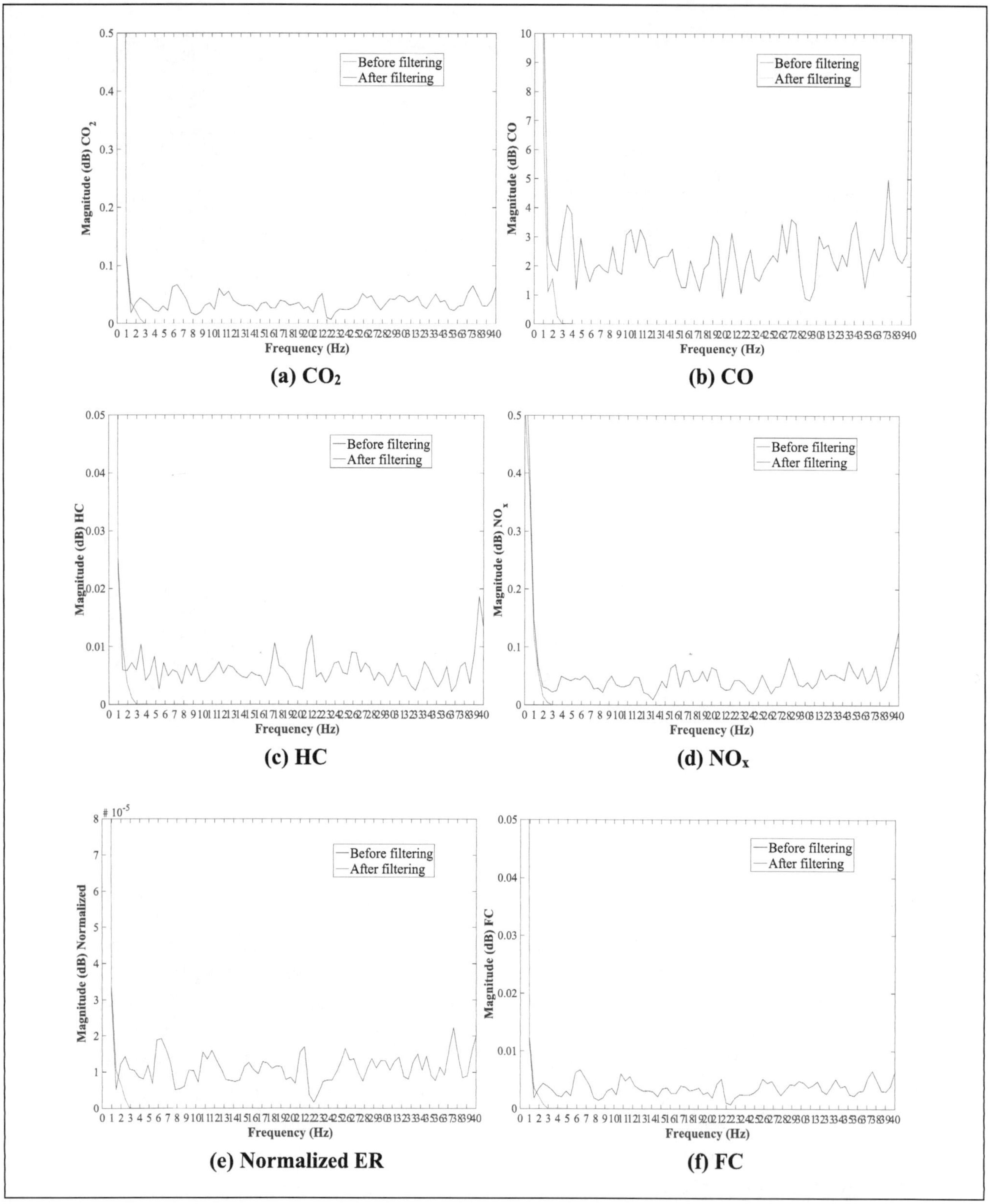

Figure 6. Original and filtered power spectral density (PSD) functions.

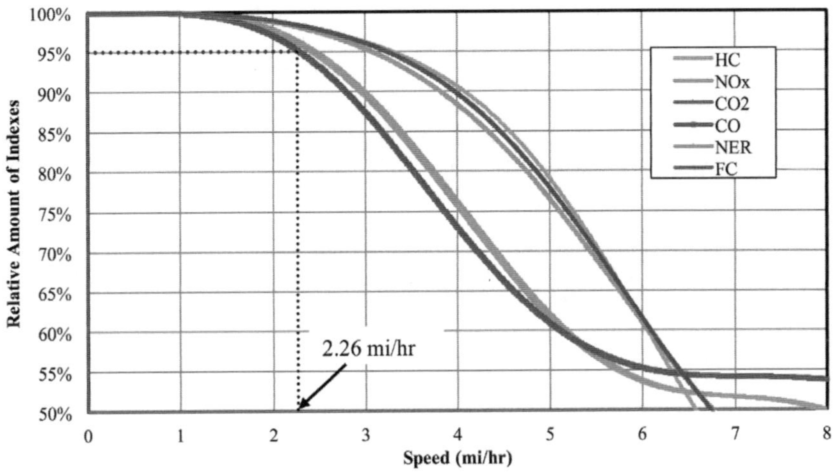

Figure 7. Curves to indicate the boundary for idle status.

Table 2. Parameters of Designed Low-Pass Infinite Impulse Response (IIR) Filter

Parameter	Option/parameter	Value
Design method	Chebyshev type I	
Order of filter	N	5
Sample frequency	F_s	1,000
Frequency pass	F_p	2 Hz
Pass of magnitude	A_p	1 dB

low-pass Chebyshev Type I filter was designed, the specifications of which are listed in Table 2.

The relevant MATLAB commends are H = fdesign. lowpass ('N, F_p, A_p', N, F_p, A_p, f_s); Ch = design (*h*, 'Cheby1'). The sorted and mean subtracted emission and FC data were filtered using the designed Chebyshev Type 1 filter. The MATLAB command is Efilter = filter (Ch, *y*). Figure 5 illustrated the sorted and mean subtracted emission and fuel consumption data series before (blue: busy lines) and after (red: less changing curve) filtering. In Figure 5, the red curve represents the trend of emission rates/fuel consumption, which is difficult to tell directly from the original blue data sets.

To scientifically exam if higher-frequency portions (the blue curves in Figure 5) are fully removed, the PSD of emission and fuel consumption rates before and after filtering are compared in Figure 6. Still, the blue curves are the PSD before filtering, while the red curves are the PSD after filtering. Obviously, all components with frequencies greater than 2 Hz are removed (in red curves) from the original data (in blue curves).

Figure 7 shows that the exhaust emission and fuel consumption rates in a small range of speeds (less than 8 mph) are close to zero. All the emission and fuel consumption rates have been normalized to 100% by dividing each of their maximum values, so that they can be placed in one figure for comparison. In Figure 7, it is seen that, beginning from a little bit more than 1 mph, the relative amount of all indexes starts to decrease. At about 2.26 mph, some of the emission components dropped to 95% of those at zero speed. If a 5% drop was chosen as the selection criteria, the driving speed (*v*) of 2.26 mph could be the boundary for an idle status. In other words, for traffic restrictions, such as traffic signals and congestions, the tested vehicle is in gear and brakes frequently to stand still or creep forward with a light or intermittent pressure on the breake pedal. When the vehicle is fully stopped or the creeping speed is equal to 2.26 mph or below, the vehicle's operational mode is defined as under an idle mode. This also defines Operating Mode Bin 1 (for speed slower than 2.26 mph) for Texas light-duty vehicles with similar features to the test vehicle.

Conclusions and Future Work

This study proposed a method to theoretically identify the boundary of an idle and non-idle mode of vehicle running based on a significant amount of on-board emission measurement and fuel consumption by a 12-year-old light-duty gasoline passenger car with automatic transmission. The method is a seven-step procedure, which includes the design of a Chebyshev filter and power spectrum analysis. The filter was designed was to characterize the low-frequent change portion of the data that had already been sorted according to speed (from zero to maximum). The power spectral density functions before and after filtering ensured that all higher-frequency components (i.e., frequencies larger than 2 Hz) had been removed. Then, all less moving curves for all emission components were plotted together. With an allowable changing percentage (say, 5%) predefined, the boundary to separate an idle and non-idle mode can be determined to be 2.26 mph.

While the identified boundary in this paper was based on a set of 10,000-mile emission tests on one vehicle in Texas, USA, the actual practical boundary should be further validated through even larger-scale tests on more vehicles of different vehicle types, fuel types, mileage, age, and other specifications.

Acknowledgments

The authors acknowledge that this research is supported in part by the National Science Foundation (NSF) under grants #1137732, and the U.S. National Tier 1 University Transportation Center (UTC) TranLIVE #DTRT12GUTC17/KLK900-SB-003. The authors appreciate Pengfei Liu, Mahreen Nabi, Mabuba Khan, and Wu Ying in data collection. The opinions, findings, and conclusions or recommendations expressed in this material are those of the author(s) and do not necessarily reflect the views of the funding agencies.

Author Contributions

The authors confirm contribution to the paper as follows: study conception and design: Qing Li, Fengxiang Qiao, Lei Yu, Shuyan Chen, Tiezhu Li; data collection: Qing Li, Fengxiang Qiao, Lei Yu; analysis and interpretation of results: Qing Li, Fengxiang Qiao, Lei Yu, Shuyan Chen, Tiezhu Li; draft manuscript preparation: Qing Li, Fengxiang Qiao, Lei Yu. All authors reviewed the results and approved the final version of the manuscript.

References

1. Caiazzo, F., A. Ashok, I. A. Waitz, S. H. L. Yim, and S. R. H. Barrett. Air Pollution and Early Deaths in the United States. Part I: Quantifying the Impact of Major Sectors in 2005. *Atmospheric Environment (Elsevier)*, Vol. 79, 2013, pp. 198–208.
2. Li, Q., F. Qiao, and L. Yu. Clustering Pavement Roughness Based on the Impacts on Vehicle Emissions and Public Health. *Journal of Ergonomics*, Vol. 6, No. 1, 2016, pp. 1000146. http://www.omicsgroup.org/journals/clustering-pavement-roughness-based-on-the-impacts-on-vehicleemissions-and-public-health-2165-7556-1000146.pdf
3. Rau, B. *Washington's Water Quality Management Plan to Control Nonpoint Sources of Pollution.* Publication no. 15-10-015. Washington State Department of Ecology, Olympia, 2015. https://fortress.wa.gov/ecy/publications/documents/1510015.pdf
4. Frey, H. C., and K. Kim. *Operational Evaluation of Emissions and Fuel Use of B20 Versus Diesel Fueled Dump Trucks.* FHWA/NC/2005-07. Department of Civil, Construction and Environmental Engineering, North Carolina State University, Raleigh, NC, 2005.
5. Dardiotis, C., G. Fontaras, A. Marotta, G. Martini, and U. Manfredi. Emissions of Modern Light Duty Ethanol Flex-Fuel Vehicles over Different Operating and Environmental Conditions. *Fuel*, Vol. 140, 2015, pp. 531–540. http://publications.jrc.ec.europa.eu/repository/handle/JRC91941
6. Li, Q., F. Qiao, and L. Yu. Will Vehicle and Roadside Communications Reduce Emitted Air Pollution? *International Journal of Science and Technology*, Vol. 5, No. 1, 2015, pp. 17–23. http://ejournalofsciences.org/archive/vol5no1/vol5no1_3.pdf
7. Li, Q., F. Qiao, and L. Yu. Implications of Wireless Communication System for Traffic Operations on Vehicle Emissions. *Proc., 107th Air & Waste Management Association (AWMA)*, Raleigh, NC, June 22–25, 2015.
8. Li, Q., F. Qiao, and L. Yu. Impacts of Vehicles to Infrastructure Communication Technologies on Vehicle Emissions. *Proc., International Conference on Environmental Science and Technology*, American Academy of Sciences, Houston, TX, June 9–13, 2014.
9. Qiao, F., Q. Li, and L. Yu. Testing Impacts of Work Zone X2V Communication System on Safety and Air Quality in Driving Simulator. *Proc., 21st ITS World Congress*, Detroit, MI, September 7–11, 2014, pp. TS64.
10. Li, Q., F. Qiao, X. Wang, and L. Yu. Impacts of P2V Wireless Communication on Safety and Environment in Work Zones through Driving Simulator Tests (paper # 26-179). *Proc., 26th Annual Conference of the International Chinese Transportation Professionals Association (ICTPA)*, Tampa, FL, May 24–26, 2013.
11. Cheng, A. M., and G. F. Hagen. An Accurate Predictive Emissions Monitoring System (PEMS) for an Ethylene Furnace. *Environmental Progress & Sustainable Energe*, Vol. 15, 1996, pp. 19–27.
12. Qiao, F., L. Yu, and M. Vojtisek-Lom. On-Road Vehicle Emission and Activity Data Collection and Evaluation in Houston, Texas. *Transportation Research Record: Journal of the Transportation Research Board*, 2005. 1941: 60–71.
13. Environmental Protection Agency (EPA). *MOVES (Motor Vehicle Emission Simulator) 2014a.* http://www3.epa.gov/otaq/models/moves/. Accessed March 13, 2016.
14. Kousoulidou, M., G. Fontaras, L. Ntziachristos, P. Bonnel, Z. Samaras, and P. Dilara. Use of Portable Emissions Measurement System (PEMS) for the Development and Validation of Passenger Car Emission Factors. *Atmospheric Environment*, Vol. 64, 2013, pp. 329–338. http://www.sciencedirect.com/science/article/pii/S1352231012009363
15. Li, Q., F. Qiao, and L. Yu. Calibrating Emission Factors for Highways Considering Pavement Roughness Information. Paper # 80. *Proc., 2016 Air Quality Measurement Methods and Technology*, Chapel Hill, NC, March 15–17, 2016.
16. Sjödin, A., and M. Jerksjö. Evaluation of European Road Transport Emission Models against On-road Emission Data as Measured by Optical Remote Sensing. *Proc., 17th International Transport and Air Pollution Conference*, Graz, Austria, 2008.
17. Environmental Protection Agency (EPA). *User's Guide to MOBILE6.1 and MOBILE6.2. Mobile Source Emission Factor Model.* EPA420-R-03-010. 2003. https://www3.epa.

gov/otaq/models/mobile6/420r03010.pdf. Accessed on July 31, 2017.
18. Qiao, F., L. Yu, and V. Soltani. Characteristics of Truck Idling Emissions under Real-World Conditions. Presented at 85th Annual Meeting of the Transportation Research Board, Washington, D.C., 2006.
19. Gaines, L. L., C.-J. B. Hartman, and M. Solomon. Energy Use and Emissions Idling-Reduction Options for Heavy-Duty Diesel Trucks: A Comparison. *Transportation Research Record: Journal of the Transportation Research Board*, 2009. 2123: 8–16.
20. Gaines, L., and A. Burnham. Emissions from Idling Heavy-Duty Trucks and Idling-Reduction Equipment. Presented at 95th Annual Meeting of the Transportation Research Board, Washington, D.C., 2016.
21. Chen, Y., and J. Borken-Kleefeld. NOx Emissions from Diesel Passenger Cars Worsen with Age. *Environmental Science & Technology*, 2016. ACS publication. https://pdfs.semanticscholar.org/c1d0/3d735c509f37abee1d4b23823713a401d89c.pdf. Accessed November 10, 2017.
22. Qiao, F., L. Yu, and L. Li. Estimating Impact of Nonrecurring Congestion on Vehicle Emissions. Presented at 86th Annual Meeting of the Transportation Research Board, Washington, D.C., 2007.
23. Redman, T. C., and A. Blanton. *Data Quality for the Information Age*. Artech House, Inc., Norwood, MA, 1997.
24. Li, Q., F. Qiao, and L. Yu. A Machine Learning Approach for Light-Duty Vehicle Idling Emission Estimation based on Real Driving and Environmental Information. *Journal of Environment Pollution and Climate Change*, Vol. 1, No.106, 2016, p. 2.
25. Fuller, J. *How Idle-Stop Systems Work*. https://auto.howstuffworks.com/fuel-efficiency/hybrid-technology/idle-stop-system.htm. Accessed November 15, 2017.
26. Nice, K. *How Torque Converters Work*. https://auto.howstuffworks.com/auto-parts/towing/towing-capacity/information/torque-converter1.htm. Accessed November 15, 2017.
27. Mitra, S. *Digital Signal Processing Applications*. http://www.cems.uvm.edu/~gmirchan/classes/EE275/Mitra_4/applications/Applications2.pdf. Accessed November 14, 2017.
28. Qiao, F., H. Liu, and L. Yu. Incorporating Wavelet Decomposition Technique to Compress TransGuide Intelligent Transportation System Data. *Transportation Research Record: Journal of the Transportation Research Board*, 2006. 1968: 63–74.
29. Stein, J. Y. *Digital Signal Processing: A Computer Science Perspective*. Wiley, Hoboken, NJ, 2000, p. 115.
30. Cerna, M., and A. F. Harvey. The Fundamentals of FFT-Based Signal Analysis and Measurement. *National Instruments, Junho*, Vol. 54, 2000. http://www.lumerink.com/courses/ece697/docs/Papers/The%20Fundamentals%20of%20FFT-Based%20Signal%20Analysis%20and%20Measurements.pdf
31. Welch, P. D. The Use of Fast Fourier Transform for the Estimation of Power Spectra: A Method Based on Time Averaging over Short, Modified Periodograms. *IEEE Transactions on Audio and Electroacoustics*, Vol. 15, No. 2, 1967, pp. 70–73.
32. Enochson, L. D., and R. K. Otnes. *Programming and Analysis for Digital Time Series Data*. U.S. Dept. of Defense, Shock and Vibration Info. Center, Washington, D.C., 1968, p. 142.
33. Smith, S. W. *The Scientist and Engineer's Guide to Digital Signal Processing. Chapter 20: Chebyshev Filters*, 1st ed. California Technical Publishing, 1997, pp. 333–342. http://www.analog.com/media/en/technical-documentation/dsp-book/dsp_book_Ch20.pdf
34. Butterworth, S. On the Theory of Filter Amplifiers. *Wireless Engineer*, Vol. 7, No. 6, 1930, pp. 536–541. http://www.gonascent.com/papers/butter.pdf
35. Franco, V., M. Kousoulidou, M. Muntean, L. Ntziachristos, S. Hausberger, and P. Dilara, Road Vehicle Emission Factors Development: A Review. *Atmospheric Environment*, Vol. 70, 2013, pp. 84–97.

The Standing Committee on Transportation and Air Quality (ADC20) peer-reviewed this paper (18-00316).

Article

TRR
JOURNAL OF THE TRANSPORTATION RESEARCH BOARD

Transportation Research Record
2018, Vol. 2672(25) 46–56

DOI: 10.1177/0361198118780826
journals.sagepub.com/home/trr
SAGE

Characteristics Analysis and Modeling of Emissions for Bus with Liquefied Natural Gas Fuel System in Real World Driving

Yingjiu Pan[1], Shuyan Chen[1], Fengxiang Qiao[2], Bin Zhang[1], and Saisai Li[1]

Abstract

To estimate accurately the emission rates of buses fueled with a liquefied natural gas (LNG) system, a method capturing the emission characteristics of LNG buses is proposed. In previous studies, vehicle specific power (VSP), a good indicator for instantaneous power demand on engines, has been mostly used in analyzing the emission characteristics of traditional vehicles such as passenger cars, diesel buses and heavy-duty trucks, with less effort on new energy buses, especially LNG buses. Based on the characteristic analyses in this study, it does not accurately describe the emission characteristics of LNG buses if the operating condition is classified only according to VSP. In this paper, we identify that the factors significantly affecting emissions are both speed and VSP using correlation analysis. Then a two-dimensional algorithm is proposed to divide operating conditions into several operating bins based on speed and VSP, and the validity of this method is verified by comparing the measured values of emissions between a training dataset and a test dataset. Based on the analysis of emission characteristics in two-dimensional operating bins, a regression model for estimating emission rates of LNG buses is established. The feasibility and accuracy of the model are validated by using a test dataset. The results indicate that all the relative errors between measured values and estimated values are within 16%, and the values of normalized mean square error (NMSE) are close to zero, indicating that the model established based on both speed and VSP can be applied to the estimation of LNG bus emissions.

With the development of urban transportation, the exhaust emissions of vehicles are increasingly becoming important factors affecting our environment. Public transit is noted not only for its large passenger carrying capacity, but also for high transport efficiency and economic competitiveness. However, there are many harmful pollutants, such as carbon monoxide (CO), nitrogen oxides (NO_X), and others in the exhaust emissions from buses in real-world driving. In recent years, liquefied natural gas (LNG) buses have been gaining popularity in many cities. Although LNG buses are more environmentally friendly than conventional diesel buses, they produce many harmful gases, as well as greenhouse gases. It is, therefore, necessary to study the emission characteristics of LNG buses, and their significance in efforts to reduce traffic pollution and to protect the environment.

In recent years, emission characteristics analyses and estimation models using vehicle specific power (VSP)-based methods have developed rapidly. VSP was proposed to describe the instantaneous engine power per unit mass of a vehicle, taking the power needed to overcome rolling and aerodynamic resistance into account (*1*). VSP contains the information of speed, acceleration, mass of vehicles, and road grade, which can be identified as an explanatory variable significantly correlated with exhaust emissions. VSP-based methods have, therefore, been well accepted and are becoming more and more popular for estimating the exhaust emissions of vehicles in real-world driving (*2*).

In fact, several studies have been carried out to assess the relationship between VSP and emission rates (*3–5*). VSP-based methods were applied not only to analyze the differences in emission characteristics for different vehicles or road conditions, but also to compare average emission rates for E85 versus gasline (*6*, *7*). However,

[1]School of Transportation, Southeast University, Jiangsu, China
[2]Department of Transportation Studies, Texas Southern University, Houston, TX

Corresponding Author:
Address correspondence to Shuyan Chen: chenshuyan@seu.edu.cn

these studies mostly focused on passenger cars (*8*), conventional diesel buses (*9*, *10*), and heavy-duty trucks (*11*), with less effort made to analyze the emission characteristics of new energy buses, especially LNG buses.

Based on previous studies, the data on the driving behavior of three LNG buses running on bus line 51 in Zhenjiang, China were collected. The key parameters significantly affecting emission rates were then identified and the relationship between them and the emissions of four pollutants is explored. After that, a method is proposed to divide the operating condition into several operating bins and analyze the emission characteristics in each bin. Furthermore, a multivariate nonlinear regression model is established based on identified key parameters to estimate the emissions. Finally, the feasibility of the estimation model is validated using a test set, indicating that the method proposed in this paper is feasible and effective in estimating exhaust emissions for LNG buses in real-world driving.

Literature Review

Many studies have been conducted on the characteristic analysis and modeling of vehicle emissions all over the world. For the research object examined in this study, our commentary focuses on the three aspects including studies on the factors that influence vehicle emissions, studies on emissions modeling of vehicles on urban roads, and studies on the emission characteristics of natural gas buses. The research results and the empirical findings on these three aspects are briefly summarized.

Studies on the Factors Influencing Vehicle Emissions

According to existing studies, the main factors influencing vehicle emissions are speed, acceleration, deceleration, idling, vehicle mass, vehicle types, and so forth. Furthermore, different traffic conditions such as road grade, road types, and the control modes of intersections could also affect vehicle emissions. Zhang et al. (*12*) evaluated the CO_2 emission characteristics of buses with different fuel systems including conventional diesel, natural gas, and diesel hybrid under different road conditions. They found that a bus's speed is the most significant factor affecting fuel consumption, and that buses with different fuel systems have significantly different emission characteristics. Boroujeni et al. (*8*) quantified the impact of road grade on the emission rates of vehicles and Wyatt et al. (*13*) analyzed the impact of road grade on CO_2 emissions from passenger cars in real-world driving and proposed a road-grade estimation methodology called light detection and ranging (LiDAR)–geographic information system (GIS). This found that over or underestimation of CO_2 emission could be caused if the road grade was neglected in real-world driving. Yu et al. (*2*) took the impact of passenger load into consideration to evaluate the emissions of transit buses. They found that the influence of passenger load on the emission rates became significant when the buses were traveling at a relatively high speed (30 km/h or above).

Based on the research results presented above, speed, acceleration, and deceleration are key parameters influencing emissions. Moreover, emissions will be over or underestimated if road grade and the passenger load factor for buses are neglected. In our study, we, therefore, obtained road-grade information by using GPS devices and acquired the passenger load factor by recording the number of passengers boarding and alighting from buses at stops.

Studies on Emissions Modeling of Vehicles on Urban Roads

As VSP contains information on vehicle speed, acceleration, road grade, and vehicle mass including both the net weight and passengers load, it is often identified as an explanatory variable for estimating vehicle emissions. Frey et al. (*3*, *5*) and Zhai et al. (*4*) evaluated the relationship between the VSP and vehicle emissions. The VSP-based method was then developed for emission estimation and the validity was verified. Wang et al. (*14*) developed a VSP-based model to estimate bus emissions based on datasets collected from China. Song et al. (*15*) proposed a mathematical model based on VSP distribution for the fuel consumption estimation. Based on this model, Wu et al. (*16*) proposed a method to estimate fuel consumption by integrating VSP and controller-area network bus technology. Wang et al. (*17*) developed a convex second-order polynomial fuel consumption model for conventional diesel and hybrid-electric buses, the model was validated by comparing the estimation with in-field measurements and predictions from CMEM, MOVES, and the concave VT-CPFM model.

Studies on the Emission Characteristics of Natural Gas Buses

To reduce the pollution to the environment caused by transportation, alternative fuels have to be found to replace hydrocarbons such as diesel and petrol. Natural gas is one alternative; it is widely available and renewable, and offers greenhouse gas reductions. It produces fewer emissions compared with other traditional and alternative fuels (*18*). Furthermore, in the transportation field, natural gas can be used in the form of compressed natural gas (CNG) or LNG.

CNG is made by compressing natural gas to a pressure of 200–250 kg/cm^2. LNG is generated by cooling natural gas to −162°C, at which point it becomes a

Table 1. Vehicle Parameters of Tested Buses

No.	Vehicle length (m)	Engine displacement (L)	Trip length (km)	Fuel type	Engine type	Emission standard
1	12	5.9	28,368	LNG	WECHAI WP6NG240E50	GB11340-2005
2	12	5.9	31,252	LNG	WEICHAI WP6NG240E50	GB11340-2005
3	12	5.9	35,586	LNG	WEICHAI WP6NG240E50	GB11340-2005

liquid. The density of LNG is 435 kg/m^3 as compared with 175 kg/m^3 for CNG at 200 bar, indicating that an LNG powered vehicle can travel up to 2.4 times the distance of a CNG vehicle for a given capacity fuel tank (*19*).

Owing to the payload restrictions as well as weight of the cylinders required for a suitable range, CNG is mainly used in light-duty vehicles. By contrast, LNG offers the possibility of using natural gas fuel for heavy-duty road transport applications such as buses due to its higher energy density, but it is only used rarely in light-duty vehicles due to its volatility (*20, 21*). To the best of the authors' knowledge, there is relatively little analysis and modeling done of emissions from LNG buses in the real-world driving, especially in the road conditions in China.

Data Description

Source of Data

The Portable Emissions Measurement System (PEMS) including SEMTECH-DS and SEMTECH-EFM3, were utilized to collect emission rates data from LNG buses. This PEMS uses a non-dispersive infrared sensor for CO and CO_2 measurement, and a nondispersive ultra-violet analyzer to measure NO and NO_2 separately and simultaneously (*2*). A global positioning system receiver model GPS 16-HVS was employed to track the route, speed, and elevation of the bus under test.

The field data collection was conducted from three LNG buses running on bus line 51 in Zhenjiang, China. Zhenjiang is a prefecture-level city in Jiangsu Province with an urban population of 3.2 million, and located in a transition zone of the plains and hills in the east of China, meaning that road grade has to be considered in this study. There were more than 1,000 buses running daily on 111 bus lines in the public transport system of the city in 2016. With the gradual promotion of new energy buses, it is important to study the emission performance of LNG buses.

Features of Data

The vehicle parameters of the buses tested are described in Table 1. The buses investigated were operated in normal mode with their air-conditioning systems switched off. The routes traveled by the buses are shown in Figure 1. The emission data were collected at the peak and off-peak hours on three business days, and the passengers were counted when boarding and alighting from the bus at stops. Furthermore, the tests were conducted under normal and similar weather conditions.

The measured data contains three parts: (a) emission data recorded by PEMS, (b) driving characteristics data recorded by GPS, and (c) passenger load data recorded by investigators. The GPS device acquires a position datum every 2 seconds, and the PEMS device records an emission datum every 2 or 3 seconds. To apply the measured data to the study of this paper effectively, the linear interpolation method is employed to get the second-by-second driving characteristics data and emissions data, and then the three parts of data are fused into one dataset.

For a better understanding of the characteristics of the measured data, the descriptive statistics are analyzed. The mean, maximum, and minimum values are calculated respectively to describe the degree of concentration of each variable, and the degree of dispersion of each variable is described using the standard deviation. Furthermore, to describe the distribution of each variable, both skewness and kurtosis values are calculated (Table 2).

Methodology

Characteristics of LNG Buses

The force analysis of the LNG bus being tested is illustrated in Figure 2. According to the definition of VSP, the calculation formula can be described as Equation 1:

$$\begin{aligned} VSP &= \frac{F_t v}{m} = \frac{(F_f + F_w + F_i + F_j)v}{m} \\ &= \frac{\left[mgf + 0.5\rho_a C_D A v^2 + mg\sin\alpha + (1+\varepsilon_i)ma\right]v}{m} \\ &= v[gf + g\sin\alpha + a(1+\varepsilon_i)] + 0.5\rho_a C_D A v^3/m \end{aligned} \tag{1}$$

where F_t is tractive force of the bus (N); v is driving speed (m/s) and m is the total mass of the bus including both the net weight and passenger load (kg); F_f, F_w, F_i, F_j,

Table 2. Descriptive Statistics Analysis of Tested Buses

Variables	Min	Max	Mean	Std Dev	Skewness		Kurtosis	
					Values	S.E.	Values	S.E.
CO (g/s)*100	0.118	29.532	10.614	3.456	0.049	0.036	1.552	0.073
CO_2 (g/s)	0.063	24.104	15.997	2.407	−3.295	0.036	16.818	0.073
HC (g/s)*1000	0.105	55.670	14.108	7.096	1.263	0.036	1.415	0.073
NO_X (g/s)*100	1.730	43.585	22.322	13.240	−0.20	0.036	−1.55	0.073
Speed (km/h)	0	52.0	16.729	13.621	0.395	0.036	−1.002	0.073
Acceleration (m/s^2)	−10.417	3.333	0.989e^{-3}	0.533	−2.159	0.036	33.554	0.073
VSP (kw/ton)	−29.011	41.633	0.563	4.582	−0.184	0.036	5.391	0.073

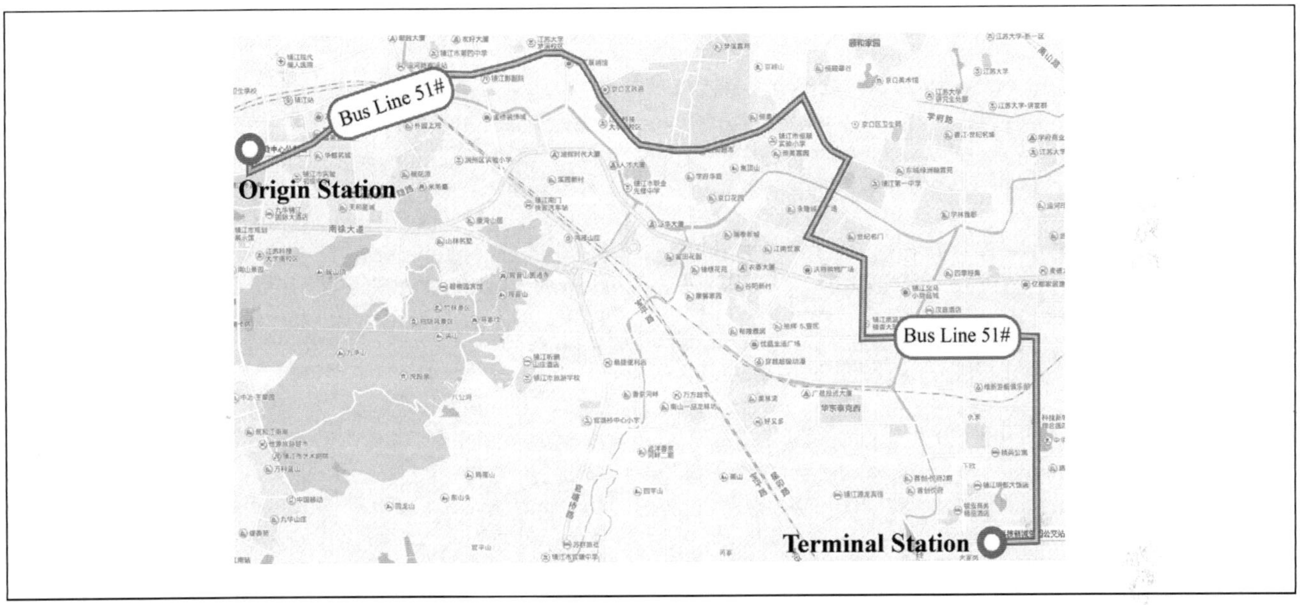

Figure 1. Experimental routes traveled by LNG buses.

represent rolling resistance, wind resistance, gradient resistance, and acceleration resistance, respectively (N); a is acceleration (m/s^2) and g is the acceleration of gravity (9.8 m/s^2); f is rolling resistance coefficient (0.00938, dimensionless) and ε_i represents mass factor (0.15); α is road grade and ρ_a is air density (1.207 kg/m^3); C_D is drag coefficient (0.6) and A is frontal area of bus (8.3 m^2).

Substituting the values of known parameters into Equation 1, the equation can be represented as below in Equation 2, where m is a variable because passenger load is constantly changing.

$$VSP = v(1.15a + 0.132 + 9.8\sin\alpha) + \frac{3.005}{m}v^3 \quad (2)$$

Due to the restriction on the conditions of the investigation, the number of passengers was recorded only when they boarded and alighted from the bus. To simplify the calculation, the passenger mass is calculated by the number of passengers multiplied by 50 kg approximately. In addition, values of road grade can be acquired by calculating the difference of elevations between two points.

The VSP distribution characteristics of testing LNG buses are shown in Figure 3. Based on the cumulative frequency of VSP obtained from LNG buses in Zhenjiang, more than 95% of the data are concentrated in the interval between −10 kw/ton and 10 kw/ton. Therefore, we mainly consider the case of VSP concentrated between −10kw/ton and 10kw/ton in this paper.

Correlation Analysis and Identification of Modeling Parameters

Qian Yu et al. (*2*) pointed out that the most fundamental factors affecting diesel bus emissions are speed, acceleration, and VSP. But for LNG buses, the key factors that

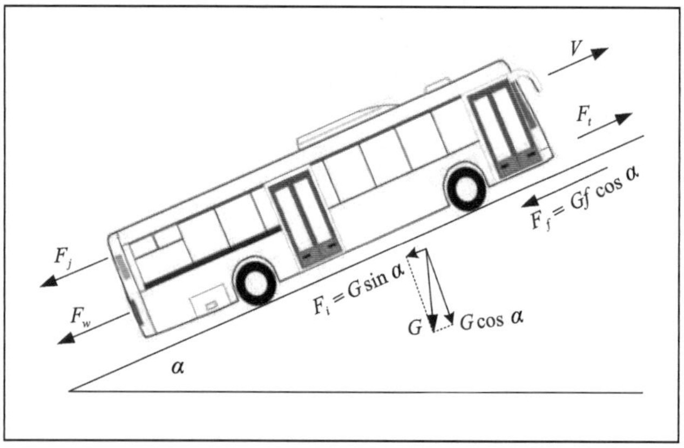

Figure 2. Force analysis of buses being tested.

could affect instantaneous emission rates have not been identified. In this section, a correlation analysis between the emission rates of four pollutants and vehicle driving parameters including speed, acceleration, and VSP is conducted.

Tables 3 and 4 show the results of Pearson and Spearman correlation analyses, respectively. The linear correlation between two variables is evaluated by Pearson correlation coefficients while the non-linear correlation between two variables is evaluated by Spearman correlation coefficients.

At the significance level of 0.05, both speed and VSP are significantly correlated with the emission rates of four pollutants. As for acceleration, there is no obvious correlation between the acceleration and the CO_2 emission rate. In addition, the Pearson correlation coefficient between acceleration and VSP is 0.697, meaning that acceleration and VSP have a strong collinearity. Therefore, speed and VSP are selected to classify operating bins and establish the emission-rates model in this paper.

Training and Test Dataset Establishment

The tested data are divided into a training dataset and a test dataset accounting for 75% and 25% of all the data, respectively. The training dataset is employed to establish the emission-rates model and the test dataset is employed to validate and evaluate the established model. To ensure that the vehicle travel data of the two datasets conform to the similar distribution, the Kolmogorov–Smirnov test (K-S test) is applied to test whether the training dataset and the test dataset are consistent with a theoretical distribution or whether there is a significant difference between the two empirical distributions.

The null hypothesis of the K-S test is that the two samples obey the same distribution. Here $F_1(x)$ and $F_2(x)$ are employed to represent the cumulative empirical distribution of the two samples, respectively. The test statistics obey the normal distribution approximately, as expressed in Equation 3.

$$Z = \max_j |D_j| \sqrt{\frac{n_1 n_2}{n_1 + n_2}} \quad (3)$$

where $D_j = F_1(X_j) - F_2(X_j)$, n_1 and n_2 are the size of the two samples, respectively. If the p-value is greater than 0.05, the null hypothesis can be accepted, which means

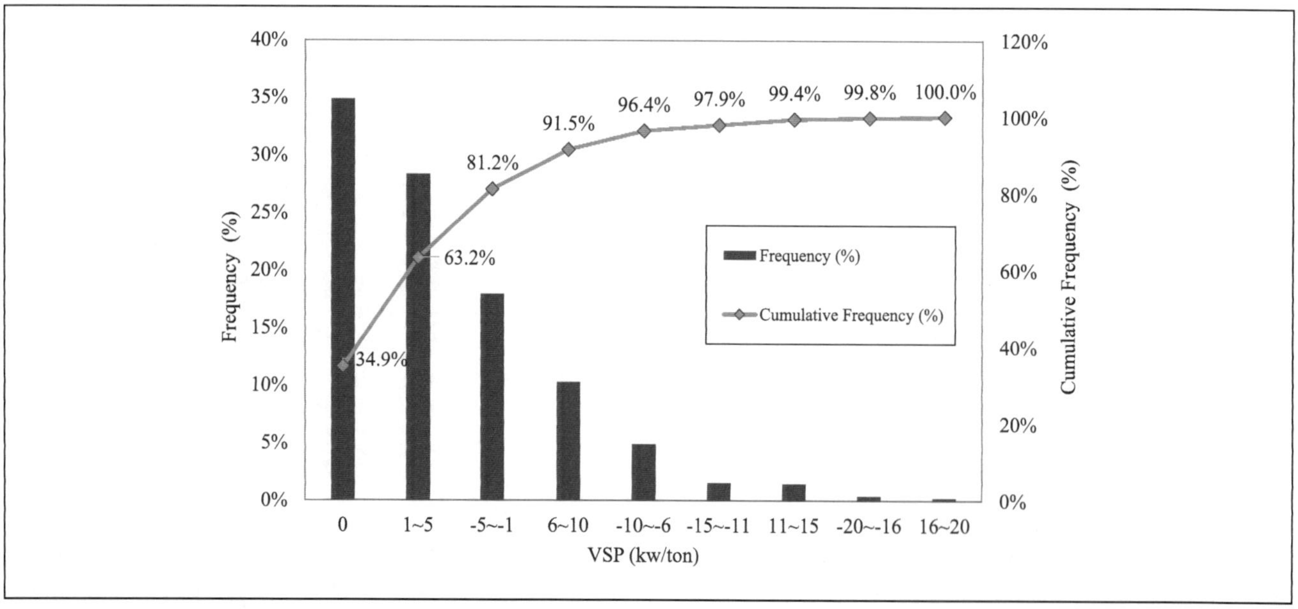

Figure 3. VSP distribution characteristics of buses being tested.

Table 3. Results of Pearson Correlation Analysis

Pollutants	Pearson	Speed	Acceleration	VSP	CO	CO_2	HC	NO_X
Speed	Correlation coefficient	1.000	0.069*	0.103*	0.058*	−0.017	−0.280*	0.605*
	p-value	–	0.000	0.000	0.000	0.258	0.000	0.000
Acceleration	Correlation coefficient	0.069*	1.000	0.697*	−0.046*	0.022	−0.073*	−0.162*
	p-value	0.000	–	0.000	0.002	0.143	0.000	0.000
VSP	Correlation coefficient	0.103*	0.697*	1.000	−0.041*	0.015	−0.101*	−0.010
	p-value	0.000	0.000	–	0.006	0.316	0.000	0.485
CO	Correlation coefficient	0.058*	-0.046*	−0.041*	1.000	0.488*	0.235*	0.034*
	p-value	0.000	0.002	0.006	–	0.000	0.000	0.021
CO_2	Correlation coefficient	−0.017	0.022	0.015	0.488*	1.000	−0.073*	0.109*
	p-value	0.258	0.143	0.316	0.000	–	0.000	0.000
HC	Correlation coefficient	−0.280*	−0.073*	−0.101*	0.235*	−0.073*	1.000	−0.430*
	p-value	0.000	0.000	0.000	0.000	0.000	–	0.000
NO_X	Correlation coefficient	0.605*	−0.162*	−0.010	0.034	0.109*	−0.430*	1.000
	p-value	0.000	0.000	0.485	0.021	0.000	0.000	–

Table 4. Results of Spearman Correlation Analysis

Pollutants	Spearman	Speed	Acceleration	VSP	CO	CO_2	HC	NO_X
Speed	Correlation coefficient	1.000	0.122*	0.160*	0.036*	0.049*	−0.359*	0.614*
	p-value	–	0.000	0.000	0.016	0.001	0.000	0.000
Acceleration	Correlation coefficient	0.122*	1.000	0.710*	−0.057*	0.007	−0.039*	−0.167*
	p-value	0.000	–	0.000	0.000	0.651	0.009	0.000
VSP	Correlation coefficient	0.160*	0.710*	1.000	−0.035*	0.038*	−0.091*	−0.036*
	p-value	0.000	0.000	–	0.017	0.011	0.000	0.016
CO	Correlation coefficient	0.036*	−0.057*	−0.035*	1.000	0.430*	0.230*	0.037*
	p-value	0.016	0.000	0.017	–	0.000	0.000	0.012
CO_2	Correlation coefficient	0.049*	0.007	0.038*	0.430*	1.000	−0.282*	0.104*
	p-value	0.001	0.651	0.011	0.000	–	0.000	0.000
HC	Correlation coefficient	−0.359*	−0.039*	−0.091*	0.230*	−0.282*	1.000	−0.439*
	p-value	0.000	0.009	0.000	0.000	0.000	–	0.000
NO_X	Correlation coefficient	0.614*	−0.167*	−0.036*	0.037*	0.104*	−0.439*	1.000
	p-value	0.000	0.000	0.016	0.012	0.000	0.000	–

the two databases obey the same distribution. Based on the results of hypothesis testing, the null hypothesis is accepted because p-values of all variables are greater than 0.05.

Method of Operating Bins Classification

Based on the results of the correlation analysis, the operating bins will be classified based on both speed and VSP in this section. As the data were collected in real-world driving, there are significant differences in emission sharing rates in both different VSP bins and speed ranges. Analytical errors may be generated when the emission model is established if the emission sharing rates of some VSP bins are too small. Therefore, the operating bins should be classified to ensure a balanced distribution of emission sharing rates.

As mentioned above, an approach for operation bins classification is proposed, which is composed of six steps as following.

Step 1: Speed interval classification. Based on the measured data, the interval of low speed, medium speed, and high speed are defined as (0, 20], (20, 40], (40, $+\infty$), respectively.

Step 2: VSP bins are shown in Table 1. The proportion of time, the emission-sharing rate, which is the ratio of emission qualities in each bin to total emission qualities, and the mean values of emission rates in each VSP bin are calculated.

Step 3: Setting the upper limit of emission-sharing rates. The upper limit of the emission-sharing rates is set to 20%. When the sum of the sharing rates between adjacent bins is greater than 20%, the two bins cannot be combined.

$$DER_{i,i+1} = |ER_i - ER_{i+1}| \tag{4}$$

$$SESR_{i,i+1} = ESR_i + ESR_{i+1} \tag{5}$$

where $DER_{i,i+1}$ is the difference of emission rates between two adjacent bins; ER_i is the emission rate of the

Table 5. Results of Dividing Operating Bins based on Both Speed and VSP

Bin	Speed range	VSP range
101	(0, 20]	[−10, −1]
102	(0, 20]	0
103	(0, 20]	[1, 4]
104	(0, 20]	[5, 10]
201	(20, 40]	[−10, −1]
202	(20, 40]	[0, 3]
203	(20, 40]	[4, 10]
301	(40, + ∞)	[−10, 10]

ith bin; ESR_i is the emission sharing rate of the ith bin; $SESR_{i,i+1}$ is the sum of emission sharing rates between the ith bin and the $(i+1)$th bin.

Step 4: Calculating the sum of emission-sharing rates and the difference in emission rates between adjacent bins respectively. If the sum of emission-sharing rates between adjacent bins is less than the upper limit and the difference of emission rates is within the acceptable range, the data between the adjacent bins can be merged into a new bin.

Step 5: Calculating the emission-sharing rate and the mean value of emission rates for the new merged bins.

$$ESR_m = ESR_k + ESR_{k+1} \quad (6)$$

$$ER_m = \frac{F_k \times ER_k + F_{k+1} \times ER_{k+1}}{F_k + F_{k+1}} \quad (7)$$

where m is the new merged bin number; F_k is the percentage of time occupied by the kth bin.

Step 6: Repeat Step 4 and Step 5 until there is no bin that can be merged.

Based on the algorithm above, a total of eight operating bins are classified in terms of VSP within three speed ranges (Table 5).

The emission characteristics of the four pollutants are analyzed according to the results of the operating bins classification (Figure 4).

Figure 4*a* shows the emission characteristics of CO in different operating bins. Although the emission sharing rate is slightly larger than the time share rate in bin 202, the emission-sharing rate is approximately equal to the time-share rate in other bins, which indicates that different values of VSP have no significant effect on CO emission rates. In low speed conditions, the emission rate of CO increases gradually with the increase of VSP, but decreases suddenly in bin 104, which may be caused by the error generated by the small data samples. In addition, the emission rates of CO at medium and high speed are generally higher than at low speed conditions, and the emission rates of CO reach the maximum at bin 202, indicating that high CO concentrations should be concerned under both medium and high speed conditions (more than 20 km/h).

Figure 4*b* shows the emission characteristics of CO_2 in different operating bins. Similar to the CO, the emission-sharing rate of CO_2 is approximately equal to the time-share rate in each bin, which indicates that different values of VSP have no significant effect on CO_2 emission rates. There is no significant variation trend about the emission rate of CO_2 in low speed conditions. Under medium speed conditions, the emission rate of CO_2 increases gradually with the increase of VSP, but at high speeds, the emission rate reduces significantly compared with medium speeds. Overall, the emissions rates of CO_2 in bin 202 and bin 203 are higher than those in other bins, indicating that high CO_2 concentrations should be concerned under medium speed conditions with VSP values greater than 0.

Figure 4*c* shows the emission characteristics of HC in different operating bins. There is a difference between the curve of the emission-sharing rate and that of time-share rate. At low speeds, the emission-sharing rate is larger than the time-share rate but at medium and high speeds, the emission-sharing rate is smaller than the time-share rate, indicating that different values of VSP have a significant effect on HC emission rates. At low speed, the emission rate curve of HC rises first and then falls with the increase of VSP, and the emission rate reaches the maximum at bin 102. At medium and high speeds, the emission-rate curve of HC is gradually declining with the increase of bins. Overall, emissions rates of HC in bin 101, 102, and 103 are higher than those in other bins, indicating that high HC concentrations should be concerned under low speed conditions with VSP values lower than 4. Therefore, bus drivers should avoid low-speed operations to reduce HC emissions in real-world driving.

Figure 4*d* shows the emission characteristics of NO_X in different operating bins. There is a significant difference between the curve of the emission sharing-rate and that of the time-share rate. At low speed, the emission-sharing rate is smaller than the time-share rate but at medium and high speeds, the emission-share rate is greater than the time-share rate, which indicates that different values of VSP have a significant effect on NO_X emission rates. At low speed, the value of the emission rate is large in bin 101 but decreases significantly in bin 102, 103, and 104. At medium speed, the values of emission rates increase significantly compared with low speeds, and there is a slight decreasing trend between bin

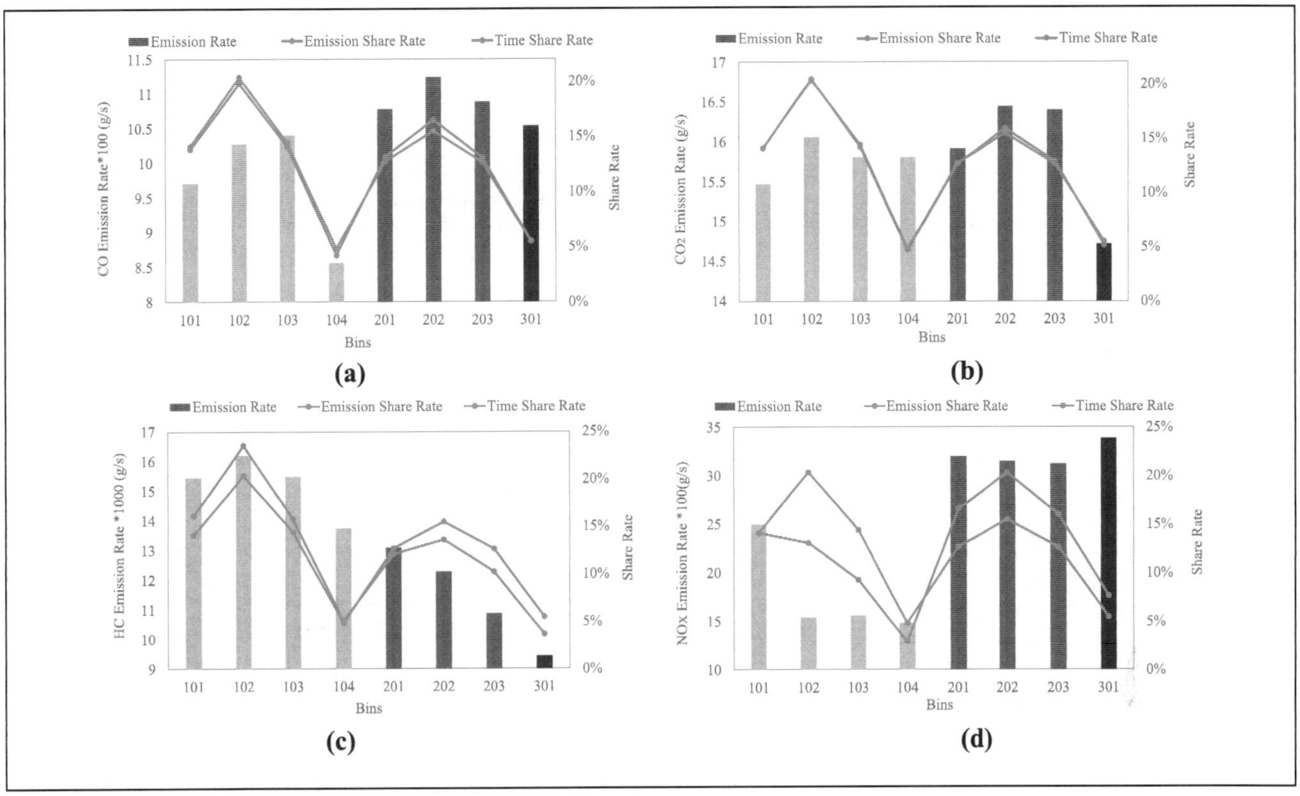

Figure 4. Emission characteristics of four pollutants: (*a*) emission characteristics of CO; (*b*) emission characteristics of CO_2; (*c*) emission characteristics of HC; (*d*) emission characteristics of NO_X.

201, 202, and 203. At high speeds, the emission rate increases slightly and reaches the maximum in bin 301. Overall, the emission rates of NO_X in bin 102, 103, and 104 are significantly lower than those in other bins, indicating that high NO_X concentrations should be concerned under low speed conditions with VSP values lower than 0 and under medium and high speed conditions.

Emission Rates Modeling

Based on the analyses above, the VSP distribution alone is not enough to describe the emission characteristics of LNG buses in real-world driving, so the emission characteristics of pollutants are analyzed according to both VSP and speed. Moreover, since the emission rates in the idling state are significantly different from those in the normal running process, the idling state should be classified as one bin separately. In this paper, the data whose speed ranges from 0 to 1 m/s are defined as the idling condition and the mean value is employed to estimate exhaust emissions in the idling condition generally. There is no significant conclusion to be drawn about emission rates in the idling condition by analyzing the emission data of the idling bin. Furthermore, the idling condition is not contained in the eight bins divided previously. Therefore, the data concerning the idling condition should be deleted when the estimation model is established.

Taking the impact of both VSP and speed on the emission rates into account, the nonlinear regression model for estimating emission rates is established based on three speed ranges which include low, medium, and high speed.

The emission-rates model at low speed, (0, 20], is shown as follows:

$$ER_{CO} = \begin{pmatrix} 153.656 - 23.551v - 2.187\,p^2 \\ + 3.338v^2 - 0.141v^3 + 0.162\,p^2 v \end{pmatrix} \times 10^{-3}$$
$$R^2 = 0.569 \tag{8}$$

$$E_{CO_2} = 16.588 + 0.072\,p - 0.131v - 3.893 \times 10^{-3}\,pv$$
$$R^2 = 0.565$$
$$+ 0.711 \times 10^{-3}\,p^2 + 4.685 \times 10^{-3} v^2 \tag{9}$$

$$E_{HC} = \begin{pmatrix} 22.284 - 5.443\,p - 2.065v + 0.803\,pv \\ + 0.189\,p^2 + 0.129\,v^2 - 0.018\,p^2 v - 0.029v^2\,p \end{pmatrix} \times 10^{-3}$$
$$R^2 = 0.666 \tag{10}$$

$$E_{NO_X} = \begin{pmatrix} 104.938 + 15.725v - 0.771\,pv \\ + 0.474\,p^2 - 0.758v^2 \end{pmatrix} \times 10^{-3} \quad R^2 = 0.853 \tag{11}$$

The emission-rates model at medium speed, (20, 40], is shown as follows:

$$ER_{CO} = \begin{pmatrix} -466.34 - 7.066\,p + 28.98v \\ -0.161\,p^2 - 0.011v^3 + 0.008v^2\,p \end{pmatrix} \times 10^{-3} \quad R^2 = 0.511 \tag{12}$$

$$ER_{CO_2} = -13.06 + 0.137\,p + 1.545v - 0.004\,pv$$

$$R^2 = 0.606 \tag{13}$$

$$-0.13\,p^2 - 0.632 \times 10^{-3} v^3 + 4.465 \times 10^{-3}\,p^2 v$$

$$ER_{HC} = \begin{pmatrix} -100.18 - 2.614\,p + 7.872v \\ -0.021\,p^2 - 0.137v^2 + 0.085vp \end{pmatrix} \times 10^{-3}$$

$$R^2 = 0.566 \tag{14}$$

$$ER_{NO_X} = \begin{pmatrix} -153.396 + 5.764\,p + 31.814v - 0.719\,p^2 \\ -0.533v^2 + 0.019\,p^3 + 0.024\,p^2 v - 0.01v^2\,p \end{pmatrix} \times 10^{-3}$$

$$R^2 = 0.764 \tag{15}$$

The emission-rates model at high speed, (40, $+\infty$), is shown as follows:

$$ER_{CO} = \begin{pmatrix} 1817 + 26.292\,p - 57.635v \\ -0.067\,p^3 + 0.01v^3 - 0.012v^2\,p \end{pmatrix} \times 10^{-3}$$

$$R^2 = 0.634 \tag{16}$$

$$ER_{CO_2} = \begin{pmatrix} 605.4 + 37.169\,p - 15.744v \\ -0.045\,p^3 + 0.003v^3 - 0.017v^2\,p \end{pmatrix} \times 10^{-1}$$

$$R^2 = 0.513 \tag{17}$$

$$ER_{HC} = \begin{pmatrix} 358.3 + 5.08\,p - 15.166v - 0.102\,pv \\ -0.745\,p^2 + 0.163v^2 - 0.013\,p^3 + 0.018\,p^2 v \end{pmatrix} \times 10^{-3}$$

$$R^2 = 0.732 \tag{18}$$

$$ER_{NO_X} = \begin{pmatrix} -3385 + 25.444\,p + 157.429v - 0.718\,pv \\ + 24.407\,p^2 - 1.65v^2 - 0.141\,p^3 - 0.555\,p^2 v \end{pmatrix} \times 10^{-3}$$

$$R^2 = 0.516 \tag{19}$$

where ER_{CO}, ER_{CO_2}, ER_{HC}, ER_{NO_X} are emission rates of four pollutants (g/s); p is Vehicle Specific Power (VSP) (kw/t); v is speed of LNG buses (km/h).

From the model presented above, it is known that all values of R^2 are greater than 0.5, indicating that the established two-dimension model presents a good fit with the increasing speed level.

Model Verification

The values of R^2 indicate that the regression model of exhaust emission rates based on both speed and VSP established previously fitted the training data well. To verify the applicability of the regression model, it is evaluated and validated based on the test dataset. First, the operation condition is divided into eight bins based on the operating bins classification method proposed previously. Second, the measured emission-rates values are acquired by calculating the mean values of second-by-second data in the test dataset. Third, when the data in each bin are substituted into the established regression model, the estimated emission rates of four pollutants can be calculated. The relative errors between measured values and estimated values are calculated and analyzed. Finally, we use NMSE (normalized mean square error) to evaluate the mean relative discrete degree of each pollutant emission rates, as shown in Equation 20.

$$NMSE = \frac{1}{n}\sum_{i}^{n}\frac{(C_{mi} - C_{ei})^2}{\overline{C_m} \cdot \overline{C_e}} \tag{20}$$

where n is the number of bins; C_{mi} and C_{ei} are the measured value and estimated value of emission rate in the ith bin, respectively; $\overline{C_m}$ and $\overline{C_e}$ are mean values of measured data and estimated data in all bins. In an accurate model, the NMSE should be close to 0, and $NMSE < 0.5$ is the acceptable limit according to previous studies (*20*).

Relative errors exist in the estimated values and measured values of exhaust emission rates, but the maximum value is 16% and most relative errors are within 10% (Figure 5). The NMSE of CO, CO_2, HC, NO_X is 0.012, 0.4×10^{-3}, 0.004, and 0.003 respectively, which indicates that the regression model above can estimate the emission rates of LNG buses effectively.

Conclusion and Discussion

This study was designed to estimate emissions of LNG buses in real-world driving. On the basis of a large amount of real world emissions and traffic data, we analyzed and evaluated the key parameters significantly affecting the emission rates of LNG buses. The method for estimating practical emission rates of LNG buses was proposed. The proposed approach in this paper can help transportation planners and decision-makers to evaluate the environmental pollution of LNG buses in road networks. It can also be a good reference for traffic management. The main findings in this study can be summarized as follows.

1. The characteristics of real-world LNG buses based on VSP bin distribution were analyzed and

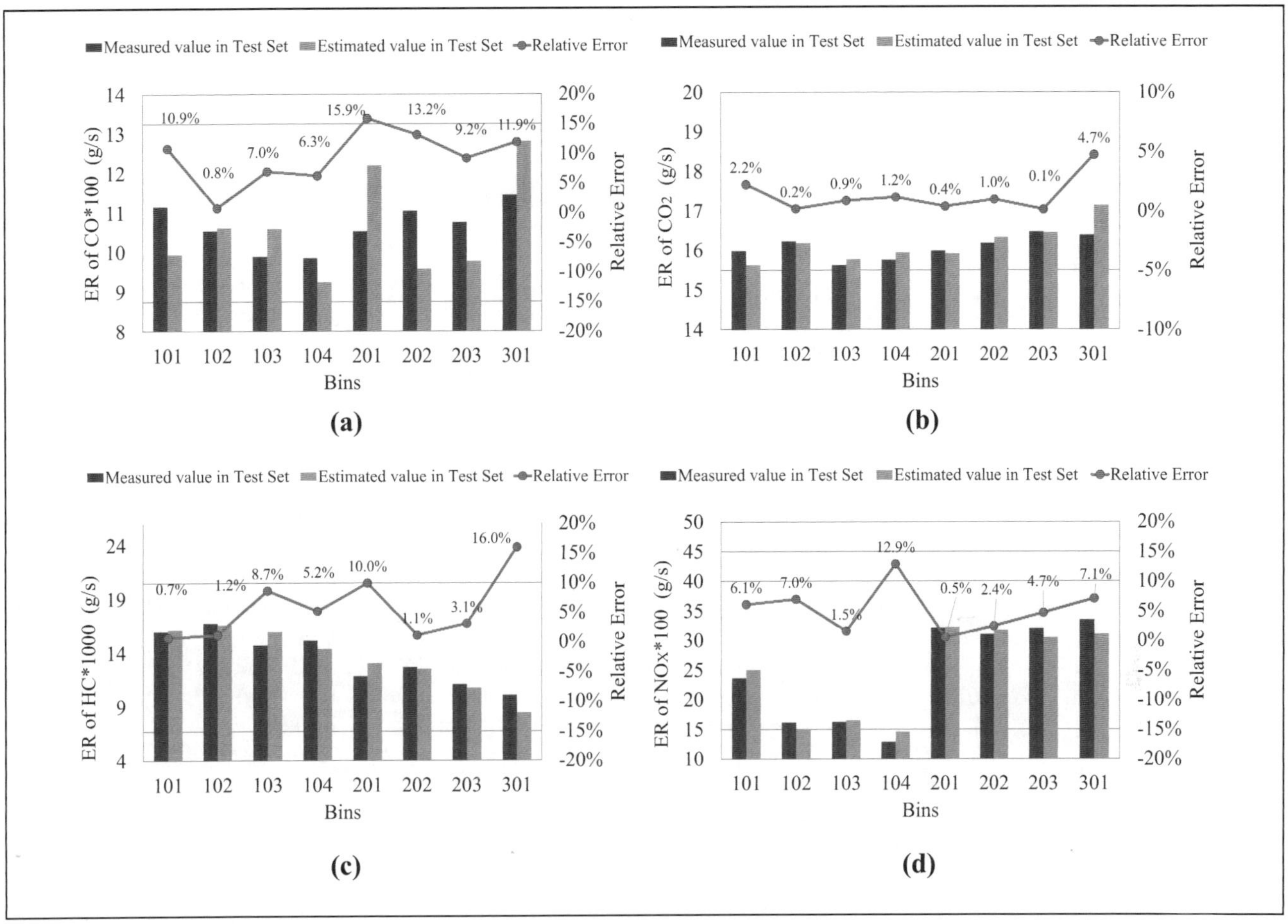

Figure 5. Comparison results between measured values and estimated values: (*a*) comparison of CO emission rates; (*b*) comparison of CO_2 emission rates; (*c*) comparison of HC emission rates; (*d*) comparison of NO_X emission rates.

showed that the values of VSP mainly range from −10 kw/ton to 10 kw/ton.

2. In the real-world driving, only the VSP bin distribution fails to reflect the emission characteristics of LNG buses accurately, both VSP and speed are significantly correlated with exhaust emissions of the four pollutants.
3. An approach of operating bins classification including six steps was proposed based on both VSP and speed. The emission characteristics of the four pollutants were analyzed and evaluated according to the classified operating bins. We found that CO emissions will be increased significantly under both medium and high speed conditions; high CO_2 concentrations should be concerned under medium speed condition with VSP values greater than 0; HC emissions will be increased significantly under low speed conditions with VSP values lower than 4; high NO_X concentrations should be concerned under low speed condition with VSP values lower than 0 and under medium and high speed conditions. Therefore, the instantaneous emission rates of the four pollutant show different relationships with both VSP and speed.
4. A regression model for emission rates estimation was built based on both VSP and speed, and the model was verified by using relative error and NMSE. All the relative errors are within 16% of measured and estimated values, and the values of NMSE are close to 0, indicating that the model can estimate the emission rates of LNG buses effectively.

Further research should focus on the following three aspects. First, the datasets used in this paper only include three LNG buses running on the same bus line. A larger sample size including different bus lines would be helpful to further verify the method proposed in this paper. Second, the values of emission rates are low and mean values are usually used to estimate the emission rates in the idling condition, so emission rates in the idling condition were not taken into account in this study. Finally, other types of buses such as diesel buses and gas-electric hybrid buses will be tested and the characteristics will be analyzed and compared between different types of bus, and a model for emissions estimation will be established for other types of buses.

Acknowledgments

This work is supported by the project of National Natural Science Foundation of China (No.61374195) and the Postgraduate Research & Practice Innovation Program of Jiangsu Province (No. KYCX17_0143). The authors would also be thankful to Prof. Zhirui Ye and Prof. Tiezhu Li who provided the data used in our experiments.

References

1. Jiménez-Palacios, J. L. *Understanding and Quantifying Motor Vehicle Emissions with Vehicle Specific Power and TILDAS Remote Sensing*. PhD dissertation. Massachusetts Institute of Technology, Cambridge, 1999.
2. Yu, Q., T. Li, and H. Li. Improving Urban Bus Emission and Fuel Consumption Modeling by Incorporating Passenger Load Factor for Real World Driving. *Applied Energy*, Vol. 161, 2016, pp. 101–111.
3. Frey, H. C., N. M. Rouphail, H. Zhai, T. L. Farias, and G. A. Gonçalves. Comparing Real-world Fuel Consumption for Diesel- and Hydrogen-fueled Transit Buses and Implication for Emissions. *Transportation Research Part D: Transport and Environment*, Vol. 12, No. 4, 2007, pp. 281–291.
4. Zhai, H., H. C. Frey, and N. M. Rouphail. A Vehicle-Specific Power Approach to Speed- and Facility-specific Emissions Estimates for Diesel Transit Buses. *Environmental Science and Technology*, Vol. 42, No. 21, 2008, pp. 7985–7991.
5. Frey, H. C., N. M. Rouphail, and H. Zhai. Link-based Emission Factors for Heavy-duty Diesel Trucks Based on Real-world Data. *Transportation Research Record: Journal of the Transportation Research Board*, 2008. 2058: 23–32.
6. Yazdani Boroujeni, B., and H. C. Frey. Road Grade Quantification Based on Global Positioning System Data Obtained from Real-world Vehicle Fuel Use and Emissions Measurements. *Atmospheric Environment*, Vol. 85, 2014, pp. 179–186.
7. Zhai, H., H. C. Frey, N. M. Rouphail, G. A. Gonçalves, and T. L. Farias. Comparison of Flexible Fuel Vehicle and Life-cycle Fuel Consumption and Emissions of Selected Pollutants and Greenhouse Gases for Ethanol 85 Versus Gasoline. *Journal of the Air & Waste Management Association*, Vol. 59, No. 8, 2009, pp. 912–924.
8. Jaikumar, R., S. M. Shiva Nagendra, and R. Sivanandan. Modeling of Real Time Exhaust Emissions of Passenger Cars Under Heterogeneous Traffic Conditions. *Atmospheric Pollution Research*, Vol. 8, No. 1, 2016, pp. 1–9.
9. Yu, Q., and T. Li. Evaluation of Bus Emissions Generated Near Bus Stops. *Atmospheric Environment*, Vol. 85, 2014, pp. 195–203.
10. Liu, Z., Y. Ge, K. C. Johnson, A. N. Shah, J. Tan, C. Wang, and L. Yu. Real-world Operation Conditions and On-road Emissions of Beijing Diesel Buses Measured By Using Portable Emission Measurement System and Electric Low-pressure Impactor. *Science of the Total Environment*, Vol. 409, No. 8, 2011, pp. 1476–1480.
11. Abdelmegeed, M. A. E., and H. Rakha. Heavy-Duty Diesel Truck Emissions Modeling. *Transportation Research Record: Journal of the Transportation Research Board*, 2017. 2627: 26–35.
12. Zhang, S., Y. Wu, H. Liu, R. Huang, L. Yang, Z. Li, L. Fu, and J. Hao. Real-world Fuel Consumption and CO_2 Emissions of Urban Public Buses in Beijing. *Applied Energy*, Vol. 113, 2014, pp. 1645–1655.
13. Wyatt, D. W., H. Li, and J. E. Tate. The Impact of Road Grade on Carbon Dioxide (CO_2) Emission of a Passenger Vehicle in Real-world Driving. *Transportation Research Part D: Transport and Environment*, Vol. 32, 2014, pp. 160–170.
14. Wang, H., L. Fu, Y. Zhou, and H. Li. Modelling of the Fuel Consumption for Passenger Cars Regarding Driving Characteristics. *Transportation Research Part D: Transport and Environment*, Vol. 13, No. 7, 2008, pp. 479–482.
15. Song, G., and L. Yu. Characteristics of Low-speed Vehicle-specific Power Distributions on Urban Restricted-access Roadways in Beijing. *Transportation Research Record: Journal of the Transportation Research Board*, 2011. 2233: 90–98.
16. Wu, Y., L. Yu, G. Song, and L. Xu. Feasibility Study of Fuel Consumption Prediction Model by Integrating Vehicle-Specific Power and Controller Area Network Bus Technology. *Transportation Research Record: Journal of the Transportation Research Board*, 2013. 2341: 66–75.
17. Wang, J., and H. A. Rakha. Convex Fuel Consumption Model for Diesel and Hybrid Buses. *Transportation Research Record: Journal of the Transportation Research Board*, 2017. 2647: 50–60.
18. Lin, W., N. Zhang, and A. Gu. LNG (Liquefied Natural Gas): A Necessary Part in China's Future Energy Infrastructure. *Energy*, Vol. 35, No. 11, 2010, pp. 4383–4391.
19. Kumar, S., H. T. Kwon, K. H. Choi, W. Lim, J. H. Cho, K. Tak, and I. Moon. LNG: An Eco-Friendly Cryogenic Fuel for Sustainable Development. *Applied Energy*, Vol. 88, No. 12, 2011, pp. 4264–4273.
20. Wang, A., Y. Ge, J. Tan, M. Fu, A. N. Shah, Y. Ding, H. Zhao, and B. Liang. On-Road Pollutant Emission and Fuel Consumption Characteristics of Buses in Beijing. *Journal of Environmental Sciences*, Vol. 23, No. 3, 2011, pp. 419–426.
21. Ab, S., and A. Bassi. Liquefied Natural Gas (LNG) as Fuel for Road Heavy Duty Vehicles Technologies and Standardization. *SAE 10th International Conference on Engines & Vehicles*, 2011.

The Standing Committee on Transportation and Air Quality (ADC20) peer-reviewed this paper (18-01370).

TRR
JOURNAL OF THE TRANSPORTATION RESEARCH BOARD

Article

Transportation Research Record
2018, Vol. 2672(25) 57–66
© National Academy of Sciences:
Transportation Research Board 2018
Article reuse guidelines:
sagepub.com/journals-permissions
DOI: 10.1177/0361198118780682
journals.sagepub.com/home/trr
SAGE

The Impact of Different Validation Datasets on Air Quality Modeling Performance

Haneen Khreis[1,2,3,4,5], Kees de Hoogh[6,7], Josias Zietsman[1], and Mark J. Nieuwenhuijsen[2,3,4]

Abstract

Many studies rely on air pollution modeling such as land use regression (LUR) or atmospheric dispersion (AD) modeling in epidemiological and health impact assessments. Generally, these models are only validated using one validation dataset and their estimates at select receptor points are generalized to larger areas. The primary objective of this paper was to explore the effect of different validation datasets on the validation of air quality models. The secondary objective was to explore the effect of the model estimates' spatial resolution on the models' validity at different locations. Annual NO_x and NO_2 were generated using a LUR and an AD model. These estimates were validated against four measurement datasets, once when estimates were made at the exact locations of the validation points and once when estimates were made at the centroid of the 100m×100m grid in which the validation point fell. The validation results varied substantially based on the model and validation dataset used. The LUR models' R^2 ranged between 21% and 58%, based on the validation dataset. The AD models' R^2 ranged between 13% and 56% based on the validation dataset and the use of constant or varying background NO_x. The validation results based on model estimates at the exact validation site locations were much better than those based on a 100m×100m grid. This paper demonstrated the value of validating modeled air quality against various datasets and suggested that the spatial resolution of the models' estimates has a significant influence on the validity at the application point.

Since it is often not possible to measure air pollution exposures for epidemiological and health impact assessments, many studies rely on less costly and more practical approaches such as exposure modeling for the large populations. Land use regression (LUR) modeling (*1*, *2*) and atmospheric dispersion (AD) modeling (*3*, *4*) are two common methods used to obtain estimates of air pollution exposures for relatively large areas and numbers of people. As discussed in depth in Khreis and Nieuwenhuijsen (*5*), these two exposure modeling methods are fundamentally different and vary in their spatial and temporal resolution, specificity to traffic, advantages, and disadvantages.

The LUR method is an empirical method. It uses least squares regression to combine air pollution measurements at certain locations with geographic information system (GIS)-based predictor variables that reflect the pollutant sources (for example road, traffic, population or building density, green space, etc.). As such, a prediction model applicable to nonmeasured locations, for example residential addresses of cohort members, is built (*5*). LUR models do not require fundamental understanding of the underlying emission and dispersion processes. AD models, on the other hand, rely on mathematical formula and an understanding of underlying processes to predict air pollution exposure estimates (*6*). The correlation between and the performance of both LUR and AD models is often similar but can vary from poor to very good (*7–9*).

An important aspect that has received less attention in evaluating the performance of both LUR and AD models is the potential influence of the validation datasets. Generally, these air quality models are validated against one measured dataset only (*9*, *10*). The validation results vary widely by model, pollutant, and study area (*9*, *10*). Yet, it is unknown whether the validation results would also vary based on the validation dataset used. Furthermore, in health impact assessments, estimates of air quality models at select receptor points are extrapolated and assumed to apply to larger areas and populations. The impact of this extrapolation on the estimates' validity at the application points is under studied.

[1]Texas A&M Transportation Institute (TTI) and Center for Advancing Research in Transportation Emissions, Energy, and Health (CARTEEH), College Station, TX
[2]ISGlobal, Centre for Research in Environmental Epidemiology, Barcelona, Spain
[3]Universitat Pompeu Fabra, Barcelona, Spain
[4]CIBER Epidemiologia y Salud Publica, Madrid, Spain
[5]Institute for Transport Studies, University of Leeds, UK
[6]Swiss Tropical and Public Health Institute, Basel, Switzerland
[7]University of Basel, Basel, Switzerland

Corresponding Author:
Address correspondence to Haneen Khreis: H-Khreis@tti.tamu.edu

Table 1. Summary Statistics of Adjusted Measured NO_2 and NO_x Concentrations at the 41 ESCAPE Sites

ESCAPE site type	Rural background	Urban background	Traffic
Definition	Measurements in the smaller towns and villages of the cohort	A site with fewer than 3,000 vehicles per day passing within a 50-m radius	A site in a major road carrying at least 10,000 vehicles per day
Number	2	24	15
Average adjusted NO_2 (μg/m^3)	16.9	24.1	29.7
Average adjusted NO_x (μg/m^3)	23.6	38.4	59.4
Average NO_2/NO_x ratio (μg/m^3)	0.72	0.63	0.50
Minimum adjusted NO_2 (μg/m^3)	16.7	17.2	19.4
Maximum adjusted NO_2 (μg/m^3)	17.0	34.1	44.9
Minimum adjusted NO_x (μg/m^3)	22.4	25.1	33.6
Maximum adjusted NO_x (μg/m^3)	24.7	59.1	110.5

Note: NO_2 = nitrogen dioxide; NO_x = nitrogen oxides; ESCAPE = European Study of Cohorts for Air Pollution Effects.

In this paper, the impact of the validation dataset selection on the validation metrics was explored using two datasets of annual NO_x and NO_2 from an existing LUR (*2*) and a newly developed AD model (*11*) in Bradford, UK. Estimated NO_x and NO_2 concentrations from both models were compared against four different validation datasets and differences in the results were explored. The validation datasets were not used to calibrate the models. The effect that the resolution of predictions has on the validation metrics was further explored.

Methods

Setting

This study is set in Bradford, a city in the North of England. In terms of population, Bradford is the fifth largest English metropolitan district, with an estimated 534,300 inhabitants (*12*). Bradford's population has a notably different structure from the other cities in England and Wales (E&W) with more people under the age of 16 (Bradford has 22.6% while E&W have 18.7%) (*13*). Based on the British government's residential area Index of Multiple Deprivation, Bradford is one of the 10% most deprived local authorities in the UK, with significant deprivation discrepancy between the different neighborhoods (*13, 14*). Another distinct characteristic of Bradford is its ethnic diversity: over 20% of the population is of South Asian origin (*14*). Bradford is also home to a longitudinal birth cohort study known as the Born in Bradford (BiB) cohort. BiB was established in 2007 in response to growing concern regarding the health impacts of air pollution and high rates of childhood morbidity and mortality in the city (*14*).

These characteristics set Bradford apart from other UK cities and offer a unique opportunity to investigate the associations between air pollution, health, and socioeconomic status. The work presented in this paper is the basis for ongoing work in Bradford assessing the childhood asthma burden due to air pollution exposures (*11*). The year of analysis was 2009, when the LUR model and the traffic model used to build the AD model were available.

Land Use Regression Modeling

The Bradford's LUR model was built as part of the European Study of Cohorts for Air Pollution Effects (ESCAPE) project (*15*). The models were based on NO_2 and NO_x measurements at 41 sites across Bradford using Ogawa passive samplers (www.ogawausa.com). The passive samplers were administered between 1 June 2009 and 15 December 2009 (*16*). The measurement sites were classified as regional background ($n = 2$), urban background ($n = 24$), and traffic sites ($n = 15$) (Table 1). Measurements were typically made at the façade of homes as the objective of the ESCAPE project was to characterize residential exposures and associated health outcomes (*16*). Therefore, air pollution levels measured were generally representative of residential exposures.

At each site, measurements were made for three 14-day periods. Each period represented a different season namely the warm, cold, and intermediate seasons. The measurements were adjusted for temporal variability using measurements obtained from a reference fixed-site monitoring station that was operated all year around (*2, 16*). The adjusted measurements were then used to calculate adjusted annual average concentrations (*16*). The summary statistics of the adjusted measurements made at these 41 sites are shown in Table 1.

Atmospheric Dispersion Modeling

The AD modeling was conducted using the commercial package Atmospheric Dispersion Modelling System – Urban (ADMS-Urban) version 3.0.0. (*17*). As inputs, the AD model used:

1. link-based traffic flows and average speeds obtained from a previously established Simulation and Assignment of Traffic to Urban Road Networks (SATURN) model (*18*),

2. NO_x exhaust emissions based on average-speed emission functions sourced from the European COmputer Programme to calculate Emissions from Road Transport (COPERT) model (*19*, *20*).

SATURN Traffic Flows and Average Speeds Data. A previously developed and validated SATURN traffic model covering the Bradford district was used to extract geographical locations of the start and end of each road link and estimate link-based traffic flows (vph) and average speeds (km/h). The SATURN model covered 4,500 road links (*21*). The model simulated three periods on an average weekday: the AM peak, inter-peak, and PM peak hour. For all hours outside the simulation periods, including hours on an average weekend, a scaling factor developed from observations of vehicle flows at 19 automatic traffic counters was applied to the model's outputs. From this, a diurnal traffic profile was derived (*20*). The estimated traffic flows were split into different vehicle classes using 2009 standard fleet compositions in Urban England (*22*). The AM inter-peak and PM peak hour speeds, as estimated by SATURN, were used to calculate emissions at those hours. For all other hours, the inter-peak speed was used to complement the derived traffic flows, as no other speed data was available.

The model was run in SATURN version 11.1.09 and independently validated at 19 automatic traffic counters with complete flow data for a neutral week in 2009, as described elsewhere (*20*). The model was found to be a simplistic schematic of the actual road network as the geographical locations of the starts and ends of road links were not very accurate and the road links were represented as straight lines, rather than curved paths.

COPERT NO_x Emissions Data. All NO_x average-speed-emission functions were sourced from the COPERT 4 version 10.0 emission model (spreadsheets are freely available at http://naei.beis.gov.uk/data/ef-transport). Based on Urban England's split of vehicle classes, EURO emission standards, weight categories, and exhaust after-treatment technologies of year 2009 (*22*), there were 167 applicable average-speed-emission functions. These were sourced from COPERT and coded onto an Excel spreadsheet (*20*). In this spreadsheet, the user is prompted to enter the SATURN link-based average traffic speed (km/h) and traffic flow (vph) for each vehicle type for its NO_x emission factor (g/km) to be calculated. For each vehicle type in the traffic fleet, the calculated NO_x emission factor (g/km) is multiplied by the estimated number of vehicles for that type. The result is NO_x emissions in g/km, at each road link, for each vehicle type. The sum of the NO_x emissions across all vehicle types over each link is the total link-based NO_x in g/km.

In line with the requirement of ADMS-Urban, the 4,500 road sources and their NO_x emission rates were converted into g/km/s, using a time conversion factor, as below:

$$\text{Road link emission rate}\left(\frac{\frac{\text{g}}{\text{km}}}{\text{s}}\right) = \Sigma(\text{vehicle class specific emission factor at link speed}\,(\text{g}/\text{km}) * \text{number of vehicles iin each dass}) * \left(\frac{1\text{ hour}}{3600\text{ seconds}}\right)$$

This process was undertaken 48 times corresponding to the 24 h in an average weekday and the 24 h in an average weekend. Thus, at each road link, and in each hour of the day, an emission rate was calculated and this data was used to develop time-varying emission factors (*23*).

ADMS-Urban Set Up and Runs. The ADMS-Urban model was populated with input details about the modeling site and meteorological conditions including the latitude, surface roughness and albedo of the dispersion site, and the surface roughness of the meteorological site. Meteorological data was entered as a series of hourly sequential data covering year 2009. The following parameters were included: time of data, temperature in C°, wind speed in m/s, wind direction in degrees (°), precipitation rate in mm/h, cloud cover in oktas and relative humidity in percent (*17*). This data is used within the model to calculate the boundary layer height and parameters that are used in the dispersion simulations (*17*). The meteorological data was obtained from Bingley Samos weather station (*24*). Road source emission rates from the 4,500 available road links were *not* calculated from traffic flow data and ADMS-Urban's in-built database of traffic emission factors (*7*, *25*). Instead, they were entered directly into the model. This bypassed the in-built emission models and was done to enable a detailed description of the traffic fleet and its emissions as the traffic flow categorization permitted within ADMS-Urban was restricted to the six following vehicle categories:

- passenger car <3.5 t;
- London taxi/Hackney carriage <3.5 t;
- light-duty vehicles <3.5 t;
- motorcycles/moped <3.5 t;
- heavy-duty vehicles ⩾3.5 t;
- buses/coaches ⩾3.5 t.

Conversely, the present study modeled 167 vehicle classes (*20*).

As only one dataset of road emissions can be entered in ADMS-Urban (it is not possible to enter the 48-h emission datasets to be modeled at once), the AM peak hour was the one selected to be directly used. A Microsoft Access

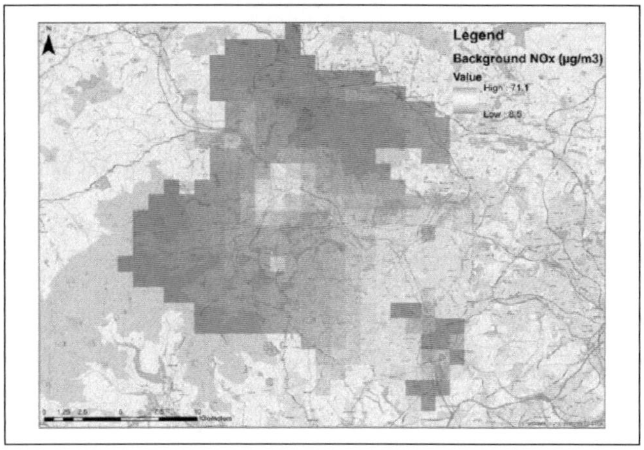

Figure 1. Annual (2010) NO_x background map at 1- × 1-km spatial resolution. Data source: Department for Environment, Food and Rural Affairs (*26*).

database, in ADMS-Urban Emission Inventory format (*17*), was prepared as an input file. The emission inventory contained the AM emission rates and other road characteristics including source names, road widths (an average of 17 m was used [7]), and geographical locations of the start and end of each road link. An additional modeling option specifying time-varying emission factors was used to enable modeling NO_x concentrations from vehicle emissions that vary by hour of the day. The time-varying emissions represented the different traffic flows and associated emissions across the different hours of the average weekday and the weekend. During the simulation, the ADMS-Urban model used the provided time-varying emission factors to multiply the AM peak hour emission rates at each road (the only input) by the appropriate factor specified for each hour.

NO_x Background Data. To account for air pollution sources besides traffic, annual average background NO_x concentrations were added to the ADMS-Urban estimates. Annual average NO_x concentrations were available either as a constant value of 38.4 $\mu g/m^3$ (Table 1) or as spatially varying values at 1km×1km grids coming from a national modeling study on emissions sources like industry, rail, domestic and aircraft (Figure 1) (*26*). The varying annual NO_x concentrations ranged from 9 to 71 $\mu g/m^3$ (mean = 14.73). Both NO_x background data sources were used to explore model performance in association with each. However, the varying background NO_x concentrations were considered more realistic.

NO_x to NO_2 Conversion Data. The proportion of NO_x that is NO_2 in exhaust emissions (termed primary NO_2) is highly uncertain and variable, with wide ranges documented in the literature; for example between 5% and 60%, across different vehicle classes, fuels, EURO emission standards, and after-treatment technologies (*17*, *27–30*). Further, secondary NO_2, produced in the atmosphere by complex photochemical processes, mainly depending on O_3 concentrations, contributes significantly to ambient NO_2 (*28*). Thus, converting NO_x to NO_2, especially at the emission estimation stage, is an uncertain process.

In this study, the conversion was undertaken for the final AD modeling outputs of NO_x concentrations. Based on the ESCAPE measurements in Bradford, the NO_2/NO_x ratio ranged from 0.39 to 0.75 with a calculated average of 0.60. This average ratio (0.60) was consistent with the average ratio of 0.59 calculated for 36 European study areas and with ratios in English cities like Manchester (0.58) and London/Oxford (0.58) (*16*). The final ADMS-Urban modeled NO_x estimates, with background NO_x added, were therefore converted to NO_2 using the average ratio of 0.60.

Validation Datasets. Four different validation datasets were available from different sources (Table 2). One dataset comes from the ESCAPE 2009 41 sites where NO_x and NO_2 were measured by Ogawa badges (Table 1); two datasets come from the local city council who administered 29 NO_2 diffusion tubes and eight continuous fixed-site monitoring stations in 2009 (*31*); and the final dataset comes from a different time period (2007/2008) when 48 NO_2 Palmes tubes were used to characterize residential exposure in preparation for the ESCAPE study (referred to as the 'de Hoogh' dataset). Therefore, the only direct source of NO_x data was ESCAPE.

Estimations Spatial Resolution. Using the LUR and AD models, annual NO_x concentrations were first modeled at the exact locations (X and Y coordinates) of the 126 validation sites (Table 2). The comparison between the modeled and the measured NO_x and NO_2 values at these exact points is referred to as "*at points*."

Subsequently, annual NO_x concentrations were modeled at 46,452 specified output points throughout the city. Each of these points was the centroid of a 100m×100m grid. At each 100m×100m grid, the centroid's modeled NO_x concentration was applied to the whole 100m×100m grid and a raster air pollution map was developed. The locations of the 126 validation sites were then intersected with the developed raster map and the NO_x values at the intersection points were extracted. The comparison between the modeled and the measured NO_x and NO_2 values using the raster intersection method is referred to as "*at raster*."

The aim of this exercise was to explore whether the accuracy of the concentration estimates differ if they were estimated at the exact location or at the raster level. This is important as in health impact assessments, estimates of air quality models at select receptor points are extrapolated and assumed to apply to larger areas and populations (*11*, *33*).

Results

Figure 2 shows the relation between the AD and the LUR models' NO_x estimates at the 46,452 points representing the

Table 2. NO_x and NO_2 Measurement Sites in Bradford used for Models' Validation

Measurement campaign and dataset (*n* = 126)	Pollutants measured	Measurement device	Year and time interval for final dataset	Locations and purpose of measurements	Reference
ESCAPE *diffusion tubes* (*n* = 41)	NO_2 and NO_x	Ogawa badges	2009 (annualized)	At the façade of homes of study subjects as the primary objective of the ESCAPE project was to characterize residential exposures and associated health	Cyrys, Eeftens (*16*)
CBMDC *diffusion tubes* (*n* = 29)	NO_2	"Diffusion tubes"	2009 (annualized)	Three sites were not close to main road while the rest were curbside sites at 0.5–5 m from the nearest road, monitoring undertaken to review and assess air quality progress	City of Bradford Metropolitan District Council (*31*) (internal document)
de Hoogh *diffusion tubes* (*n* = 48)	NO_2	Palmes tubes	Four 2-week periods during 2007–2008	Close to the front door of 48 homes of study subjects from the Born in Bradford cohort to characterize their residential exposures and compare with future ESCAPE work	Smith (*32*)
CBMDC *fixed-site monitoring* (*n* = 8)	NO_2	Automatic urban network chemiluminescence	2009 (annualized)	Two sites were classified as urban background whereas the rest were curbside sites at 1.5–2 m from the nearest road, monitoring undertaken to review and assess air quality progress	City of Bradford Metropolitan District Council (*31*) (internal document)

Note: CMBDC = City of Bradford Metropolitan District Council; NO_2 = nitrogen dioxide; NO_x = nitrogen oxides; ESCAPE = European Study of Cohorts for Air Pollution Effects.

centroids of the 100m×100m grids. A linear model between NO_x estimates from both models only captured 0.25 of the variability (Pearson $r = 0.50$), indicating that there is only a moderate correlation between the models. At 37,548 (81% of the) specified output points, the AD model estimated lower NO_x than the LUR model (ranging from −0.0007% to −296.33%, on average by −57.5%). At the remaining 8,904 specified output points, the AD model estimated higher NO_x than the LUR by +0.0109% to +100% (on average by +76.9%).

As shown on the bottom left side of Figure 2, the AD model estimated NO_x values between 10 μg/m^3 and 50 μg/m^3 when the LUR estimated almost 0 μg/m^3 NO_x. This trend had to do with the fact that the LUR model equations resulted in negative values at some output points in rural areas where traffic was very low and green space was high. These negative values were set to the minimum NO_x estimated by the LUR: 0.0006 μg/m^3. The removal of these points (6,101 with negative estimated NO_x) improved the correlation between the two models, bringing R^2 up by almost 0.10 to 0.34.

Table 3 shows the results of the different models' validation against the four validation datasets described in Table 2. The LUR models had a good performance against the (internal) dataset measured at the 41 ESCAPE sites. This data set, however, represented the measurements that were used to develop the model in the first place. The R^2 of the LUR against the ESCAPE measurements was 0.58 and 0.54 for NO_x and NO_2, respectively. When the estimates were made at a raster level, the predictive power of the model dropped by 0.23, in both the NO_x and NO_2 validation. At the 48 NO_2 diffusion tubes from the de Hoogh dataset, the NO_2 LUR model performed similarly well, with an R^2 of 0.61. The predictive power of the model dropped by 0.29 when the estimates were a made at a raster level. When the LUR model estimates were compared with the Bradford's council measurements, however, the model performed significantly worse with an R^2 of 0.21 and 0.38 in comparison with the NO_2 diffusion tubes and the NO_2 fixed-site measurement data, respectively. The stark difference was that the LUR model could not estimate higher NO_2 values recorded by the council (data available from the authors). This was a reasonable finding as the LUR model's prediction range is bound by the measured lower and upper pollutant values underlying the model. Like the other datasets, the predictive power of the model dropped by 0.15% when the estimates were a made at a raster level.

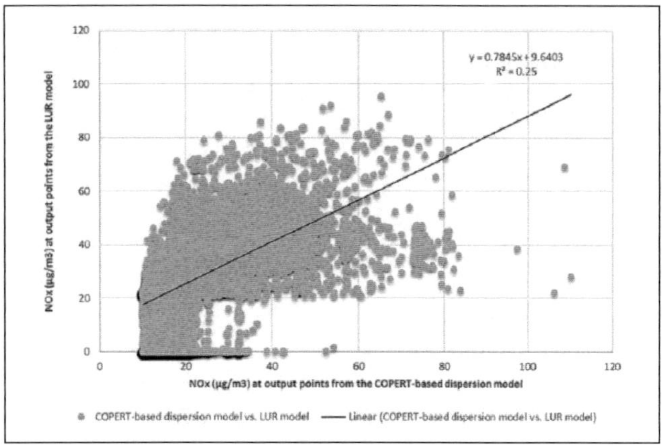

Figure 2. COPERT-based dispersion modeling vs. LUR modeling annual average NO_x estimates (µg/m^3) at 46,452 specified output points centering each 100m×100m grid.

In comparison to the LUR models, the AD model performed worse. Using varying background NO_x concentrations, the R^2 of the AD model at the 41 ESCAPE sites was 0.23 and 0.30 for NO_x and NO_2, respectively. Overall, using constant background levels resulted in worse performance (Table 3). When the estimates were made at a raster level, R^2 slightly decreased (by 0.02 and 0.05). This was in line with the LUR observations above, but with a lesser decrease in R^2. At the council's diffusion tubes and continuous fixed-site measurement sites and using varying background levels, the AD model had an R^2 of 0.23 and 0.28, respectively. When the estimates were made at a raster level, R^2 dropped by 0.04 to 0.26. Overall, the AD models with the varying background NO_x concentrations performed better than those with constant background.

Trends in the validity of the estimates at points and at raster suggested that for the LUR model, the validity was consistently better when estimates were made at points. For the AD model, this trend was also apparent but was less strong. This, alongside manual oversight of the SATURN network (see section on "SATURN Traffic Flows and Average Speeds Data"), was thought to indicate a possible issue with the traffic links' inaccurate geolocations.

The ESCAPE campaign was the only direct source of NO_x data and therefore the only dataset allowing direct comparison with the AD model estimates. Further analysis showed that measured NO_x at the 41 ESCAPE sites was generally higher than AD model estimates. This is apparent when inspecting the Bland–Altman agreement plot shown in Figure 3, in which most of the points fell above the zero line, and the ESCAPE's measurement minus the AD model's estimate was greater than zero. This suggested that background NO_x concentrations at most of these locations were underestimated or that traffic-related air pollution was underestimated due to, for example, low vehicle-emission factors, or both. The AD models underestimated NO_x at 35 out of the 41 measurement sites by 1.5% to 72.1%, or on average by 14.7 µg/m^3 NO_x (31.7 %). Fourteen of these sites were traffic sites whereas 20 were urban background sites.

There were two traffic ESCAPE sites (circled in red in Figure 3) that were considered as potential outliers. At these two sites, the difference between the measured and the AD-modeled NO_x was highest. These two points were influential on the AD models' validation and their removal substantially improved the models' validation increasing R^2 from 0.23 (Table 3: COPERT dispersion model NOx at points: varying background) to 0.49. One of these points was indeed explained and treated as an outlier in a relevant previous analysis (*2*). Similarly, the removal of these two points increased the LUR's R^2 from 0.58 to 0.73 (NO_x validation).

Finally, as mentioned above, a concern was that the poorer performance of the AD models was, in part, related to the inaccurate links geolocations in the original SATURN network. In an attempt to overcome this issue, a stepwise user-specific conditioned snapping procedure (*34*) was undertaken. This was done to snap the SATURN road links closer to the real roads locations as identified by Ordnance Survey Open Roads Maps. The aim of the snapped SATURN model was to increase the accuracy of the links geolocations. The snapped SATURN model was run again in ADMS-Urban. The validation results of this model, excluding the two outliers identified above, showed that R^2 went up from 0.49 to 0.60.

Discussion and Conclusions

In this paper, LUR and AD model estimates were validated against four different validation datasets. The validation metrics varied substantially, based on which model (combination) and which validation dataset was used. The LUR model performed better with the ESCAPE and the de Hoogh diffusion tubes, whereas the AD model performed better with council's fixed monitoring sites (when constant background NO_x was used) and with the de Hoogh diffusion tubes (when varying background NO_x was used). The performance of both models was similar with the council's diffusion tubes. The validation results based on the actual points' locations were generally much better than when the estimates were a made at a raster level (100m×100m grid). The estimates from the LUR and AD model had a moderate correlation. The AD model underestimated NO_x by 31.7%, on average. This under estimation was more prominent at the traffic sites.

The higher correlation between the LUR estimates and the de Hoogh measurements may be explained by the fact that both datasets came from tube measurements outside residences, thereby ensuring similar conditions and potentially similar air pollution variability. On the other hand, the council's diffusion tubes tended to be placed closer to roads, indicating that both the LUR and AD models did not capture roadside variations of NO_x so well. Across all metrics, the

Table 3. COPERT-based Dispersion Models and LUR Model Validation against Different Datasets

Model combinations	ESCAPE NO_x diffusion tubes (n = 41)	ESCAPE NO_2 diffusion tubes (n = 41)	CBMDC NO_2 diffusion tubes (n = 29)	de Hoogh NO_2 diffusion tubes (n = 48)	CBMDC NO_2 fixed-site monitoring (n = 8)
	Validation dataset				
LUR models					
NO_x LUR estimates *at points*	$R^2 = 0.58$				
NO_x LUR estimates *at raster*	$R^2 = 0.35$				
NO_2 LUR estimates *at points*		$R^2 = 0.54$	$R^2 = 0.21$	$R^2 = 0.61$	$R^2 = 0.38$ ($r = 0.62$)
NO_2 LUR estimates *at raster*		$R^2 = 0.31$	$R^2 = 0.06$	$R^2 = 0.32$	$R^2 = 0.38$ ($r = -0.61$)
COPERT-based dispersion model					
COPERT dispersion model NO_x *at points* (constant background)	$R^2 = 0.13$				
COPERT dispersion model NO_x *at points* (varying background)	$R^2 = 0.23$				
COPERT dispersion model NO_x *at raster* (constant background)	$R^2 = 0.16$				
COPERT dispersion model NO_x *at raster* (varying background)	$R^2 = 0.21$				
COPERT dispersion model NO_2 *at points* (constant background)		$R^2 = 0.17$	$R^2 = 0.27$	$R^2 = 0.34$	$R^2 = 0.56$
COPERT dispersion model NO_2 *at points* (varying background)		$R^2 = 0.30$	$R^2 = 0.23$	$R^2 = 0.50$	$R^2 = 0.28$
COPERT dispersion model NO_2 *at raster* (constant background)		$R^2 = 0.17$	$R^2 = 0.21$	$R^2 = 0.15$	$R^2 = 0.01$
COPERT dispersion model NO_2 *at raster* (varying background)		$R^2 = 0.25$	$R^2 = 0.19$	$R^2 = 0.30$	$R^2 = 0.02$

Note: CMBDC = City of Bradford Metropolitan District Council; COPERT = COmputer Programme to calculate Emissions from Road Transport; LUR = land use regression; NO_2 = nitrogen dioxide; NO_x = nitrogen oxides; ESCAPE = European Study of Cohorts for Air Pollution Effects.

AD model estimates had a slightly better correlation with the council's diffusion tube measurements. This is thought to indicate that AD may better capture the variability in air pollution concentrations from the roads as the vehicle sources were explicitly modeled. The differences between the point estimates and the raster estimates are likely explained by measurement error, as the resolution in the prediction point reduces significantly when using a raster. The magnitude of the effect (a halving of R^2), however, was rather large. This observation has implications for health impact assessment studies that usually use air quality model estimates at select receptor points to extrapolate air pollution concentrations to larger areas and populations.

A few studies which validated the ADMS-Urban model were found in the literature and are in line with the underestimation and validation metrics documented here. For example, Briant et al. (*35*) measured a summerly monthly mean value of 22.5 μg/m^3 NO_2 at 62 diffusion tube sites in Paris, compared with a 9.6 μg/m^3 NO_2 as modeled by ADMS-Urban. In the winter campaign, differences were higher with a measured monthly mean of 35.15 μg/m^3 NO_2 compared with a 19.4 μg/m^3 NO_2 as modeled by ADMS-Urban. Model validation against one dataset resulted in an R^2 of 0.55 to 0.62, depending on the season. Peace et al. (*36*) set up and validated an ADMS-Urban model for Greater Manchester. The validation was undertaken with one validation dataset from 12 continuous fixed-site monitoring stations. The results showed that the model underestimated NO_x and NO_2 concentrations but that R^2 equaled 0.88. Dėdelė and Miškinytė (*37*) used ADMS-Urban to model NO_2 concentrations in Kaunas city and validated the modeled concentration against measurements from 41 Ogawa passives samplers operated as part of ESCAPE. Differently from the other studies, the vehicle fleet was assigned an age of 14 years to calculate emissions. Overall, the ADMS-Urban estimates were higher than the average measured NO_2. However, the model tended to underestimate the maximum concentrations and overestimate the minimum concentrations. The R^2 with 40 of the available diffusion tubes equaled 0.75 to 0.79, depending on the season. In their follow-up study, Dėdelė and Miškinytė (*38*) compared modeled and validated NO_2 concentrations with another validation dataset from four continuous air

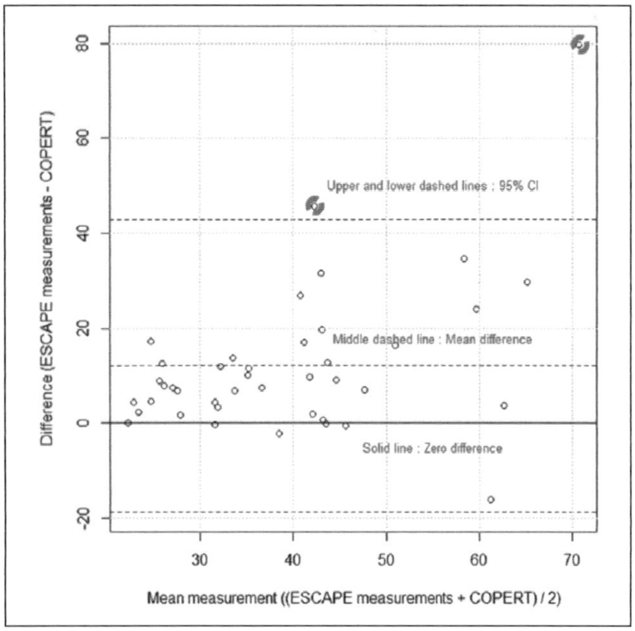

Figure 3. Bland–Altman agreement plot for ESCAPE measurements versus COPERT-based dispersion modeling NO_x estimates.

quality monitoring stations in Kaunas city and this resulted in an R^2 of 0.56 to 0.91. At the two traffic stations and the background station, the average modeled NO_2 was lower than the observed, whereas the opposite trend was observed at the residential site.

Finally, in Bradford, de Hoogh et al. (*7*) used ADMS-Urban to model NO_2 concentrations and compare the modeled estimates with measurements from 40 Ogawa passives samplers operated as part of ESCAPE (one influential outlier was excluded). The authors found that the median and minimum modeled concentrations were underestimated (measured vs. modeled median = 25.5 µg/m³ vs. 19.8 µg/m³, and minimum 16.7 µg/m³ vs. 13.0 µg/m³), whereas the maximum concentration was overestimated, but to a lower extent (measured vs. modeled maximum = 36.7 µg/m³ vs. 38.0 µg/m³). The measured and modeled NO_2 concentrations correlated well with a median R^2 (range) of 0.55 (0.01 to 0.74) (*7*).

The novelty of this study lies in the comparison of the LUR versus the AD models and the comparison of all the models' validation metrics against different measured datasets. Recent literature of similar studies is limited (*7–9*). Beelen et al. (*9*) compared the performance of 100m×100m grid LUR and AD models used to estimate NO_2 levels at 69,975 receptor output points in a large Dutch urban area. The authors showed moderate correlations between NO_2 levels estimated from the LUR and the AD models ($r = 0.55$, compared with 0.50 with the COPERT-based model in this study). The authors also showed that when modeled concentrations were compared with measured concentrations at 18 independent validation sites, the AD model performed better than the LUR model with $r = 0.77$ compared with 0.47 for the LUR model. This latter observation was not confirmed in the current study in which validation performance was contingent on the validation dataset used. Linked to the relevance of the validation dataset selection, de Nazelle et al. (*8*) showed that the performance of the ESCAPE LUR model is better when validated against (internal) measurements from the ESCAPE sites, whereas the R^2 dropped by 0.17 to 0.18 when the model was validated against external, independent sites. A similar pattern was demonstrated for other LUR models when applied to external validation datasets. The worsening of performance was suggested to be due to a combination of over-fitting and differences in the sampling campaigns (for example years and site selection) (*8*). The results of this study are in line with these observations but also show that the validation dataset selection affects the validity parameters for AD models as well. A new addition was to show that producing air pollution estimates at a raster or a point level makes a difference to the validation.

The strengths of this study are the well-developed LUR and AD models used, the availability of multiple validation datasets, and the extensive analyses undertaken. The limitations are in the limited validation data for NO_x, which was the only pollutant directly modeled with the AD model. Furthermore, the AD model tended to underestimate NO_x. This underestimation is likely to be due to the combination of: 1) unrealistic low vehicle-emission factors sourced from COPERT (*39*), which are derived from conservative laboratory rather than higher real-world emission measurements (*27, 40*); 2) overestimated average speeds from the SATURN model, which generally undermine the proportions of congestion, speed variation, and stop-start driving (*20, 21*); and 3) the fact that the SATURN road network did not model all roads in the study area, specifically smaller roads with cold-starts emissions. The contribution of each of these factors to the overall underestimation of the AD models warrants further research. Finally, due to the unavailability of required data, NO_2 was generated by conversion, which is a simplistic procedure and one that conceals spatial variability.

In conclusion, this study showed the value of validating modeled air quality estimates against various datasets to obtain a better understanding of the different models' performance. The results showed that the models' performance was contingent on the validation dataset used and that there is value in reporting validation metrics against multiple measurements, when available. The work provided estimates of the decrease in the models' predictive power, when predictions are made with a raster rather than an exact location approach. This decrease in predictive power has implications for health impact assessment studies that typically estimate air quality at select receptor points and extrapolate these estimates to larger areas and populations.

References

1. Eeftens, M., R. Beelen, K. de Hoogh, T. Bellander, G. Cesaroni, M. Cirach, C. Declercq, A. Dedele, E. Dons, A. de Nazelle, and K. Dimakopoulou. Development of Land Use Regression Models for PM2. 5, PM2. 5 Absorbance, PM10 and PMcoarse in 20 European Study Areas; Results of the ESCAPE Project. *Environmental Science & Technology*, Vol. 46, No. 20, 2012, pp. 11195–11205.
2. Beelen, R., G. Hoek, D. Vienneau, M. Eeftens, K. Dimakopoulou, X. Pedeli, M. Y. Tsai, N. Künzli, T. Schikowski, A. Marcon, and K. T. Eriksen. Development of NO_2 and NO_x Land Use Regression Models for Estimating Air Pollution Exposure in 36 Study Areas in Europe–the ESCAPE Project. *Atmospheric Environment*, Vol. 72, 2013, pp. 10–23.
3. Rancière, F., N. Bougas, M. Viola, and I. Momas. Early Exposure to Traffic-related Air Pollution, Respiratory Symptoms at 4 Years of Age, and Potential Effect Modification by Parental Allergy, Stressful Family Events, and Gender: A Prospective Follow-up Study of the PARIS Birth Cohort. *Environmental Health Perspectives*, Vol. 125, 2017, pp.737–745.
4. Yamazaki, S., M. Shima, T. Nakadate, T. Ohara, T. Omori, M. Ono, T. Sato, and H. Nitta. Association Between Traffic-related Air Pollution and Development of Asthma in School Children: Cohort Study in Japan. *Journal of Exposure Science and Environmental Epidemiology*, Vol. 24, No. 4, 2014, pp. 372–379.
5. Khreis, H., and M. J. Nieuwenhuijsen. Traffic-Related Air Pollution and Childhood Asthma: Recent Advances and Remaining Gaps in the Exposure Assessment Methods. *International Journal of Environmental Research and Public Health*, Vol. 14, No. 3, 2017, p. 312.
6. Nieuwenhuijsen, M. J. *Exposure Assessment in Environmental Epidemiology*. Oxford University Press, New York, NY, 2015.
7. de Hoogh, K. et al., Comparing land use regression and dispersion modelling to assess residential exposure to ambient air pollution for epidemiological studies. *Environment International*, 2014. 73: p. 382-392.
8. de Nazelle, A., I. Aguilera, M. Nieuwenhuijsen, R. Beelen, M. Cirach, G. Hoek, K. de Hoogh, J. Sunyer, J. Targa, B. Brunekreef, and N. Künzli. Comparison of Performance of Land Use Regression Models Derived for Catalunya, Spain. *Atmospheric Environment*, Vol. 77, 2013, pp. 598–606.
9. Beelen, R., M. Voogt, J. Duyzer, P. Zandveld, and G. Hoek. Comparison of the Performances of Land Use Regression Modelling and Dispersion Modelling in Estimating Small-scale Variations in Long-term Air Pollution Concentrations in a Dutch Urban Area. *Atmospheric Environment*, Vol. 44, No. 36, 2010, pp. 4614–4621.
10. Khreis, H., C. Kelly, J. Tate, R. Parslow, K. Lucas, and M. Nieuwenhuijsen. Exposure to Traffic-related Air Pollution and Risk of Development of Childhood Asthma: A Systematic Review and Meta-analysis. *Environment International*, Vol. 100, 2017, pp. 1–31.
11. Khreis, H., K. De Hoogh, and M. J. Nieuwenhuijsen. Full-Chain Health Impact Assessment of Traffic-Related Air Pollution and Childhood Asthma. *Environment International*, Vol. 114, 2018, pp. 365–375.
12. City of Bradford Metropolitan District Council. *Population*. https://www.bradford.gov.uk/open-data/our-datasets/population/. Accessed July 20, 2017.
13. Fielding, T. *Health Impact Assessment of Low Emission Zone proposals in Bradford and Leeds*, 2012. Methodology, baseline assessment and evidence review.
14. Wright, J., N. Small, P. Raynor, D. Tuffnell, R. Bhopal, N. Cameron, L. Fairley, D. A. Lawlor, R. Parslow, E. S. Petherick, and K. E. Pickett. Cohort Profile: The Born in Bradford Multi-ethnic Family Cohort Study. *International Journal of Epidemiology*, Vol. 42, No. 4, 2013, pp. 978–991.
15. European Study of Cohorts for Air Pollution Effects. FP7 - ESCAPE - European Study of Cohorts for Air Pollution Effects (project description), http://www.escapeproject.eu/. Accessed March 30, 2016.
16. Cyrys, J., M. Eeftens, J. Heinrich, C. Ampe, A. Armengaud, R. Beelen, T. Bellander, T. Beregszaszi, M. Birk, G. Cesaroni, and M. Cirach. Variation of NO_2 and NO_x Concentrations Between and Within 36 European Study Areas: Results from the ESCAPE Study. *Atmospheric Environment*, Vol. 62, 2012, pp. 374–390.
17. Cambridge Environmental Research Consultants Ltd. *ADMS-Urban, An Urban Air Quality Management System, User Guide, Version 3.0*. Cambridge Environmental Research Consultants Ltd, Cambridge, UK, 2010.
18. Van Vliet, D. S. A. T. U. R. N. (1982). SATURN-a modern assignment model. *Traffic Engineering & Control*, 23(HS-034 256).
19. Samaras, S., D. Tsokolis, S. Toffolo, A. Garcia-Castro, C. Vock, L. Ntziachristos, and Z. Samaras. Limits of Applicability of COPERT Model to Short Links and Congested Conditions. *Proc., 20th International Transport and Air Pollution Conference 2014*. Graz, Austria, 2014.
20. Khreis, H., and J. Tate. Alternative Methods for Vehicle Exhaust Emission Modelling and Impact on Local Road Transport Emission Inventories: The Case Study of Bradford, UK. *Transportation Research Part D: Transport and Environment*, Vol. 5, 2017, S50.
21. Steer Davies Gleave, S. PT and Highway LMVR. *Steer Davies Gleave and JMP*, 2009.
22. National Atmospheric Emissions Inventory. *Ricardo Energy and Environment*. Vehicle Fleet Composition Projections (Base 2013). National Atmospheric Emissions Inventory, UK, 2014.
23. Cambridge Environmental Research Consultants Ltd. *Environmental Software: ADMS-Urban Model*, http://www.cerc.co.uk/environmental-software/ADMS-Urban-model.html. Accessed May 12, 2014.
24. *Bingley SAMOS Observations Map*. http://www.metoffice.gov.uk/public/weather/observation/map/gcwdjeczy#?zoom=9&lat=54.10&lon=-4.44. Accessed March 8, 2017.
25. City of Bradford Metropolitan District Council. *Bradford Low Emission Zone Feasibility Study*, 2013.
26. Department for Environment, Food and Rural Affairs. *Background Concentration Maps: User Guide*, 2016.
27. Carslaw, D., S. Beevers, E. Westmoreland, M. Williams, J. Tate, T. Murrells, J. Stedman, Y. Li, S. Grice, A. Kent, and I. Tsagatakis. *Trends in NOx and NO_2 emissions and ambient measurements in the UK*. Defra, London, 2011.
28. Mavroidis, I., and A. Chaloulakou. Long-term Trends of Primary and Secondary NO_2 Production in the Athens Area. Variation of the NO_2/NOx Ratio. *Atmospheric Environment*, Vol. 45, No. 38, 2011, pp. 6872–6879.

29. Rhys-Tyler, G. *Road Vehicle Exhaust Emissions: An age of uncertainty, in Dispersion Modellers User Group 2017.* Holiday Inn, Kensington, London, 2017.
30. Sjödin, A. M. Jerksjö. *Evaluation of European Road Transport Emission Models Against On-road Emission Data as Measured by Optical Remote Sensing*, 2008.
31. City of Bradford Metropolitan District Council. *Air Quality Progress Report for Bradford.* City of Bradford Metropolitan District Council, Bradford, 2010, p. 49.
32Smith, R. B. *Assessment and Validation of Exposure to Disinfection By-products During Pregnancy, in an Epidemiological Study Examining Associated Risk of Adverse Fetal Growth Outcomes*. Imperial College London, 2011.
33. Mueller, N., D. Rojas-Rueda, X. Basagaña, M. Cirach, T. Cole-Hunter, P. Dadvand, D. Donaire-Gonzalez, M. Foraster, M. Gascon, D. Martinez, and C. Tonne. Urban and Transport Planning Related Exposures and Mortality: A Health Impact Assessment for Cities. *Environmental Health Perspect*, Vol. 125, 2017, pp. 89–96.
34. ESRI. *GIS Dictionary: Snapping*, http://support.esri.com/other-resources/gis-dictionary/term/snapping. Accessed November 7, 2016.
35. Briant, R., C. Seigneur, M. Gadrat, and C. Bugajny. Evaluation of Roadway Gaussian Plume Models with Large-scale Measurement Campaigns. *Geoscientific Model Development*, Vol. 6, No. 2, 2013, p. 445.
36. Peace, H., B. Owen, and D. Raper. Comparison of Road Traffic Emission Factors and Testing by Comparison of Modelled and Measured Ambient Air Quality Data. *Science of the Total Environment*, Vol. 334, 2004, pp. 385–395.
37. Dėdelė, A., and A. Miškinytė, Estimation of inter-seasonal differences in NO_2 Concentrations Using a Dispersion ADMS-Urban Model and Measurements. *Air Quality, Atmosphere & Health*, Vol. 8, No. 1, 2015, pp. 123–133.
38. Dėdelė, A., and A. Miškinytė. The Statistical Evaluation and Comparison of ADMS-Urban Model for the Prediction of Nitrogen Dioxide with Air Quality Monitoring Network. *Environmental Monitoring and Assessment*, Vol. 187, No. 9, 2015, p. 578.
39. Williams, M., R. Barrowcliffe, D. Laxen, and P. Monks. *Review of Air Quality modelling in DEFRA*, 2011. http://uk-air.defra.gov.uk/assets/documents/reports/cat20/1106290858_DefraModellingReviewFinalReport.pdf. Accessed September 22, 2014.
40. Khreis, H. Critical Issues in Estimating Human Exposure to Traffic-related Air Pollution: Advancing the Assessment of Road Vehicle Emissions Estimates. Presented at the World Conference on Transport Research - WCTR 2016, Shanghai, 10–15 July 2016, Transportation Research Procedia.

The Standing Committee on Transportation and Air Quality (ADC20) peer-reviewed this paper (18-01950).

TRR
JOURNAL OF THE TRANSPORTATION RESEARCH BOARD

Article

Transportation Research Record
2018, Vol. 2672(25) 67–78

DOI: 10.1177/0361198118796070
journals.sagepub.com/home/trr
SAGE

Effect of Light Duty Vehicle Performance on a Driving Style Metric

Shams Tanvir[1], H. Christopher Frey[2], and Nagui M. Rouphail[2]

Abstract

Eco-driving involves alterations to driving style to improve energy efficiency. The observed driving style reflects the combined effects of roadway, traffic, driver, and vehicle performance. Although the effect of roadway and traffic characteristics can be inferred from microscale driving activity data, the effect of vehicle performance on driving style is not properly understood. This paper addresses two questions: (1) how different is an individual driver's driving style when operating vehicles with differences in performance?; and (2) how dissimilar are the driving styles of different drivers when operating vehicles that have similar performance? To answer these questions, we have gathered microscale vehicle activity measurements from 17 controlled real-world driving schedules and two years of naturalistic driving data from five drivers. We also developed a metric for driving style termed "envelope deviation," which is a distribution of gaps between microscale activity (1 Hz) and fleet average envelope. We found that there is significant inter-driver heterogeneity in driving styles when controlling for vehicle performance. However, no significant inter-vehicle heterogeneity was present in driving styles while controlling for the driver. Findings from this study imply that the choice of vehicle does not significantly alter the natural driving style of a driver.

The transportation sector in the U.S. consumed about 36 quadrillion Btu of energy in the form of petroleum in 2016 (*1*). In addition to producing greenhouse gases (GHG), petroleum dependence makes the economy vulnerable to price shocks (*2*). Many fuel consumption reduction measures have been attempted such as improving vehicle technology, reducing travel demand, and increasing vehicle occupancy and transit use, among others (*3*). However, the effectiveness of these countermeasures has shown mixed results because of the expense and inconvenience involved in their adoption (*4*). In contrast, "eco-driving," which involves changing the behavior or the style of driving to achieve greater energy efficiency is less costly than other options; however, it is more difficult to administer. The estimated benefit from eco-driving is reported as ranging from 5% to 40% (*5*, *6*). High benefit values are generally estimated for a very idealistic condition where drivers can choose the cruising speed and the level of acceleration and deceleration. However, in a naturalistic environment, drivers are constrained by congestion on their routes, road geometry, traffic control, weather, and vehicle performance. In this paper, we have explored the effect of vehicle performance on driving style.

Driving style is a generic term which is sometimes used interchangeably with the term "driver aggressiveness." Generally, driving style refers to accumulated microscale driving activity, such as speed or acceleration over a single or a set of synthetic or real-world drive cycle/s gathered from the same driver–vehicle combination. Synthetic drive cycles such as the Federal Test Procedure (FTP) and Highway Fuel Economy Driving Schedule (HWFET) are mainly used for certification purposes in chassis dynamometer tests. In contrast, real-world or naturalistic drive cycles are manifested as the driving style reacting to both in-vehicle and out-of-vehicle environments.

Vehicle performance is the capability of a vehicle to accelerate or decelerate under particular speed conditions. Malliaris et al. (*7*) found an inverse power relation between vehicle time to accelerate from 0 to 60 mph and its power-to-weight ratio. Vehicle performance delineates the limiting conditions on the driving style. The effect of power-to-weight ratio, as a surrogate for vehicle performance, has been studied in previous research (*8*, *9*).

[1]Institute for Transportation Research and Education (ITRE), North Carolina State University, Raleigh, NC
[2]Department of Civil, Construction, and Environment Engineering (CCEE), North Carolina State University, Raleigh, NC

Corresponding Author:
Address correspondence to Shams Tanvir: stanvir@ncsu.edu

Most studies examined vehicle performance and driving styles in a disjointed way, not considering the effect of driver–vehicle interaction on the specified driving style metric. In other words, heterogeneity in driving styles across different drivers was not considered while measuring the effect of vehicle performance. Absence of consideration of driver–vehicle interaction may have led to wide variability in the measured effect of vehicle performance on driving style. Moreover, characterization of this driver–vehicle interaction can feed into recently developed agent based traffic simulators (*10*, *11*) to test the effects of different policies on energy use.

In the next section, we present a review of the literature on the measures of driving style and methods of testing the effect of vehicle performance on driving style. Methods and data for the paper are discussed in the subsequent sections. Next, we discuss the results and, in the final section, the conclusions from this analysis.

Review of Driving Style Measures

Several approaches have been proposed to characterize driving style or the "aggressiveness" of driving styles. The most common way to characterize driving style is by characterizing microscale driving activity. Descriptive statistics of instantaneous speed and acceleration over the drive cycle such as average speed, standard deviation of speed, average positive acceleration, standard deviation of positive acceleration, percentage trip time accelerating over a threshold, and jerk frequency, among others, have been used to characterize driving style (*12*, *13*). However, aggregate cycle based characteristics are not very useful in comparing multiple driving styles in terms of fuel consumption. Another mode of characterizing driving style is to use a normalized energy consumption measure. In this approach, fuel efficiency (measured in miles per gallon) or the ratio of cycle average speed to energy consumption (*14*) is used to describe driving style. A major shortfall of this approach is that the direct measurement of fuel consumption is needed. In addition, aggregate measures do not consider the differences in drivers' reactions to different dynamics which exist at different speed levels.

To account for vehicle speed–acceleration dynamics, joint distribution of speed and acceleration has been suggested as a characteristics of driving style. EPA (*15*) used the fraction of speed–acceleration instances falling outside the range of standard drive cycles such as FTP or HWFET to characterize driver aggressiveness. Berry (*16*) proposed the use of an "Aggressiveness Factor" from microscale driving activity data by calculating the amount of extra work relative to steady-state driving. She defined three separate "aggressiveness factors" based on trip average speeds to distinguish neighborhood (below 20 mph), city (between 20 and 45 mph), and highway (above 45 mph) driving. With the same principles, Stichter (*17*) developed a "proposed aggressivity metric" using average positive and negative acceleration frequencies for trucks. These single number driving style metrics overlook the distribution of deviations of observed microscale activities from the standardized cycles or steady-state driving style. To explain the dynamics of deviation, Liu and Frey (*18*) suggested using "activity envelopes" for light duty vehicles. They defined activity envelope as the 95% frequency range of acceleration for each of 15 speed bins from 0 to 75 mph in intervals of 5 mph, and one bin for speeds greater than 75 mph. This discretized way of portraying driving style provided more information than a single number metric. However, the entire distribution of acceleration activities in relation to the standardized cycles was not explicitly considered in their metric.

Review of Experimental Designs

The method for experimenting with driving styles and vehicle performance varies substantially across the literature. In most cases, the objective of the research was not to measure the effect of vehicle performance on driving style; rather the focus was on finding the joint effect of driving style and vehicle characteristics on energy consumption or emissions while controlling for other factors such as route and congestion. Common approaches to gather microscale driving activity are through engine dynamometers across predetermined drive cycles (*19*) and microscopic vehicle simulation (*16*). However, these simulated driving styles cannot always accurately represent real-world characteristics of individual drivers (*20*, *21*). Frey et al. (*22*) used portable emissions measurement systems along with vehicle on-board diagnostic (OBD) information to characterize real-world driving style and associated emissions. Experiments controlling for routes, time of day, road grade, and vehicles (*23*, *24*) have been conducted to examine the resulting fuel use and emissions. However, these studies did not examine the variation in individual driver driving style at different levels of vehicle performance. In contrast to controlled experiments, naturalistic driving studies can provide long-term information on a driver's driving style. In naturalistic driving, personal vehicles are equipped with microscale driving activity recorders that collect data during regular driving. Due to the expense involved in equipping the vehicles, naturalistic driving studies are typically conducted in randomized assignments of drivers for a long period or, in case of factorial design trials, multiple driver–vehicle combinations are monitored in short phases. In the first case, even with a long-term study, the true impact of vehicle performance on driving style cannot be properly characterized due to heterogeneities in driving style in the driver population. In short-duration studies, observed driving styles do not stabilize due to variabilities in the choice of routes and departure times.

On the basis of the review of literature presented above, three main limitations can be gleaned: (1) lack of an effective driving style metric, (2) lack of controlled experiments to investigate the impact of vehicle performance on individual driver driving style, and (3) even fewer long-term factorial design driver–vehicle experiments have been conducted.

Research Questions

The key research questions addressed in this paper are: (1) How different is a single driver's driving style when operating vehicles with different performances? and (2) How different are the driving styles of different drivers operating vehicles of similar performance in the context of both controlled experiments and naturalistic studies?

Methods

The methods of this study include a proposed approach to characterize driving style, a procedure to compare different driving styles, and a plan to classify vehicle performance.

Characterizing Driving Style

The speed–acceleration joint distribution provides a comprehensive signature of driving style for a driver. All the driving style measures discussed in the literature review section are abstractions of the speed–acceleration joint distribution. Liu and Frey (*18*) developed vehicle activity envelopes for 100 light duty vehicles including conventional passenger cars, passenger trucks, and hybrid electric vehicles. Vehicle activity envelopes were defined as the 97.5th percentile and 2.5th percentile acceleration values at each of the 16 speed bins. The envelope values were examined across factors such as type of vehicle, type of transmission, road grade, engine horsepower, and power-to-weight ratio. Since the acceleration envelope values were found to be robust across all the factors considered, the combined activity envelope can be considered as the benchmark of real-world driving. Liu and Frey corrected the instantaneous accelerations with the measured road grades as shown in Equation 1 (*25*).

$$a_e = a + 35.66 \times \frac{RG}{100} \quad (1)$$

a_e = effective acceleration, km/h per second
a = acceleration, km/h per second

Table 1. Activity Envelope Values for Real-World Vehicle Fleet (*18*)

	Bounds of acceleration envelopes by speed bin	
Speed bin, $\bar{\bar{v}}$ (mph)	Effective 97.5th percentile acceleration, $E_{\bar{\bar{v}},1}$ (mph/s)	Effective 2.5th percentile acceleration, $E_{\bar{\bar{v}},0}$ (mph/s)
0–5	1.69	−2.97
5–10	5.70	−6.04
10–15	5.48	−6.01
15–20	4.87	−5.77
20–25	4.75	−5.21
25–30	4.50	−4.75
30–35	3.84	−4.12
35–40	3.08	−3.39
40–45	2.37	−2.68
45–50	2.12	−2.20
50–55	1.89	−1.94
55–60	1.75	−1.85
60–65	1.66	−1.62
65–70	1.38	−1.44
70–75	1.35	−1.22
75+	1.44	−1.05

RG = percentage increase in elevation per unit distance (%)

The adjustments made it possible to consistently compare microscale driving activities while accounting for grade. The activity envelope values for effective acceleration of a real-world fleet is listed in Table 1. Envelopes for standard drive cycles such as FTP, HWFET, SCO3, and US06 were compared with the envelope of a real-world vehicle fleet. However, no numerical measure of comparison was defined in their work (*25*). In this paper we define a new metric termed "envelope deviation" to quantitatively characterize the deviation of joint speed–acceleration distribution of drive cycles in relation to the real-world vehicle fleet envelope. Instantaneous envelope deviation is defined as the difference of absolute values of instantaneous acceleration, and the absolute values of envelope acceleration for the mode, that is, positive acceleration or negative acceleration/deceleration (Equation 2).

$$d_t = |a_{e,t}| - \left|E_{\bar{\bar{v}},m}\right| \quad (2)$$

d_t = instantaneous envelope deviation at time t (mph/s)

$E_{\bar{\bar{v}},m}$ = effective envelope acceleration for speed bin $\bar{\bar{v}}$ and mode m (mph/s)

$a_{e,t}$ = instantaneous effective acceleration at time t

$\bar{\bar{v}}$ = speed bins as left-open right-closed (mph; 16 from 0 to 75 + mph) at time t

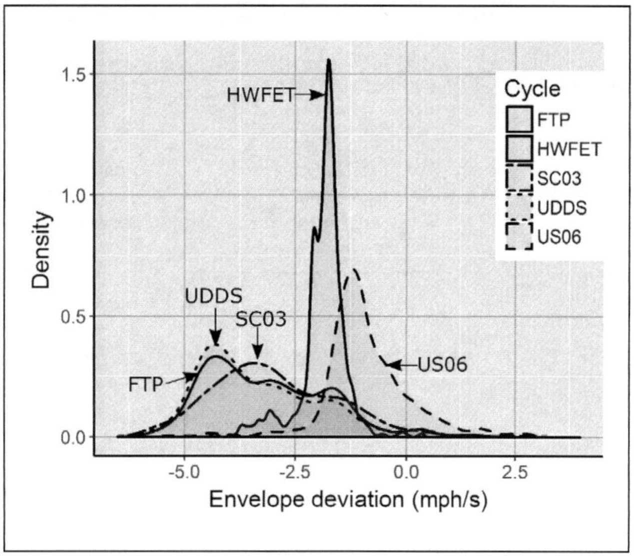

Figure 1. Distributions of envelope deviation for standard driving cycles.

m = logical at time t; 1 if $a_{e,t} \geq 0$ (for acceleration), else 0 (for deceleration)

For example, if a driver has a positive acceleration ($m = 1$) while driving at 33 mph ($\bar{\bar{v}}$ = 30–35 mph), then the effective envelope acceleration would be the 97.5th percentile acceleration for the speed bin ($E_{30-35,1}$ = 3.84 mph/s). In contrast, if a driver has a negative acceleration ($m = 0$) while driving at 50 mph ($\bar{\bar{v}}$ = 50–55 mph), then the effective envelope acceleration would be the 2.5th percentile acceleration for the speed bin ($E_{50-55,0}$ = −1.94 mph/s).

The empirical distribution of d_t provides aggressiveness characteristics of the drive cycles. Negative deviation values mean the acceleration or deceleration response for that instance was within the real-world vehicle fleet envelope. In contrast, positive deviation values suggests the instantaneous acceleration and deceleration values are more aggressive than the envelope. If the distribution of d_t were drawn for the accumulated real-world vehicle fleet, 95% of the distribution would fall on the negative side and 5% of the distribution would fall on the positive side. Distributions of envelope deviation for standard driving cycles are shown in Figure 1.

The FTP cycle has about 3.4% of the envelope deviation distribution on the positive side and the mass of the distribution is flatter compared with the other drive cycles. The HWFET cycle has only 0.52% in the positive envelope deviation zone with a sharp peak. The UDDS and SC03 cycles follow the FTP distribution closely with 3.5% and 2.9% in the positive envelope deviation zones respectively. The US06 driving cycle has 15.3% in the positive zone. The peaks of HWFET and US06 cycles are closer to zero, that is, not much different from the envelope compared with the rest of the standard cycles. US06 is a well-known aggressive cycle, where FTP and UDDS (UDDS is a part of the FTP cycle) are mild cycles, that is, acceleration is limited to around 3.5 mph/s. Hence, the envelope deviation analysis of standard drive cycles meets expectation.

Comparing Driving Styles

Driving styles are compared by comparing the empirical distributions of d_t based on 1 Hz data. No assumptions regarding the underlying distribution of envelope deviation were made. The two-sample Kolmogorov–Smirnov test (K–S test) was selected as the nonparametric method to compare two distributions, as it has sensitivity to differences in both the location and shape of the empirical distributions of the two samples. The test statistic for two-sample K–S test, D, represents the maximum distance between the two empirical cumulative density functions under consideration. Although the test statistic D is robust in presence of ties in values of underlying distributions and large sample sizes, the p-value is not. Therefore, even if the null hypothesis is rejected, that is, p-value < significance level (α) it does not always infer that the two samples came from different populations. In this paper, comparison of driving styles will be conducted by calculating K–S test statistics, D of the two microscale activity samples, termed as "Distance." Only when the null hypothesis is accepted, that is, p-value > α, is the condition reported.

Classifying Vehicle Performance

Vehicle performance depends on multiple factors such as engine size, fuel type, transmission type, weight, and engine horse power, and can depend on mileage, age, and other factors. In the context of understanding the effect of vehicle performance on driving style, only the factors which directly affect microscale activity are considered. The vehicle power-to-weight ratio has been found to be inversely correlated with vehicle acceleration potential, that is, time for 0 to 60 mph. However, very few studies verified the impact of vehicle power-to-weight ratio on real-world driving style. Andre, 2006 (*26*) found that vehicles with power-to-weight ratio greater than 0.037 hp/lb enabled higher aggressiveness than vehicles with less power-to-weight ratio. Berry, 2010 (*16*) found that drivers of vehicles with power-to-weight ratio between 0.04 hp/lb and 0.05 hp/lb have significantly higher aggressiveness than those operating vehicles of power-to-weight ratio greater than 0.05 hp/lb or less than 0.04 hp/lb. This finding was consistent across different drivers and different makes and models of vehicles.

Table 2. Specifications of the Vehicles Analyzed in the Controlled Experiment

Driver ID	Test ID[a]	Body type	Model year	Make and model	Engine disp. (L)	Power (hp)	Curb weight (lbs)	Power-to-weight (hp/lb)
1	1-1-1-13	Minivan	2006	Dodge Caravan	3.3	180	4,350	0.033
1	1-1-2-13	SUV	2013	GMC Terrain	2.4	182	3,853	0.037
1	1-2-3-13	Sedan	2012	Ford Fusion	2.5	175	3,285	0.039
1	1-2-4-13	Sedan	2012	Dodge Avenger	2.4	173	3,400	0.038
1	1-2-5-13	Minivan	2012	Toyota Sienna	3.5	266	4,310	0.044
1	1-2-6-13	Sedan	2012	VW Passat	2.5	170	3,166	0.038
1	1-3-7-13	Sedan	2008	Chevrolet Impala	3.5	211	2,655	0.059
2	2-2-8-13	Sedan	2012	Ford Fusion	2.5	175	3,285	0.039
2	2-2-9-13	Sedan	2007	Honda Accord	2.4	166	3,321	0.040
2	2-2-10-16	Minivan	2017	Kia Sedona	3.3	276	4,443	0.045
2	2-3-11-16	Sedan	2013	Chevrolet Impala	3.6	301	3,555	0.067
3	3-1-12-13	Sedan	2003	VW Jetta	2.0	115	2,934	0.030
3	3-2-13-16	Sedan	2007	Volvo S60	2.5	208	3,501	0.046
4	4-3-14-16	Sedan	2016	Chevrolet Impala	3.6	305	3,841	0.064
4	4-3-15-16	SUV	2017	Kia Sorento	3.3	290	4,369	0.053
5	5-2-16-13	SUV	2008	Nissan Xterra	4.0	261	4,387	0.048
5	5-2-17-16	Sedan	1998	Nissan Maxima	3.0	190	3,025	0.044

[a]Test ID represents four parts: Driver ID-Vehicle Class ID-Vehicle ID-Test Year.

Similar findings were made by Johansson et al. (*27*), however, the reason was not reported.

Since a complete range of vehicle power-to-weight ratio cannot be tested in both the controlled experiments and naturalistic studies conducted in this study, each vehicle has been assigned to one of three classes of power-to-weight ratio. Vehicles with power-to-weight ratio less than 0.037 hp/lb were classed as low performance, medium performance vehicles had power-to-weight ratio between 0.037 hp/lb and 0.05 hp/lb, while vehicles with power-to-weight ratio greater than 0.05 hp/lb were classed as high performance vehicle. No super-high-performance vehicles such as sports cars (usually power-to-weight > 0.06 hp/lb) and supercars were used in this study.

Study Data

Both controlled experiments and naturalistic driving studies were performed to understand the effect of vehicle performance on driving style. Microscale driving activity and vehicle dynamics information on predetermined routes were gathered from controlled experiments. In contrast, the naturalistic study focused on a smaller number of drivers over a longer time period.

Controlled Experiment Data

The detail methodology of the controlled experiment including selection of vehicles, drivers, route, time of day, and quality control process is described in Frey et al. (*23*). A similar approach was adopted in conducting the experiments in 2013 and 2016. Field measurements of microscale vehicle activity and tailpipe emissions were conducted for five drivers tagged randomly to preserve anonymity. The measurements resulted in 17 driving trials with Drivers 1 and 2, operating seven and four different vehicles, respectively, and Drivers 3, 4, and 5 each operating two different vehicles. Driver and vehicle attributes were recorded at the beginning of each experiment. The specifications of the selected vehicles are shown in Table 2.

Two origin–destination (O-D) pairs were selected as the real-world routes to conduct the controlled experiment: one from North Carolina State University campus to North Raleigh and the other from North Raleigh to Research Triangle Park. The routes were selected to include a wide range of variability in the regular traveling conditions. Both of the selected routes included a combination of roads with different functional categories such as freeways, ramps, major arterials, minor arterials, and feeder/collector streets. Also, the operating conditions of the selected routes captured a broad range of engine loads, which was suitable for studying the effect of vehicle performance on driving style.

Instantaneous vehicle speed was recorded from the CANbus system of the vehicles through the OBD port. An OBD link scan tool was used as a recorder of the engine data. Instantaneous acceleration (1 Hz) was calculated from the speed by subtracting the previous speed instance from the current instance. Locations of the vehicles were measured using Global Positioning System

Table 3. Specifications of the Selected Vehicles in the Naturalistic Driving Experiment

Driver ID	Test ID[a]	Engine disp. (L)	Power (hp)	Curb weight (lbs)	Power-to-weight (hp/lb)
A	A2	5.3	320	5,400	0.048
A	A1	1.8	126	2,700	0.035
B	B2-1	2.5	165	3,140	0.039
B	B2-2	2.0	148	2,696	0.041
C	C1	2.2	136	2,998	0.035
D	D1	2.0	146	2,809	0.037
E	E2	3.0	200	3,340	0.045

[a]Test ID represent two parts: Driver ID and Vehicle Class ID.

(GPS) receivers. The height information from the GPS integrated with data from a barometric altimeter provided estimation of the road grade (*28*).

Tagging for the Test ID field in Table 2 is done in such a way that the first part of the Test ID contains driver ID, the second part consists of the vehicle performance class ID (1, 2, 3 represents low performance, medium performance, and high performance vehicles, as defined above), the third part contains a unique vehicle ID, and the fourth part represents the test year (YY format).

Naturalistic Driving Dataset

Naturalistic driving data for this study were collected using an on-board logging system called i2D (*29*). The system consists of an on-board unit (OBU), a mobile communications network (via M2M protocols), and a secure cloud database. The OBU connects to the vehicle OBD-II interface, and includes a GPS sensor along with a three-dimensional accelerometer and a barometric altimeter. Multiple engine and vehicle dynamics measures are acquired from the OBU at high resolution (1 Hz) and transmitted to the cloud database using mobile communications every 23 seconds. An illustration of the device is shown in Figure 2.

Participation in this study was voluntary and 35 drivers were recruited randomly in Raleigh, NC. The study participants were anonymized and were not provided with any specific instructions during the study. The observed vehicles used were personal vehicles of the participants. Participants were not required to disclose specific information about their vehicles in order to maintain anonymity. Participants were mostly university staff and students with ages ranging from 20 years to 68 years. Out of the 35 participants, 25 were male and 10 were female. A total of 24 months of data were collected. Data collection started in April 2014 and ended in March 2016.

Participants were asked whether they would volunteer to disclose their vehicle type and usage information through a questionnaire. Out of the 35 participants, we received responses from only five drivers; of those only two used multiple vehicles during the study. To protect the privacy of the questionnaire respondents, only generic information regarding their vehicles was asked for, such as fuel type, engine size, horse power, and gross vehicle weight.

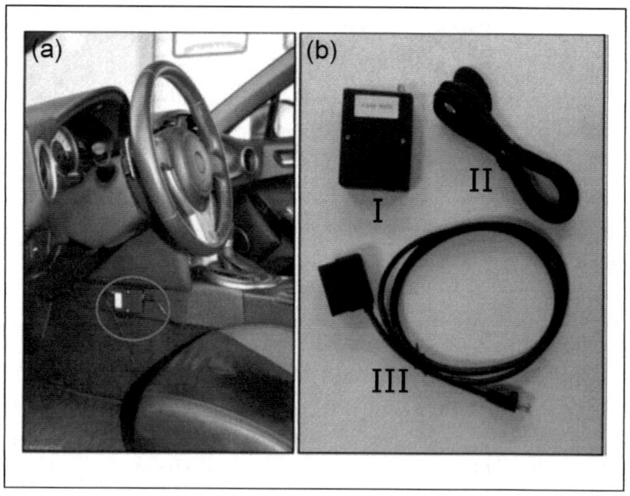

Figure 2. (*a*) i2D device connected to the vehicle. (*b*) I. the device II. antenna III. OBD-II connector cable.

Table 3 includes information of the seven driver–vehicle combinations studied from the naturalistic driving experiment based on the drivers who provided sufficient information. Tagging for the Test ID is done in a way which preserves both the vehicle and the driver information. The first letter part indicates the driver and the second numeric part represents the vehicle performance class.

Results and Discussion

Aggregate driving activity statistics for different Test IDs were analyzed. The joint speed–acceleration distributions across different groups were compared. The driving style

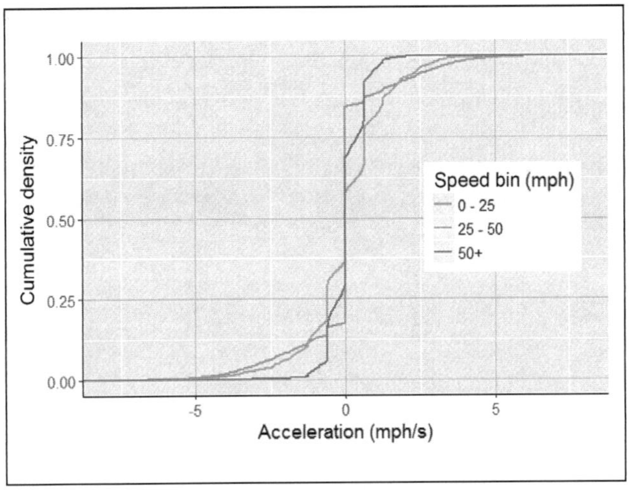

Figure 3. Cumulative density function of acceleration for Dodge Caravan (1-1-1-13) at different speed bins (n=15,450).

measure "envelope deviation" was calculated for each driving style and the effect of different vehicle power-to-weight classes on the estimated measure discussed.

Aggregate Microscale Driving Statistics

The distribution of acceleration, regardless of speed, for most driving styles appears to be surprisingly similar, with sharp peaks around 0 mph/s (Figure 3). Even after adjusting for idling time, acceleration distribution peaked sharply on the mean (mode and mean are similar because of symmetry in distribution), which is consistent with previous studies (*15*). Another feature to notice is the discrete nature of acceleration; as speed is recorded in the naturalistic study as discrete integers, there are a finite number of accelerations possible for the finite combinations of speeds.

Joint Speed–Acceleration Distributions

The joint speed–acceleration distribution for the *same driver* of the controlled experiment operating two vehicles with varying performance classes is shown in Figure 4. The black lines represent the real-world fleet activity envelope derived from Liu and Frey (*18*) to provide context to the joint distribution (as shown in Table 1). The visual representation of the distribution gives indication of the range of performance from these vehicles.

The overlapping nature of the points is shown in the heat-map style binned density plot. There are more extreme points in the high performance vehicle class than for low performance vehicles. For example, instances of deceleration greater than 5 mph/s in speeds over 40 mph (values shown with blue lines) were found in the high performance vehicle class. However, those points are sporadic and random, suggesting that even though there is the capability to perform high acceleration and deceleration events in high performance vehicles, the driver is rarely taking advantage of them.

Similar comparisons were done for other drivers for both controlled and naturalistic studies using multiple vehicle performance classes. The speed–acceleration joint distributions were found to be similar, with varying levels diffusion of the points near the edge of the distributions.

Envelope Deviation in Controlled Experiments

Descriptive statistics such as the percentage of area in positive, mean, and median of the developed measure

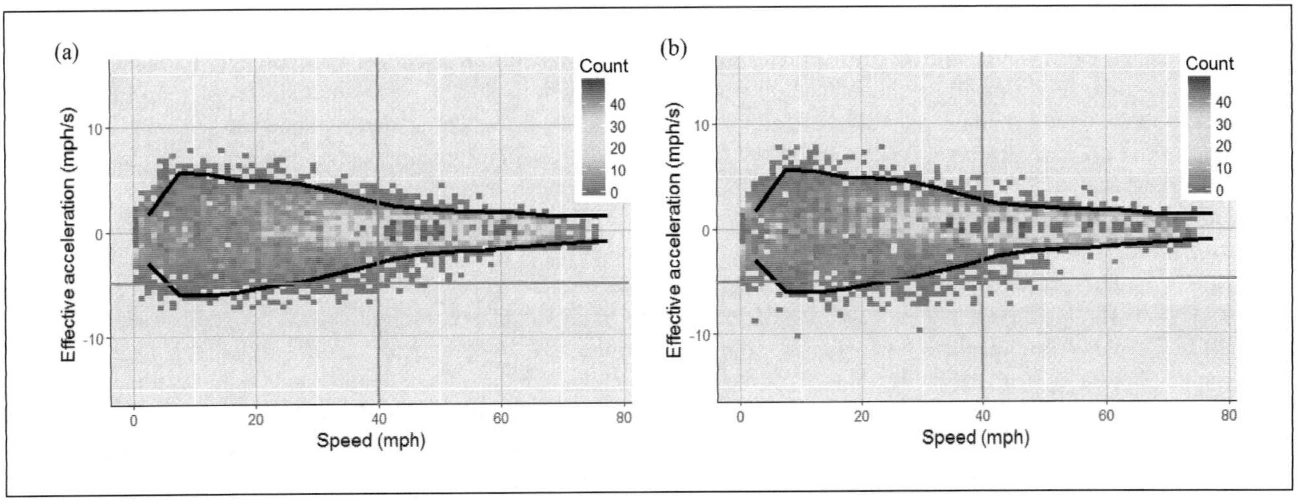

Figure 4. Joint speed–acceleration distribution binned density plots for low performance vehicle (0.033 hp/lb) (*a*) and high performance vehicle (0.059 hp/lb) (*b*): (*a*) Dodge Caravan (1-1-1-13) and (*b*) Chevrolet Impala (1-3-7-13). Black lines represent real-world vehicle activity envelope.

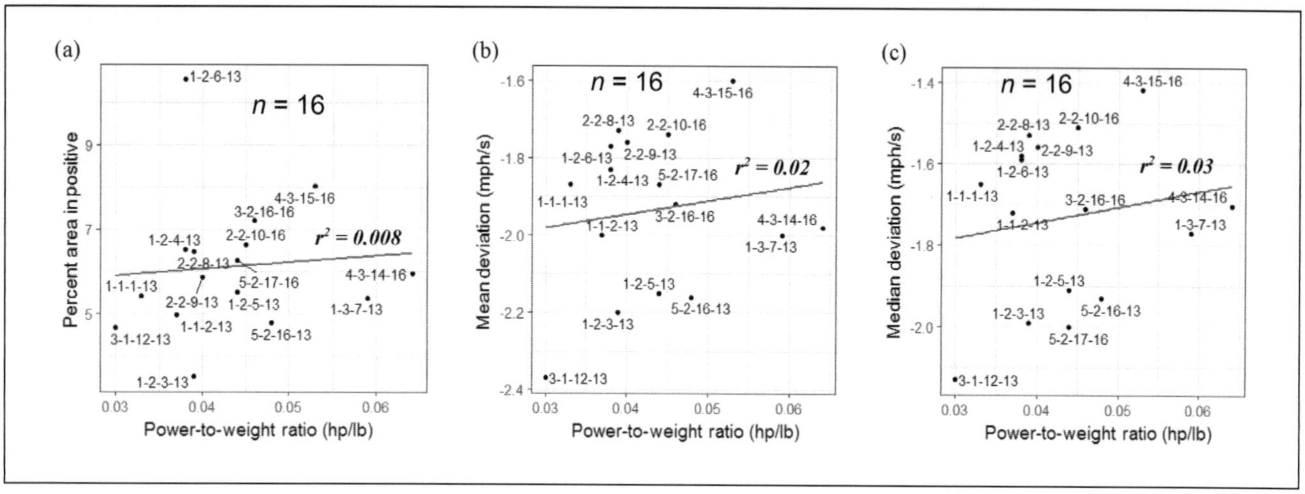

Figure 5. Measured variabilities in the summaries of envelope deviation distributions under varying power-to-weight ratios: (*a*) percentage of area in positive, (*b*) mean deviation (mph/s), and (*c*) median deviation (mph/s).

envelope deviation were calculated. By plotting all these variables against vehicle power-to-weight ratio, a very weak relationship with considerable variability was observed (Figure 5). Also, the slopes of the fitted lines in Figure 5 do not imply significant relationships.

The common expectation is that the more powerful the vehicle, the more it will generate deviations that exceed the real-world activity envelope, that is, by greater than 5%. In this study, 12 out of 16 cycles surpassed the 5% threshold. There was no indication of high performance vehicles exceeding the threshold more than the low performance vehicles; in fact medium performance vehicles generated both the highest and lowest aggressive cycles. The most aggressive cycle was from a 2012 VW Passat and the least aggressive cycle was from a 2012 Ford Fusion, which had power-to-weight ratios of 0.038 hp/lb and 0.039 hp/lb respectively. In addition, the lack of samples from very low and very high performance vehicle categories may have contributed to the insignificance of the relationship.

Pairwise comparisons of the cycles using all data for each cycle under consideration were done using the two-sample K–S test statistic, D. Table 4 includes a sample of the comparison between selected cycles from Driver 1 and Driver 2. Comparison of cycles from Driver 1 with the cycles from Driver 2 showed high D values ranging from 0.11 to 0.18. However, no pattern was detected which shows cycles from medium performance vehicles in Driver 1 had greater similarity with those from Driver 2. The cycles recorded with Driver 1 showed high variability in observed D, ranging from 0.04 to 0.15. It shows that the true signal of a driving style may become unclear from the noise induced by the traffic conditions (even the controlled experiments cannot control for all the nuances of the testing routes).

Distances among the envelope deviation distributions for the same driver were found to be in the range of 0.002 to 0.13. This is the *inter-vehicle heterogeneity* component of driving style controlling for the driver. The mean distances among the vehicles used by each driver (found from averaging all the shaded items in Table 4) were: Driver 1 0.09, Driver 2 0.002, Driver 3 0.03, Driver 4 0.05, and Driver 5 0.03. The average of the mean driver distance was 0.044.

To compare drivers, assuming vehicle performance is the same, distances among the empirical envelope deviation distributions for the same vehicle performance class were calculated. This distance is a measure of *inter-driver heterogeneity* of driving style when controlling for vehicle performance classes (Figure 6*a*). Inter-driver heterogeneity was found in the range of 0.002 to 0.19. The mean distance among the drivers driving low performance vehicles was 0.08, among drivers driving medium performance vehicles 0.11, and among drivers driving high performance vehicles 0.12. The average K–S test statistics representing the inter-vehicle component of driving style was 0.1, which is about 2.5 times higher than the case when controlling for the driver but varying vehicle performance.

Both of the previous measures were calculated while controlling for either vehicles or for the driver. However, there were conditions in this experiment where the same driver–vehicle combination was replicated multiple times. If hypothetical "true" styles for each driver–vehicle combination were present, then multiple observations would help us to identify them. In this case, multiple observations were used to quantify the variability of the measured statistic, D. This variability is termed as the *intra-driver-vehicle variability*.

For example, the intra-driver-vehicle variability for Driver 1 in medium vehicle performance group was 0.10.

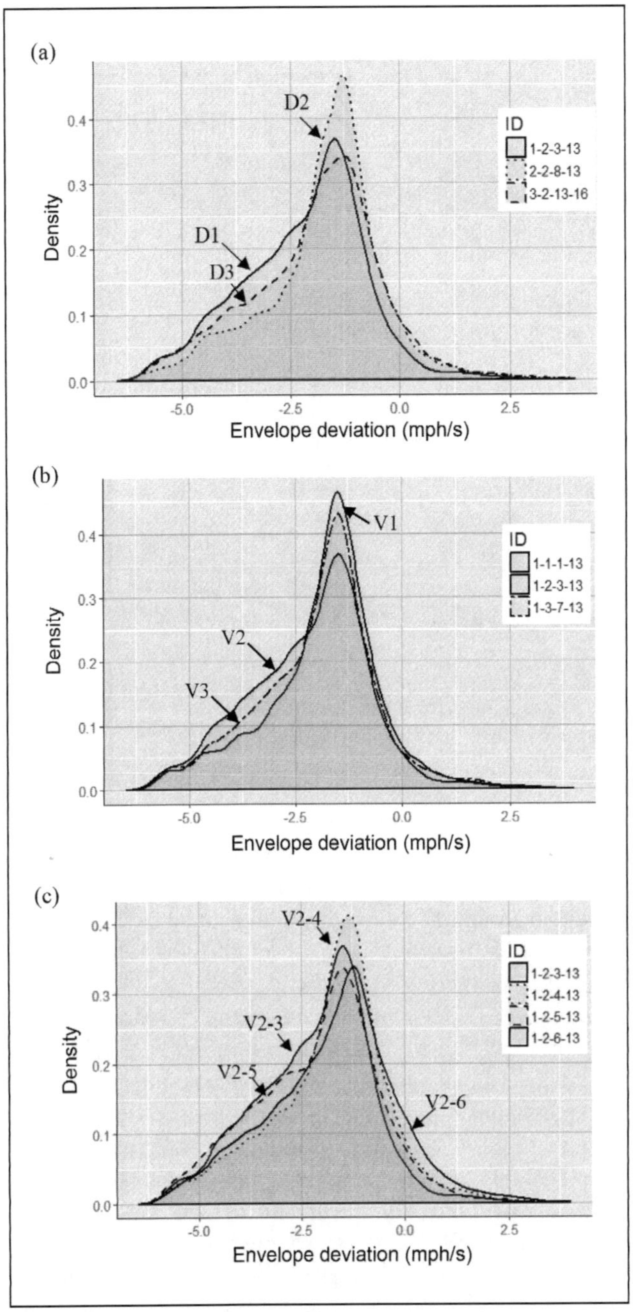

Figure 6. Comparison of envelope deviation distributions for the controlled experiment: (*a*) Drivers 1, 2, and 3 tested in medium performance vehicles (0.037 hp/lb < power-to-weight < 0.05 hp/lb), (*b*) Driver 1 tested for three vehicle performance classes, and (c) Driver 1 tested with multiple medium performance vehicles.

Intra-driver-vehicle variability for Driver 2 in medium vehicle performance group was small, at 0.002. The high variability in the first case may be due to unobserved variation by congestion along the test route on a particular day. The only observable congestion indicators on the test routes were average speed on the routes, which were 27.8 mph (1-2-3-13), 30.1 mph (1-2-4-13), 29.3 mph (1-2-5-13), and 31.3 mph (1-2-6-13). Intra-driver–vehicle variabilities did not show any correlation to the observed cycle average speed.

Envelope Deviation in Naturalistic Studies

The envelope deviation distributions for the naturalistic study for Drivers A and B are plotted together in Figure 7. The amount of variability in distances among the seven driver–vehicle combinations were found to be similar to that of the controlled experiment driving cycles. However, the percentage of the area in positive in the naturalistic observations was in the range of 5% to 7% which is narrower than the range for controlled experiments (4% to 10%).

Both Driver A and Driver B had each generated long-term microscale activity data from two different vehicles. One of the vehicles used by Driver A (ID = A2) had very different engine characteristics than the other. However, the K–S test statistic between A1 and A2 were found to be 0.05, which is comparable to the inter-vehicle heterogeneities estimated for controlled experiments. The two vehicles used by Driver B were from the medium vehicle performance class. The distance between B2-1 and B2-2 was found to be 0.04.

Driver E was found to be the most aggressive driver in the naturalistic vehicle fleet. This may be attributed either to the higher performance of the vehicle relative to the vehicles of the fleet or to the inherently "aggressive" nature of the driver. Such conclusions cannot be drawn without collecting more data. Although more data from other drivers are available from the naturalistic study, only a subset of the drivers ($N = 5$) were studied in this current research. We are currently working on a revised experimental protocol to enhance the sample size of the naturalistic study.

Conclusions and Future Work

The proposed envelope deviation distribution is found to be an effective way of describing driving style. It is essentially a transformation of the joint speed–acceleration distribution of a cycle with respect to the typical real-world driving cycle. The typical real-world drive cycle was defined by the vehicle activity envelope developed from a collection of 100 light duty vehicles. Hence, envelope deviation distribution is a metric of driver aggressiveness which quantifies the amount of deviation of driving style from the typical vehicle activity.

The standard drive cycles such as FTP, HWFET, SC03, US06, and UDDS were compared using envelope deviation distribution. The findings suggest consistency of the envelope deviation with other established

Table 4. Example Two-Sample K–S Test Statistics, D for Pairwise Comparison of Deviation Distributions from Baseline in Controlled Experiments

ID	1-1-1-13	1-2-3-13	1-2-4-13	1-2-5-13	1-2-6-13	1-3-7-13	2-2-8-13	2-2-9-13	3-2-13-16
1-1-1-13	0*	*0.13*	*0.05**	*0.12*	*0.09*	*0.06*	0.17	0.16	0.10
1-2-3-13		0*	**0.13**	**0.04**	**0.15**	*0.08*	0.17	0.18	0.16
1-2-4-13			0*	**0.11**	**0.05**	*0.09*	0.14	0.15	0.21
1-2-5-13				0*	**0.12**	*0.06*	0.15	0.16	0.18
1-2-6-13					0*	*0.12*	0.18	0.19	0.12
1-3-7-13						0*	0.11	0.11	0.06
2-2-8-13							0*	0.002*	0.26
2-2-9-13								0*	0.25
3-2-13-16									0*

Note: * = failure to reject the null hypothesis. Bold cells represent intra-driver-vehicle variability for Driver 1Vehicle Performance Class 2. Bold and italicized cells in combination represent inter-vehicle heterogeneity for Driver 1.

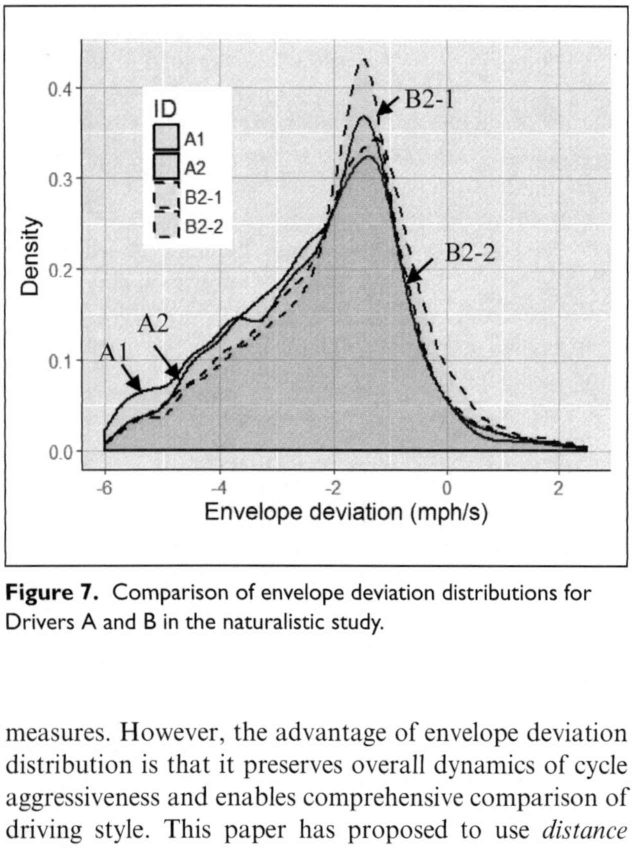

Figure 7. Comparison of envelope deviation distributions for Drivers A and B in the naturalistic study.

measures. However, the advantage of envelope deviation distribution is that it preserves overall dynamics of cycle aggressiveness and enables comprehensive comparison of driving style. This paper has proposed to use *distance* (term given to the two-sample K–S test statistic, D) among driving cycles to compare and understand variabilities in driving styles.

To understand variabilities in driving styles with vehicle performance, each controlled and naturalistic study driving cycle was assigned to one of the three vehicle performance classes, that is, low, medium, and high. It was found that difference in driving styles measured as distance can be decomposed into three components: inter-driver heterogeneity, inter-vehicle heterogeneity, and intra-driver-vehicle variability. Inter-driver heterogeneity is variabilities in estimated distances among drive cycles assuming the vehicle performance class was kept the same, whereas, inter-vehicle heterogeneity is variabilities in measured distances while controlling for the driver. Intra-driver-vehicle variability is the measure of repeatability in estimated driving style metrics for the same vehicle performance class and driver combination.

The decomposition of driving style helped us to answer both the research questions asked previously. Significant differences in a driver driving style at different vehicle performance classes can be implied if the inter-vehicle heterogeneity is significantly higher than the repeatability of the driving cycle measurements, that is, intra-driver-vehicle variability. In contrast, significance in differences among driving styles of different drivers operating vehicles of similar performance can be confirmed when the inter-driver heterogeneity is significantly higher than the intra-driver-vehicle variability component.

In both the controlled experiments and the limited scope of the naturalistic study (not all vehicle classes were available in naturalistic data), it was found that the inter-vehicle heterogeneity is not significantly higher than the intra-driver-vehicle variability. However, inter-driver heterogeneity was significantly higher than the intra-driver-vehicle variability. These findings imply that operating vehicles of different performance classes does not result in significantly different driving styles. This implication suggests that an eco-driving scheme can be directly tied to the driver and it will not require information regarding the vehicle unless exact estimation of the amount of fuel savings is necessary. Moreover, this is useful information for designing eco-driving schemes where microscale driving data is passively collected from devices such as GPS-enabled smartphones, and navigation units, among others. While the passive mode of retrieving driving style information helps substantially to increase participation in eco-driving schemes, vehicle-related information from the participants is seldom available due to privacy concerns. With the assumption that driving style can be characterized without vehicle-related information, information and incentives pertaining to driving style improvements can be provided to the driver directly.

Generalization of the observations from this research is not possible until a significant amount of replications are available for a multitude of driver–vehicle combination groups. In the future the experimental dataset will be expanded to include more samples of low and high performance vehicle classes. Additional vehicle performance indicators will be explored to examine the high variability of driving styles in the medium vehicle performance class.

Acknowledgments

This research is funded partially by the US Department of Energy (DOE) Advanced Research Project Agency - Energy (ARPA-E). The authors acknowledge the students of Frey Research Group for collecting and processing real-world microscale vehicle activity data. Thomas Chase and Behzad Agdashi from the Institute for Transportation Research and Education have helped with the collection of naturalistic driving data.

Author Contributions

The authors confirm contribution to the paper as follows: study conception and design: Shams Tanvir, H. Christopher Frey; data collection: Shams Tanvir, H. Christopher Frey; analysis and interpretation of results: Shams Tanvir, H. Christopher Frey, Nagui Rouphail; draft manuscript preparation: Shams Tanvir. All authors reviewed the results and approved the final version of the manuscript.

References

1. U.S. Energy Information Administration. *Monthly Energy Review*, DOE/EIA-0035(2017/6). U.S. Department of Energy, Washington, D.C., 2017.
2. Kilian, L., and C. Park. The Impact of Oil Price Shocks on the US Stock Market. *International Economic Review*, Vol. 50, No. 4, 2009, pp. 1267–1287.
3. Bandivadekar, A., L. Cheah, C. Evans, T. Groode, J. Heywood, E. Kasseris, M. Kromer, and M. Weiss. Reducing the Fuel Use and Greenhouse Gas Emissions of the US Vehicle Fleet. *Energy Policy*, Vol. 36, No. 7, 2008, pp. 2754–2760.
4. Graham-Rowe, E., S. Skippon, B. Gardner, and C. Abraham. Can We Reduce Car Use and, if so, how? A Review of Available Evidence. *Transportation Research Part A: Policy and Practice*, Vol. 45, No. 5, 2011, pp. 401–418.
5. Onoda, T. IEA policies—G8 Recommendations and an Afterwards. *Energy Policy*, Vol. 37, No. 10, 2009, pp. 3823–3831.
6. Bin, S., and H. Dowlatabadi. Consumer Lifestyle Approach to US Energy use and the Related CO_2 Emissions. *Energy Policy*, Vol. 33, No. 2, 2005, pp. 197–208.
7. Malliaris, A. C., H. Hsia, and H. Gould. *Concise Description of Auto Fuel Economy and Performance in Recent Model Years*. SAE Technical Paper 760045. Society of Automotive Engineers, Detroit, 1976.
8. Brundell-Freij, K., and E. Ericsson. Influence of Street Characteristics, Driver Category and Car Performance on Urban Driving Patterns. *Transportation Research Part D: Transport and Environment*, Vol. 10, No. 3, 2005, pp. 213–229.
9. Fontaras, G., N. Zacharof, and B. Ciuffo. Fuel Consumption and CO_2 Emissions from Passenger Cars in Europe – Laboratory Versus Real-World Emissions. *Progress in Energy and Combustion Science*, Vol. 60, 2017, pp. 97–131.
10. Zhou, X., S. Tanvir, H. Lei, J. Taylor, B. Liu, N. M. Rouphail, and H. C. Frey. Integrating a Simplified Emission Estimation Model and Mesoscopic Dynamic Traffic Simulator to Efficiently Evaluate Emission Impacts of Traffic Management Strategies. *Transportation Research Part D: Transport and Environment*, Vol. 37, 2015, pp. 123–136.
11. Tanvir, S., N. Karmakar, N. M. Rouphail, and B. J. Schroeder. Modeling Freeway Work Zones with Mesoscopic Dynamic Traffic Simulator: Validation, Gaps, and Guidance. *Transportation Research Record: Journal of the Transportation Research Board*, 2016. 2567: 122–130.
12. Evans, L. Driver Behavior Effects on Fuel Consumption in Urban Driving. In *Proceedings of the Human Factors Society Annual Meeting*, Sage Publications, Los Angeles, Calif., 1978, pp. 437–442.
13. Liu, J., A. Khattak, and X. Wang. A Comparative Study of Driving Performance in Metropolitan Regions Using Large-Scale Vehicle Trajectory Data: Implications for Sustainable Cities. *International Journal of Sustainable Transportation*, Vol. 11, No. 3, 2017, pp. 170–185.
14. Ericsson, E. Variability in Urban Driving Patterns. *Transportation Research Part D: Transport and Environment*, Vol. 5, No. 5, 2000, pp. 337–354.
15. USEPA. *Federal Test Procedure Review Project: Preliminary Technical Report*. EPA 420-R-93-007. U.S. Environmental Protection Agency, Washington, D.C., 1993.

16. Berry, I. M. *The Effects of Driving Style and Vehicle Performance on the Real-World Fuel Consumption of US Light-Duty Vehicles*. Doctoral dissertation. Massachusetts Institute of Technology, 2010.
17. Stichter, J. S. *Investigation of Vehicle and Driver Aggressivity and Relation to Fuel Economy Testing*. Masters thesis. The University of Iowa, 2012.
18. Liu, B., and H. C. Frey. Measurement and Evaluation of Real-World Speed and Acceleration Activity Envelopes for Light-Duty Vehicles. *Transportation Research Record: Journal of the Transportation Research Board*, 2015. 2503: 128–136.
19. Sharer, P., R. Leydier, and A. Rousseau. *Impact of Drive Cycle Aggressiveness and Speed on HEVs Fuel Consumption Sensitivity*. SAE Technical Paper 2007-01-0281. Society of Automotive Engineers, Detroit, 2007.
20. Song, G., L. Yu, and Y. Zhang. Applicability of Traffic Micro-Simulation Models in Vehicle Emission Estimations: A Case Study of VISSIM. *Transportation Research Record: Journal of Transportation Research Board*, 2012. 2270: 132–141.
21. Anya, A., N. Rouphail, H. Frey, and B. Schroeder. Application of AIMSUN Microsimulation Model To Estimate Emissions on Signalized Arterial Corridors. *Transportation Research Record: Journal of the Transportation Research Board*, 2014. 2428: 75–86.
22. Frey, H. C., A. Unal, N. M. Rouphail, and J. D. Colyar. On-Road Measurement of Vehicle Tailpipe Emissions Using a Portable Instrument. *Journal of the Air & Waste Management Association*, Vol. 53, No. 8, 2003, pp. 992–1002.
23. Frey, H. C., K. Zhang, and N. M. Rouphail. Fuel Use and Emissions Comparisons for Alternative Routes, Time of Day, Road Grade, and Vehicles Based on In-use Measurements. *Environmental Science & Technology*, Vol. 42, No. 7, 2008, pp. 2483–2489.
24. Van Mierlo, J., G. Maggetto, E. Van de Burgwal, and R. Gense. Driving Style and Traffic Measures-Influence on Vehicle Emissions and Fuel Consumption. *Proceedings of the Institution of Mechanical Engineers, Part D: Journal of Automobile Engineering*, Vol. 218, No. 1, 2004, pp. 43–50.
25. Bachman, W. H. *A GIS-Based Modal Model of Automobile Exhaust Emissions*. Epa/600/R-98/097 (NTIS PB98-165145). U.S. Environmental Protection Agency, Research and Development, National Risk Management Research Laboratory, Washington, D.C., 1998.
26. Andre, M., R. Joumard, R. Vidon, P. Tassel, and P. Perret. Real-World European Driving Cycles, for Measuring Pollutant Emissions from High- and Low-Powered Cars. *Atmospheric Environment*, Vol. 40, No. 31, 2006, pp. 5944–5953.
27. Johansson, H., P. Gustafsson, M. Henke, and M. Rosengren. Impact of EcoDriving on Emissions. *Proc., International Scientific Symposium on Transport and Air Pollution*, Avignon, France, 2003.
28. Yazdani Boroujeni, B., and H. C. Frey. Road Grade Quantification Based on Global Positioning System Data Obtained from Real-World Vehicle Fuel Use and Emissions Measurements. *Atmospheric Environment*, Vol. 85, 2014, pp. 179–186.
29. Livedrive. *i2D Telematics: How Does it Work*. www.i2d.co/i2dpubportal/portal/tec.xvw Accessed August 1, 2017.

The Standing Committee on Transportation and Air Quality (ADC20) peer-reviewed this paper (18-02023).

TRR
JOURNAL OF THE TRANSPORTATION RESEARCH BOARD

Article

Transportation Research Record
2018, Vol. 2672(25) 79–88

DOI: 10.1177/0361198118796710
journals.sagepub.com/home/trr
SAGE

An Investigation of the Uncertainty of Handbook of Emission Factors for Road Transport (HBEFA) for Estimating Greenhouse Gas Emissions: A Case Study in Beijing

Hongyu Lu[1], Guohua Song[1], Qi Zhao[2], Jingyi Wang[1], Weinan He[3], and Lei Yu[4,5]

Abstract

The reliability of estimation of emissions under the Handbook of Emission Factors for Road Transport (HBEFA) model has not been thoroughly studied and the mechanics of the model have not been fully investigated. The HBEFA structure of typical driving cycles is analyzed in this study regarding its uncertainty for estimating greenhouse gas (GHG) emissions. The key parameter of the HBEFA model, relative positive acceleration (RPA), is analyzed for its sensitivity to GHG emissions. The results indicate that uncertainties in estimation of GHG emissions do exist in the HBEFA model. The uncertainties are derived from the adoption of selected driving cycles and from the parameter of RPA. Given the uncertainties of estimating the GHG emission factors of driving cycles, it might be difficult for the selected driving cycles to represent all traffic situations. Uncertainties also exist in the selection of the top-20 driving cycles. For the selected samples of driving cycles, the sensitivity of RPA is found to be insufficient in describing the emission differences, due to the insensitivity of RPA to acceleration and deceleration. Different driving characteristics may still yield the same RPA, which indicates the potential inability of RPA to capture specific traffic scenarios such as intersections, ramps, and congestion.

The Handbook of Emission Factors for Road Transport (*1*), or HBEFA, is one of the common emission models mainly applied in Europe. Errors have been found in the estimates of emissions under the HBEFA model, while the uncertainty of the emission estimation of HBEFA remains to be investigated. As an emission model based on typical driving cycles and relative positive acceleration (RPA), the model mechanics of the HBEFA model have not been thoroughly investigated. Are the typical driving cycles capable of describing the full details of the driving behaviors of the entire traffic situation? Is the emission-sensitivity of RPA sufficient to reflect various driving activities, especially changes of acceleration? To answer these questions, the coherent mechanics of HBEFA need to be analyzed, with in-depth analysis of the typical driving cycles and the key variable, RPA. The analysis targets not only the average accuracy of the model, but also the model's uncertainty when it is required to describe the full details of the emission estimation. The aim of this study is to investigate the uncertainty of GHG emission estimation under the HBEFA model.

HBEFA is a commonly used emission model mostly applied in Austria, Germany, Switzerland, and Sweden (*2*). HBEFA provides the emission factors (EFs) of a certain category of vehicle for particular traffic situations. Local vehicle trajectories are needed for HBEFA to develop typical driving cycles and a localized HBEFA model of Beijing, China, is developed by He et al. (*3*).

Errors have been found in the emission estimation of the HBEFA model, including overestimation (*4–6*),

[1]MOE Key Laboratory for Urban Transportation Complex Systems Theory and Technology, School of Traffic and Transportation, Beijing Jiaotong University, Beijing, P.R. China
[2]School of Electrical and Control Engineering, Beijing Key Laboratory of Urban Intelligent Traffic Control Technology, North China University of Technology, Beijing, P.R. China
[3]Department of Mobile Source Emission Research, Beijing Transport Institute - Energy & Environment Center, Beijing, P.R. China
[4]Texas Southern University, Houston, TX
[5]Xuchang University and Beijing Jiaotong University, P.R. China

Corresponding Author:
Address correspondence to Guohua Song: GhSong@bjtu.edu.cn

underestimation (*7–11*), or "puzzling coincidence that cannot be explained" (*12*). However, few previous studies have aimed to examine the uncertainty of the emissions estimates under the HBEFA model.

HBEFA is based on typical driving cycles and uses RPA as the key parameter. The mechanics of HBEFA are not fully understood, especially their effects on the uncertainty of emission estimations. It has been found that the use of single typical driving cycles may cause inaccuracy in the emission estimation (*13–15*). One trajectory might not be sufficient to describe the vehicle activities of the entire traffic situation. For the HBEFA model, it has been found that the representativeness of the typical driving cycle is crucial to the accuracy of the emission estimation. Some "typical" driving cycles used in HBEFA studies might not reflect the traffic reality (*16*).

The parameter RPA is one of the three key indices used to determine the typical driving cycles of the HBEFA model. However, the correlation between RPA and emissions has yet to be fully understood and the emission-sensitivity of RPA remains to be studied.

In comparison with RPA, vehicle specific power (VSP) is a widely used emission-sensitive parameter, applied in emission models such as MOtor Vehicle Emission Simulator (MOVES) (*17*) and International Vehicle Emissions (IVE) (*18*). It has been found that the VSP distributions are able to describe different traffic characteristics (*19*, *20*). An emission model to derive EFs from speed-specific VSP distributions has been established by Song et al. (*21*).

The aim of this study is to analyze the uncertainty of the estimations of GHG emissions using the HBEFA model. By investigating the representativeness of the HBEFA driving cycles and the GHG emission-sensitivity of RPA, the mechanics of the HBEFA model are studied, with VSP distributions as the comparable variable.

Methodology

This study is organized as follows. First, the field data of a large number of vehicle trajectories and vehicle GHG emission data are described. Second, the speed-specific VSP distributions and the GHG EFs based on VSP are derived. Third, the HBEFA driving cycles are obtained and the top-20 driving cycles are selected, with RPA, average speed, and stop fraction as the indices. Fourth, the uncertainty derived from the driving cycles of the HBEFA model is investigated, as well as the uncertainties of the top-20 driving cycles. Finally, the uncertainty derived from using RPA as a parameter is studied by analyzing the GHG emission-sensitivity of RPA.

Data Preparation

Two types of field data are used in this study (*20*). The first is the second-by-second vehicle activity data of light-duty vehicles (LDVs), and the second is the GHG emission data of gasoline-powered LDVs. The first type of data is used to develop the VSP distributions and the HBEFA driving cycles. The second is used to obtain the average emission rates.

The vehicle activity data were collected in Beijing using portable GPS devices. The activity data includes 7,547,355 records of second-by-second vehicle trajectories, provided by 178 LDVs. The data was collected from April 25, 2004 to April 16, 2013. A map-matching algorithm was employed to match the vehicle trajectories to the geo-information system (GIS) map of the road network of Beijing, in order to obtain information on road type for each record of the vehicle activity data.

The vehicle GHG emission data was collected using a portable emissions monitoring system instrument on May 8, 2009, in Beijing. The tested vehicle is a manual transmission type of Hyundai Sonata with a 1.6 L conventional internal combustion engine and gross weight of 1,435 kg. The tested vehicle was a 2007 model with a mileage of 31,060 km. A total of 10,870 second-by-second records of the mass emission of GHG from the tailpipe exhaust and the corresponding vehicle speed were collected.

Emission Model Based on VSP

The GHG EFs based on VSP are calculated with the following procedures. First, micro-trips are derived from the field vehicle activity data. Second, speed-specific VSP distributions are obtained by categorizing the micro-trips by average travel speed. Third, average emission rates within each VSP bin are calculated. Finally, the emission factor of each average speed bin is developed. The development of the emission model based on VSP is consistent with the previous studies (*19–21*).

Micro-Trips for Expressway, Major Arterial, and Collector Roads. The vehicle trajectories are categorized by three types of road: expressway, major arterial and collector (minor arterial and branch) roads. The trajectories for each type of road are then divided into micro-trips. Each micro-trip of the expressway consists of 60 records of a continuous second-by-second trajectory, and each micro-trip of the major arterial/collector roads consists of 180 records of a continuous second-by-second trajectory. The micro-trips are classified according to the average speed at intervals of 2 km/h, as in Equations 1 and 2:

$$Average\,Speed = \frac{Distance\,Traveled}{Time\,Spent} = 3.6 \times \frac{\sum_{i=1}^{N} v_i}{N} \quad (1)$$

$$\forall : Average\,Speed \in [n, n+2), \quad Speed\,Bin\# = n, (n \text{ is an integer}) \quad (2)$$

where v is the instantaneous speed (m/s), N is the records of the micro-trip (60 for the expressway and 180 for the major arterial and collector roads), and *Speed Bin#* is the ID of the average speed bin.

Speed-Specific VSP Distribution. The VSP is defined by a typical equation for LDVs as shown in Equation 3:

$$VSP = v \times [1.1 \times a + 9.81 \times grade + 0.132] + 0.000302 \times v^3 \quad (3)$$

where v is the speed (m/s), a the acceleration (m/s^2), and *grade* (%) the vertical rise divided by the slope length, which is assumed to be zero in this study after excluding the data on gradient of roads. The VSP of each record of the trajectories is binned up at an interval of 1 kW/t, as shown in Equation 4:

$$\forall : VSP \in [i - 0.5, i + 0.5), \quad VSP\,\text{Bin\#} = i \quad (i\,\text{is an integer from} - 20\,\text{to}\,20) \quad (4)$$

where *VSP Bin#* is the ID of the VSP bin.

The VSP distribution of the average speed bin of 30 to 32 km/h on the expressway is presented as an example in Figure 1. More distributions could be shown but are omitted from the paper due to the length limit and can be found in Song et al. (*19*).

Emission Factor Based on VSP. The average GHG emission rates (in the unit of g/s) of the LDVs are calculated by employing the VSP binning method, as shown in Figure 1.

The EF of each average speed bin based on VSP is obtained as Equation 5:

$$EF_v = 3600/v \times \sum_{i=-20}^{20} ER_i \cdot VSPfraction_i \quad (5)$$

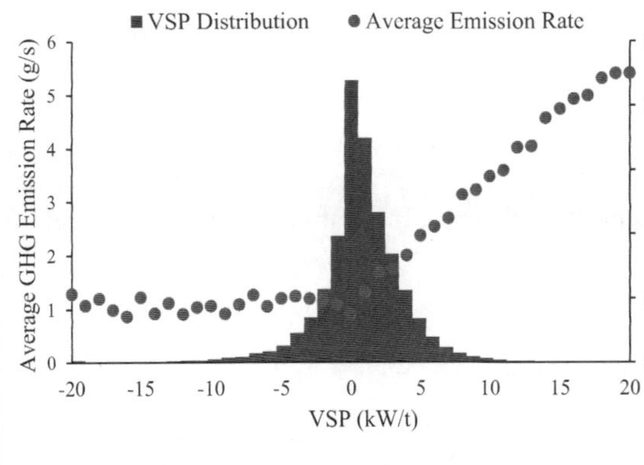

Figure 1. Average emission rates and example of speed-specific VSP distributions.

where EF_v is the emission factor (g/km) based on VSP at the average speed of v, ER_i is the average emission rate of the VSP bin i, and $VSPfraction_i$ is the time fraction of the VSP bin i.

Selection of Driving Cycles of HBEFA

According to the HBEFA model, the selection of the typical driving cycles includes three steps (*22*):

1. Driving cycles are categorized by road types, and then divided into traffic situations (in this case, levels of service, LOS).
2. The 20 most "concentrated" driving cycles are selected as the top-20 cycles, with the normalized indices of RPA, average speed, and stop fraction.
3. The typical driving cycle for each LOS is "reasonably selected" manually.

Since the aim of this study is not to investigate the average GHG EFs of the HBEFA model, in this study, the third step is not adopted and the analysis of the uncertainty is focused on all the driving cycles and the top-20 driving cycles.

Driving Cycles Categorized by Level of Service. The LDV trajectories are categorized by the three types of road and then divided into driving cycles. Each driving cycle has a minimum distance traveled of 2,000 meters and a maximum duration of 400 seconds and in total 8,267 driving cycles are distinguished. The average speed of each driving cycle is calculated as in Equation 1.

The driving cycles are classified into five LOSs according to the average speed and type of road, as shown in Table 1.

Relative Positive Acceleration, Average Speed, and Stop Fraction. RPA is the key parameter used in the HBEFA model. The RPA of each driving cycle is calculated by Equation 6:

$$RPA = \frac{\sum_{i=1}^{n} a_i \cdot v_i \cdot \Delta t}{s} \quad (6)$$

where a_i is the instantaneous acceleration (m/s^2) of record i, v_i the instantaneous speed (m/s) of record i, Δt the time interval of the records (1 s in this study), n the record count of the driving cycle and s the distance traveled (m) of the driving cycle.

Three parameters are normalized and used as the indices in selecting the top-20 driving cycles for HBEFA (*3*), which are RPA, average speed, and stop fraction. The normalization is conducted as Equations 7 to 9:

Table 1. Driving Cycles of the Levels of Service of Different Road Types

Road type	LOS	Speed range (km/h)	Driving cycle count	Total time (h)	Average of top-20 driving cycles		
					Average speed (km/h)	RPA (m/s^2)	Stop fraction
Expressway	Free flow	$v > 65$	800	22.4	71.1	0.106	0.00%
	Saturated traffic	$50 < v \leq 0 <$	1,352	47.3	57.1	0.124	0.00%
	Heavy traffic	$35 < v \leq 50$	670	31.3	42.4	0.142	0.09%
	Stop and go	$20 < v \leq 35$	591	43.9	27.1	0.154	2.62%
	Heavy stop and go	$v \leq 20$	395	43.5	12.6	0.159	23.54%
Major arterial	Free flow	$v > 40$	298	11.7	48.9	0.132	0.00%
	Saturated traffic	$30 < v \leq 40$	292	17.1	34.1	0.178	11.95%
	Heavy traffic	$20 < v \leq 30$	564	46.1	24.3	0.197	27.09%
	Stop and go	$15 < v \leq 20$	271	29.5	17.6	0.218	41.74%
	Heavy stop and go	$v \leq 15$	483	53.7	9.2	0.219	61.35%
Collector	Free flow	$v > 35$	340	15.9	40.4	0.139	0.87%
	Saturated traffic	$25 < v \leq 35$	512	35.2	28.7	0.166	7.18%
	Heavy traffic	$15 < v \leq 25$	936	94.7	18.6	0.184	21.63%
	Stop and go	$10 < v \leq 15$	580	64.4	12.3	0.187	36.36%
	Heavy stop and go	$v \leq 10$	598	66.4	5.7	0.185	63.83%

$$Z_v = \frac{v - \mu_v}{\sigma_v} \tag{7}$$

$$Z_{RPA} = \frac{RPA - \mu_{RPA}}{\sigma_{RPA}} \tag{8}$$

$$Z_{SGF} = \frac{v - \mu_{SGF}}{\sigma_{SGF}} \tag{9}$$

where Z_v, Z_{RPA}, and Z_{SGF} are the normalized indices for average speed, RPA, and stop fraction, μ_v, μ_{RPA}, and μ_{SGF} are the mean values of average speed, RPA, and stop fraction, σ_v, σ_{RPA}, and σ_{SGF} are the standard deviations of average speed, RPA, and stop fraction.

Top-20 Driving Cycles. Among all the driving cycles, 20 potential typical driving cycles are selected for each type of road and each LOS. The selected 20 driving cycles are the most concentrated cycles among all the driving cycles regarding the three indices. An index Z is calculated to describe the extent of concentration of the driving cycle as in Equation 10. For each LOS, 20 driving cycles with the lowest Zs are selected as the top-20 driving cycles:

$$Z = \sqrt{(Z_v - Z_{RPA})^2 + (Z_{RPA} - Z_{SGF})^2 + (Z_V - Z_{SGF})^2} \tag{10}$$

According to the HBEFA model, for each LOS, a typical driving cycle is selected from the top-20 driving cycles. The aim of this study is not to evaluate the accuracy of the HBEFA GHG EFs on an average level, but to investigate the possible uncertainty of the GHG EFs. Hence the typical driving cycles are not selected in this study. Instead, the uncertainty of the top-20 driving cycles, which are the potential typical driving cycles, are analyzed.

Emission Factors of HBEFA Driving Cycles

In the original HBEFA model, a passenger car and heavy-duty vehicle emission model (PHEM) is applied to generate the EFs (*23*). With the input of the typical driving cycles, PHEM uses a field-calibrated engine map to simulate the EFs of the driving cycles.

Since the technical details of PHEM are not yet publicly known, the GHG EFs of the driving cycles in this study are obtained by assigning average emission rates to the trajectories of the driving cycles. The GHG emission of each second-by-second record is assigned with the average GHG emission rate of the corresponding VSP bin. The EF of each driving cycle is calculated as in Equation 11:

$$EF_{GHG} = \frac{1000 \times \sum_{i=1}^{n} ER_i}{s} \tag{11}$$

where EF_{GHG} is the EF of GHG of the driving cycle (g/km), s the distance traveled of the driving cycle (m), ER_i the assigned average emission rate of record i (g/s), and n the number of records of the driving cycle.

It should be noted that the emission assigning method does not influence the accuracy of this investigation. The emission rate serves as a control variable in this study. The EFs derived from both HBEFA and VSP are calculated based on the same emission rates.

Uncertainty/Relative Uncertainty of GHG EF under the HBEFA Model

For all the driving cycles (without the selection of top-20 driving cycles), two parallel boundaries of GHG EFs are obtained to determine the uncertainty. The boundaries

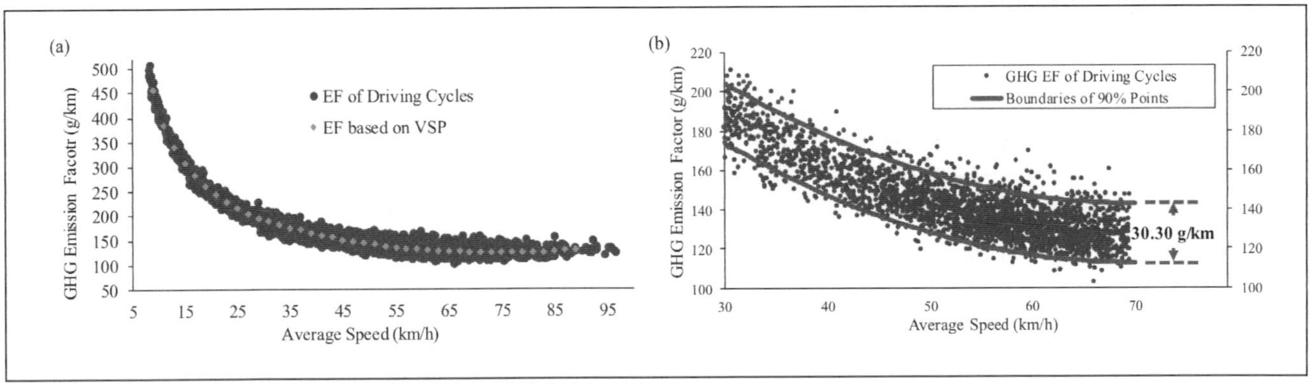

Figure 2. GHG EFs and uncertainty of all the driving cycles: (*a*) GHG EFs of all the driving cycles; (*b*) uncertainty of GHG EFs of all the driving cycles.

are calculated so that 90% of the GHG EFs fall within them. The uncertainty of the GHG EFs of the driving cycles is defined as the minimum possible difference between the boundaries. Thus 10% of the unselected driving cycles are excluded so that extreme driving cycles are not analyzed.

For the top-20 driving cycles, the uncertainty is defined as the range of the EFs, that is, the difference between the maximum GHG EF and the minimum GHG EF. The top-20 driving cycles, selected by the three indices, are supposed to be more representative than the unselected driving cycles, which is why no top-20 driving cycle is excluded in the uncertainty analysis.

The relative uncertainty of the GHG EF of the HBEFA model is defined as the quotient of the uncertainty divided by the average GHG EF, as in Equation 12:

$$RU = \frac{Uncertainty}{GHGEF_{Average}} \times 100\% \tag{12}$$

where RU is the relative uncertainty, $Uncertainty$ is the uncertainty of the GHG EF (g/km), and $GHG\ EF_{Average}$ is the average GHG EF (g/km). For the top-20 driving cycles, the average GHG EF refers to the average GHG EF of the 20 driving cycles in the LOS. For the unselected driving cycle of a certain speed bin, the average GHG EF refers to the GHG EFs of all the driving cycles in the speed bin.

Discussion

Uncertainty Derived from Use of Driving Cycles

The data on field trajectories are divided into 8,682 driving cycles and the GHG EFs of the driving cycles are calculated. The top-20 driving cycles are selected, as shown in Table 1, and the uncertainties are analyzed. The driving cycles for the expressway are introduced in this part as examples, but similar results were found for the major arterial and collector roads.

Uncertainty of All Driving Cycles. The driving cycles for the expressway are illustrated in Figure 2*a* as an example of the uncertainty of all the driving cycles. Compared with the GHG EF based on VSP, the GHG EFs of the driving cycles have more volatility, meaning they differ from one another and a considerable range exists between the maximum and the minimum EF in similar speed bins. The volatility is especially considerable in the average speed from 30 km/h to 70 km/h. For the lower volatility in the lower/higher speed range, a possible explanation is that, under conditions of either heavy congestion or free flow, drivers tend to drive less casually due to the limits of the traffic flow or the high speed of the vehicle.

The GHG EFs for driving cycles at 30 to 70 km/h are shown in Figure 2*b*. By illustrating the parallel boundaries of 90% of driving cycles, it is found that the uncertainty of estimation for speed range of 30 to 70 km/h on the expressway is 30.30 g/km.

Given the analysis above, the following conclusions are drawn:

- It is difficult for single typical driving cycles to describe precisely the full features of all the driving cycles, considering the variety of the driving cycles. The use of "typical" driving cycles might cause uncertainty in the GHG emission estimation.
- Using typical driving cycles to represent the various driving cycles may lead to an insufficient use of the remainder of the driving cycles in the data set.
- It is difficult for users of the model reasonably to select typical driving cycles.

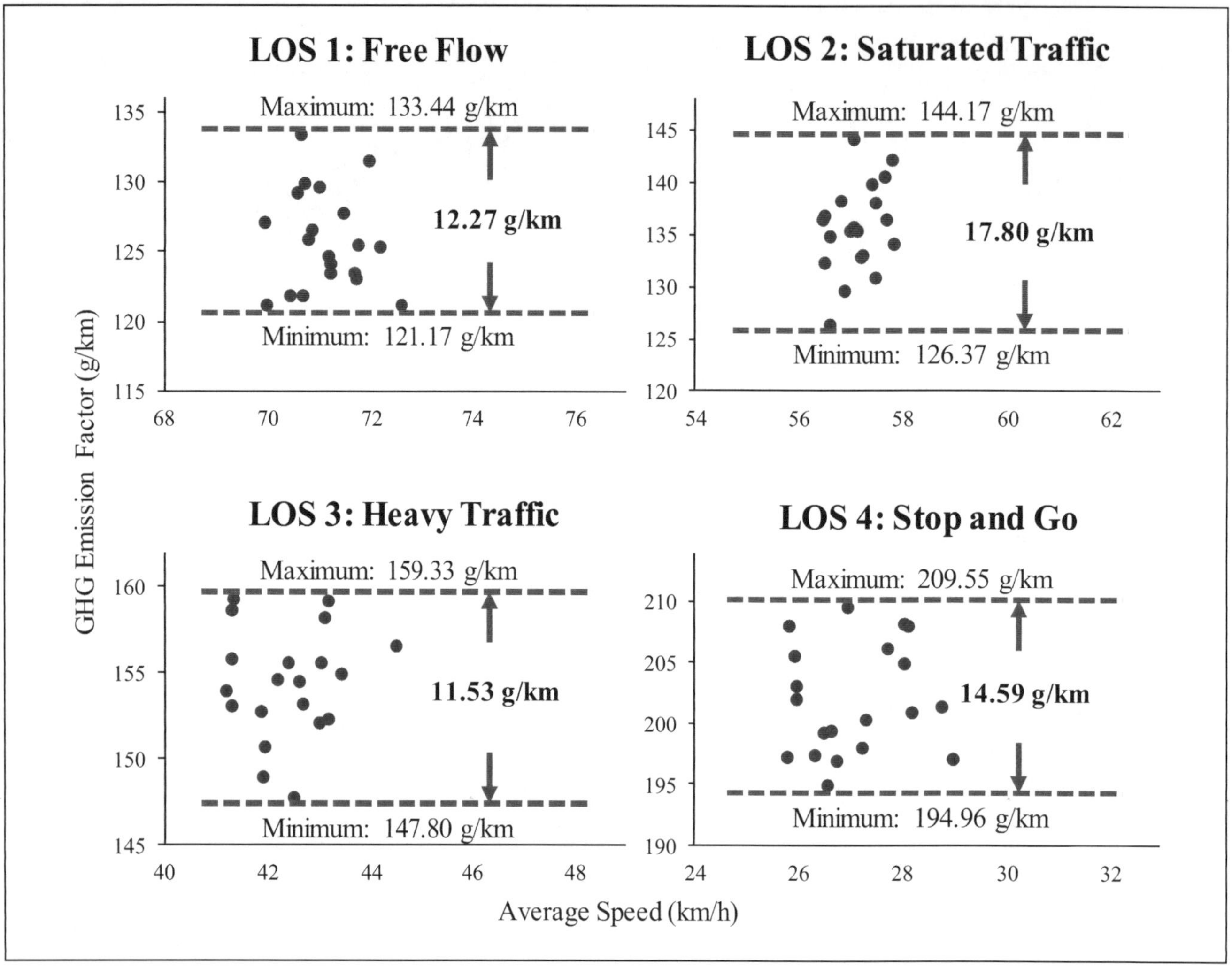

Figure 3. Uncertainties of the top-20 driving cycles.

Uncertainty of Top-20 Driving Cycles. The uncertainty of GHG emission estimation for the top-20 driving cycles is defined as the difference between the maximum GHG EF and the minimum GHG EF. Four LOSs of expressway driving are shown as an example in Figure 3. The uncertainty of the four LOSs are 12.27 g/km, 17.80 g/km, 11.53 g/km and 14.59 g/km, respectively. The relative uncertainties of the four LOSs are 9.75%, 13.12%, 7.47%, and 7.22%.

The findings of this part of the study indicate that the uncertainty of GHG emission estimation exists even after the selection of the top-20 driving cycles. Two points emerge based on the findings above:

- The three indices of RPA, average speed, and stop fraction do not guarantee a precise estimation of the GHG emissions.
- It is difficult to make sure the typical driving cycles describe the driving behaviors of the various LOSs more precisely than the remaining 19 driving cycles. It may be difficult to select the most typical driving cycle, given that the 20 most concentrated driving cycles still have a relative uncertainty of around 10%.

Uncertainty Derived from RPA

In this section, the relationship between RPA and GHG emissions is investigated. The other two indices, average speed and stop fraction, serve as control variables in this part. The driving cycles with average speed of 70 to 72 km/h are selected as examples so that the studied driving cycles have little difference in their average speed. Driving cycles with positive stop fractions are excluded to make sure all the investigated driving cycles have zero stop fractions.

There is a positive correlation between RPA and EF, as shown in Figure 4*a*. The R^2 is only 0.6482, however,

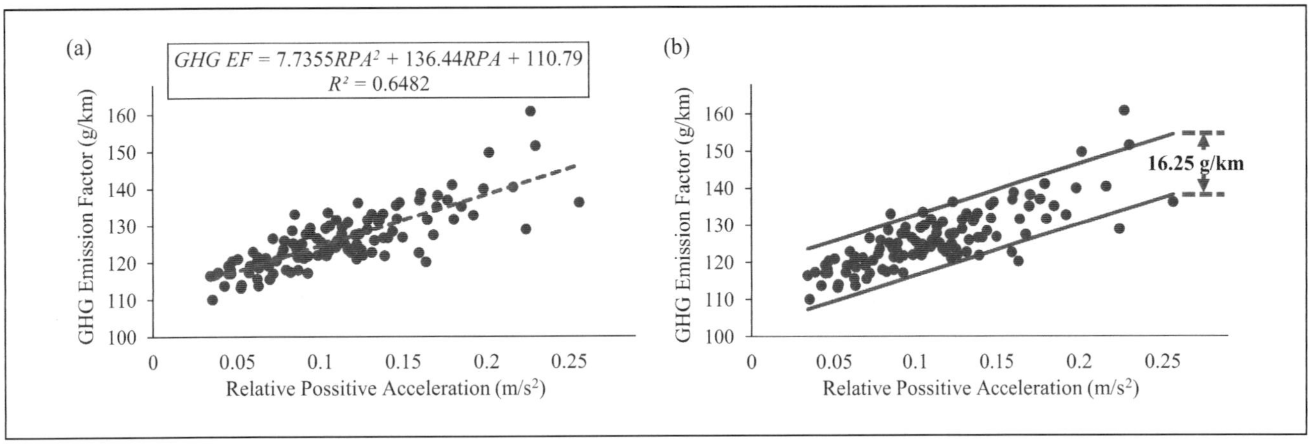

Figure 4. GHG EFs of driving cycles at 70 to 72 km/h: (*a*) correlation between RPA and GHG emissions; (*b*) uncertainty of the driving cycles.

Table 2. Samples of the Driving Cycles

Sample	ID	Road type	LOS	Average speed (km/h)	RPA	Stop fraction	EF (g/km)	Total distance (m)	Record count
I	5522	Expressway	Free flow	71.58	0.0843	0.00%	115.96	2008.2	101
II	3946	Expressway	Free flow	70.47	0.0848	0.00%	133.01	2016.3	103

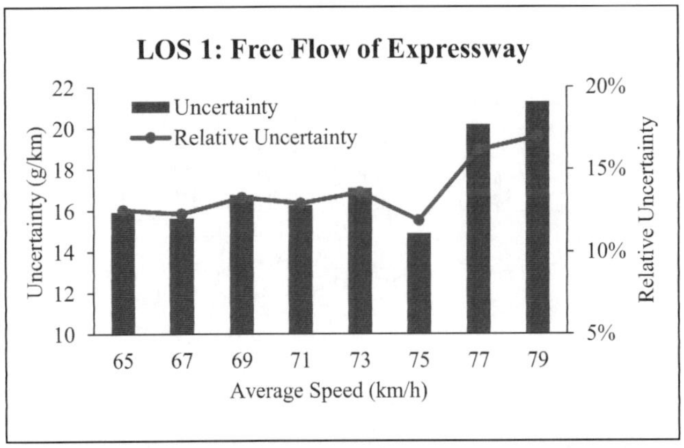

Figure 5. Uncertainties/relative uncertainties of LOS I on the expressway.

indicating the volatility of the correlation. Two parallel boundaries are illustrated in Figure 4*b* so that 90% of the EFs fall within the boundaries. The uncertainty, that is, the minimum difference between the boundaries, is 16.25 g/km. The relative uncertainty of estimation for driving cycles with average speed of 70 to 72 km/h is 12.91%.

The uncertainties and relative uncertainties of the LOS of free flow of the expressway is shown as an example in Figure 5, including the driving cycles with average speed of between 65 and 79 km/h. The relative uncertainties of the example are all greater than 11%.

To further investigate the GHG emission-sensitivity of RPA, two driving cycles with average speeds between 70 and 72 km/h are selected as the samples. The two sample driving cycles have similar average speeds and similar RPAs, and have the same (zero) stop fraction, as shown in Table 2.

The two sample driving cycles have similar RPAs while the difference between the GHG EF of the samples is 17.05 g/km, which is relatively 12.82% and 14.70% of their GHG EFs. This indicates that using RPA may cause uncertainty in the estimation of GHG emissions.

The speed sequences of the sample trajectories are investigated, followed by analysis of the acceleration distributions. The speed sequences of the sample driving cycles are shown in Figure 6.

It is found that the accelerations and decelerations in driving cycle sample II (red), which has the larger GHG EF, are more aggressive than those in driving cycle sample I (blue). The acceleration distributions of the two samples are shown in Figure 7. The three major areas of difference between the accelerations of the two samples are marked with green squares in the figure.

The rather aggressive driving behavior of sample II is also observed in the VSP distributions in Figure 7. The

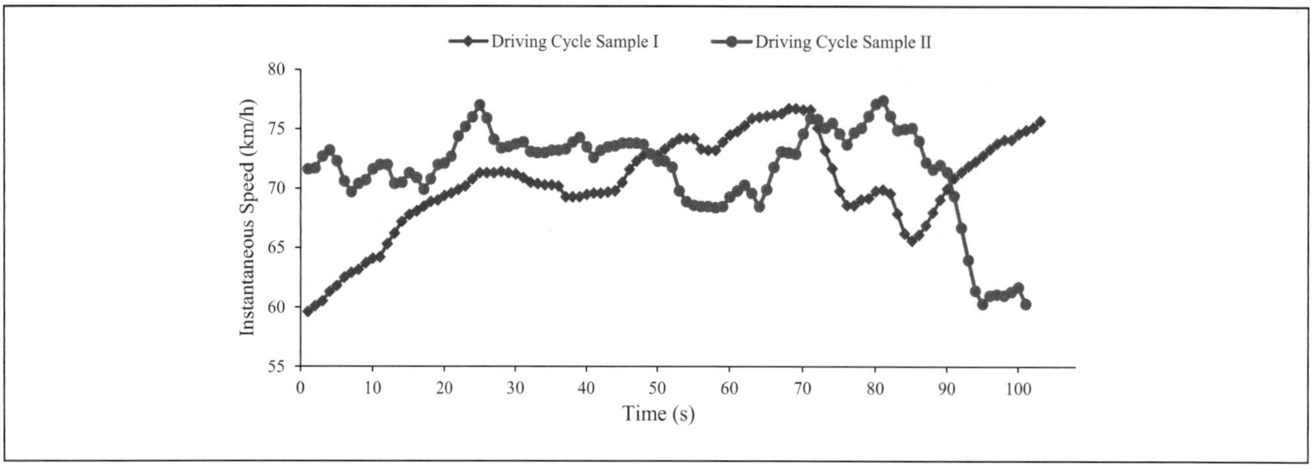

Figure 6. Speed sequences of the samples of driving cycles.

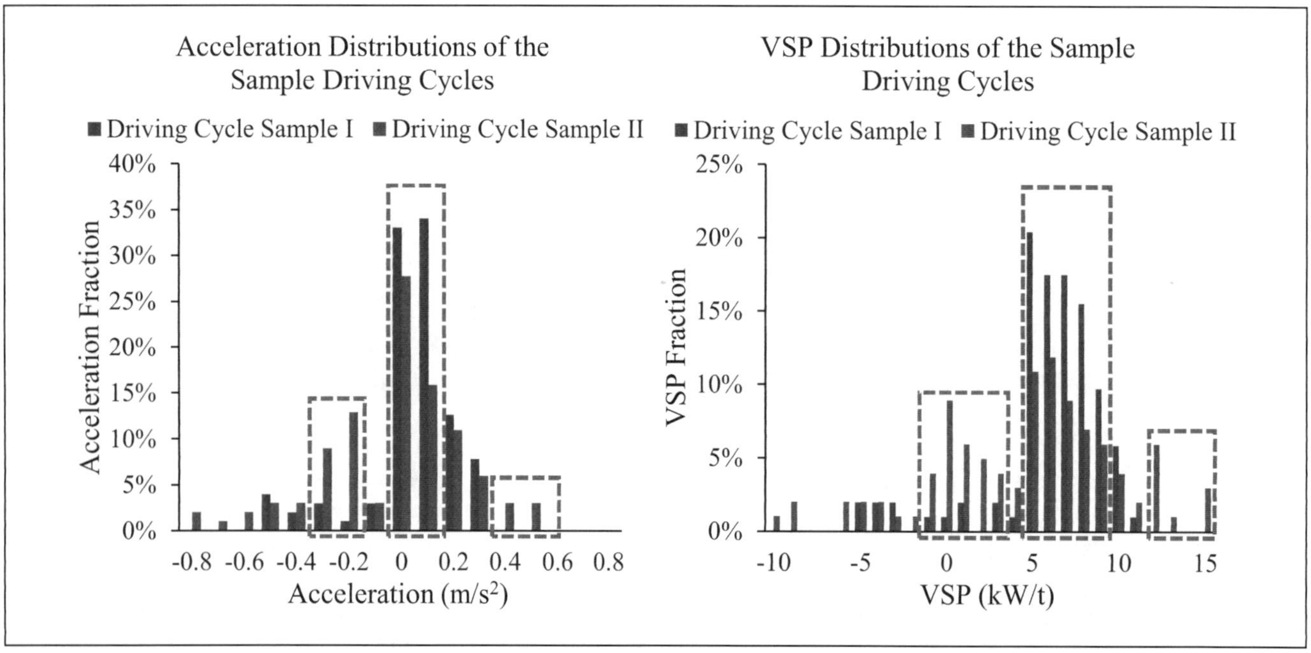

Figure 7. Acceleration and VSP distributions of the samples of driving cycles.

difference in RPA of the two samples is only 0.0005 (relatively less than 0.6%), while the difference in GHG EFs is 17.05 g/km (relatively greater than 12%). The samples indicate that, even with the same RPA, average speed, and stop fraction, the acceleration features of the driving cycles may differ. In comparison with RPA, the VSP distributions are better at describing the differences of the driving behaviors, and thus the emissions. To conclude, the samples indicate that the sensitivity of RPA is not sufficient in precisely describing the GHG EF differences of the two samples, due to the acceleration-insensitivity of RPA as a measure.

It should be noted that samples might be special, but not rare. Special traffic scenarios such as intersections, ramps of the expressway, and traffic congestion may all cause volatility in acceleration/deceleration. Using RPA to capture such complex traffic scenarios causes uncertainty in the GHG emission estimation.

Two possible assumptions regarding the sources of the acceleration-insensitivity of RPA are as follows. One is that RPA is an aggregated parameter rather than an instantaneous one, which might fail to distinguish the second-by-second characteristics of driving behaviors. Second, RPA does not take deceleration (negative acceleration) into consideration, which may cause errors in the emission estimation.

Though the second-by-second trajectories are necessary for the calculation of both RPA and VSP, RPA is not as of high resolution as VSP. For RPA, the aggregation of positive accelerations may cause the compensation of the detailed features, and thus lead to loss of estimation accuracy. Hence RPA might not be able to capture the instantaneous characteristics of the driving cycles. The outcomes of the HBEFA model might not be as high resolution as the input data, since the second-by-second details are lost in the aggregation.

Furthermore, given that the HBEFA model uses one typical driving cycle to describe each LOS, only a small portion of the second-by-second trajectories are made use of effectively in the development/calibration of the HBEFA model. In this case, fewer than 5,000 records (the exact number varies depending on which of the top-20 driving cycles is selected as the typical driving cycles) of the second-by-second data are used, out of the total of 7,547,355 records. The records effectively utilized represent less than 0.06% of the entire data, if the typical driving cycles are to be selected.

Summary and Recommendation

The study investigates the uncertainty of GHG emission estimation of the HBEFA model. The driving cycles of HBEFA are developed based on a massive data set of vehicle trajectories collected in the field. The top-20 driving cycles for each LOS of each road type are selected, with RPA, average speed, and stop fraction as the three indices. The GHG EFs of the driving cycles are obtained by assigning field emission rates to the second-by-second trajectories. The uncertainties of the GHG EFs of all the driving cycles are investigated, as well as the uncertainties of the GHG EFs of the top-20 driving cycles. The GHG emission-sensitivity of RPA is studied with two samples of the driving cycles. The conclusions are as follows:

1. Using individual driving cycles to describe GHG emissions leads to uncertainty in the estimation. It is found that the uncertainty for average speeds of 30 to 70 km/h on the expressway is 30.30 g/km. It is difficult to use typical driving cycles to represent the various traffic characteristics of all the driving cycles.
2. For the top-20 driving cycles, the uncertainties of the four LOSs of the expressway are 12.27 g/km, 17.80 g/km, 11.53 g/km, and 14.59 g/km, representing 9.75%, 13.12%, 7.47%, and 7.22% of the average EF. This indicates that uncertainties exist regarding the GHG emissions even for the selected driving cycles which are supposed to be representative.
3. It is found that RPA is not sufficiently sensitive to describe the difference between the two sample driving cycles. The GHG emission-insensitivity of RPA results from its insensitivity to acceleration. In comparison with RPA, VSP distributions are better at describing the characteristics of the driving behaviors, especially the various second-by-second details.
4. Both the adoption of the driving cycle mechanics and RPA as the key parameter cause uncertainties in estimation of GHG emissions. Besides, only a small portion of the data can be made sufficient use of, due to selection of the typical driving cycles and the insensitivity of RPA. Though the input data of the HBEFA model is required to be of high resolution, the outcomes of the model may not be of the same precision.

The case study was conducted in Beijing, China. Investigation in other parts of the world would contribute more to the study of the mechanics of the HBEFA model. The sensitivity of RPA is analyzed by a set of samples in this study. A more solid conclusion could be made if a more general investigation were conducted. Two assumptions are made in this study regarding the possible sources of the acceleration-insensitivity of RPA. A thorough analysis is suggested to study the sources of the insensitivity. A more thorough understanding of both RPA and driving cycle mechanics may provide a better understanding of how to estimate vehicle emissions more accurately.

Acknowledgments

This research is supported by the Natural Science Foundation of China (NSFC) # 51678045 and # 51578052.

Author Contributions

The authors confirm contribution to the paper as follows: study conception and design: GS, HL, and LY; data collection: JW and WH; analysis and interpretation of results: GS, HL, and QZ; draft manuscript preparation: HL and JW. All authors reviewed the results and approved the final version of the manuscript.

References

1. Keller, M., S. Hausberger, and P. Wüthrich. *HBEFA Version 3.3.* MKC Consulting GmbH, IVT Institute for Internal Combustion Engines and Thermodynamics of TU Graz, and INFRAS, Berne, Switzerland, 2017.
2. iCET. *Calculating Urban Transportation Emissions: Private Vehicles—Lessons from Existing Tools, Development, Design and Implementation.* Innovation Center for Energy and Transportation, Beijing, China, 2015. http://www.icet.org.cn/english/admin/upload/2015101235945161.pdf.
3. He, W., Y. Liu, S. Sun, and Y. Cheng. Greenhouse Gas Emissions Model for Urban Transportation Based on HBEFA—The Case of Beijing Localization Model.

Journal of Transportation Systems Engineering and Information Technology, Vol. 14, No. 4, 2014, pp. 222–229 (in Chinese).
4. Colberg, C. A., B. Tona, W. A. Stahel, M. Meier, and J. Staehelin. Comparison of a Road Traffic Emission Model (HBEFA) with Emissions Derived from Measurements in the Gubrist Road Tunnel, Switzerland. *Atmospheric Environment*, Vol. 39, 2005, pp. 4703–4714.
5. Fontaras, G., V. Franco, P. Dilara, G. Martini, and U. Manfredi. Development and Review of Euro 5 Passenger Car Emission Factors Based on Experimental Results over Various Driving Cycles. *Science of the Total Environment*, Vol. 468–469, Jan. 2014, pp. 1034–1042.
6. Maneva, K., and F. Skácel. Correlation of Highway Traffic and NOx Concentration Level. The Environmental Technology and Innovations. *Proc., 1st International Conference on Environmental Technology and Innovations*, Ho Chi Minh City, Vietnam, 2016, pp. 71–75.
7. Borken-Kleefeld, J. Real World Emissions from Road Vehicles: Results from Remote Sensing, Emission Measurements and Models. *Proc., 20th International Transport and Air Pollution Conference*, Graz, Austria, 2014, pp. 1–5.
8. Diegmann, V., L. Neunhäuserer, and G. Gässler. A Monitoring System To Control Effects and Effectiveness of Traffic Measures in Urban Areas. *Proc., Transport and Air Pollution*, Empa, Dübendorf, Switzerland, 2010.
9. Elkafoury, A., A. M. Negm, M. H. Aly, M. F. Bady, and T. Ichimura. Develop Dynamic Model for Predicting Traffic CO Emissions in Urban Areas. *Environmental Science and Pollution Research*, Vol. 23, No. 16, 2016, pp. 15899–15910.
10. Sjödin, Å., and M. Jerksjö. Evaluation of European Road Transport Emission Models against On-Road Emission Data as Measured by Optical Remote Sensing. *Proc., 17th International Conference: Transport and Air Pollution*, Graz, Austria, 2008.
11. Tsanakas, N., J. Ekström, and J. Olstam. Reduction of Errors when Estimating Emissions Based on Static Traffic Model Outputs. *Transportation Research Procedia*, Vol. 22, 2017, pp. 440–449.
12. Chen, Y., and J. Borken-Kleefeld. Real-driving Emissions from Cars and Light Commercial Vehicles—Results from 13 Years Remote Sensing at Zurich/CH. *Atmospheric Environment*, Vol. 88, 2014, pp. 157–164.
13. Frey, H., N. Rouphail, and H. Zhai. Speed- and Facility-Specific Emission Estimates for On-Road Light-Duty Vehicles on the Basis of Real-World Speed Profiles. *Transportation Research Record: Journal of the Transportation Research Board*, 2006. 1987: 128–137.
14. Smit, R., A. L. Brown, and Y. C. Chan. Do Air Pollution Emissions and Fuel Consumption Models for Roadways Include the Effects of Congestion in the Roadway Traffic Flow? *Environmental Modelling & Software*, Vol. 23, 2008, pp. 1262–1270.
15. Alam, A., and M. Hatzopoulou. Deriving Local Operating Distributions to Estimate Transit Bus Emissions across an Urban Network. *Transportation Research Record: Journal of the Transportation Research Board*, 2016. 2570: 57–68.
16. Borge, R., I. De Miguel, D. De Paz, J. Lumbreras, J. Pérez, and E. Rodríguez. Comparison of Road Traffic Emission Models in Madrid (Spain). *Atmospheric Environment*, Vol. 62, 2012, pp. 461–471.
17. *MOVES2014a User Guide*. EPA-420-B-15-095. Office of Transportation and Air Quality, U.S. Environmental Protection Agency, Washington, D.C., 2015.
18. Davis, N., J. Lents, M. Osses, N. Nikkila, and M. Barth. Development and Application of an International Vehicle Emissions Model. *Transportation Research Record: Journal of the Transportation Research Board*, 2005. 1939: 155–165.
19. Song, G., L. Yu, and Z. Tu. Distribution Characteristics of Vehicle-Specific Power on Urban Restricted-Access Roadways. *Journal of Transportation Engineering*, Vol. 138, No. 2, 2012, pp. 202–209.
20. Zhai, Z., G. Song, H. Lu, W. He, and L. Yu. Validation of Temporal and Spatial Consistency of Distributions for Emission Estimation: A Case Study in Beijing, China. *Journal of the Air & Waste Management Association*, 2017, pp. 1–9.
21. Song, G., L. Yu, and Y. Wu. Development of Speed Correction Factors Based on Speed-Specific Distributions of Vehicle Specific Power for Urban Restricted-Access Roadways. *Journal of Transportation Engineering*, Vol. 142, No. 3, 2016, p. 4016001.
22. Shengyang, S., D. Bongardt, U. Eichhorst, M. Schmied, P. Wüthrich, and M. Keller. *Modelling Energy Consumption and GHG Emissions of Road Transport in China: Technical Paper on GIZ CRTEM/HBEFA- China Model*. Deutsche Gesellschaft für Internationale Zusammenarbeit (GIZ) GmbH, Eschborn, Germany, 2014.
23. Hausberger, S., M. Rexeis, M. Zallinger, and R. Luz. *Emission Factors from the Model PHEM for the HBEFA Version 3*. Institute for Internal Combustion Engines and Thermodynamics, Graz, Austria, 2009.

The Standing Committee on Transportation and Air Quality (ADC20) peer-reviewed this paper (18-03799).

Article

TRR
JOURNAL OF THE TRANSPORTATION RESEARCH BOARD

Transportation Research Record
2018, Vol. 2672(25) 89–98

DOI: 10.1177/0361198118796939
journals.sagepub.com/home/trr
SAGE

Application and Evaluation of the Reinforcement Learning Approach to Eco-Driving at Intersections under Infrastructure-to-Vehicle Communications

Junqing Shi[1], Fengxiang Qiao[2], Qing Li[2], Lei Yu[3,4], and Yongju Hu

Abstract

Eco-driving behavior is able to improve vehicles' fuel consumption efficiency and minimize exhaust emissions, especially with the presence of infrastructure-to-vehicle (I2V) communications for connected vehicles. Several techniques such as dynamic programming and neural networks have been proposed to study eco-driving behavior. However, most techniques need a complicated problem-solving process and cannot be applied to dynamic traffic conditions. Comparatively, reinforcement learning (RL) presents great potential for self-learning to take actions in a complicated environment to achieve the optimal mapping between traffic conditions and the corresponding optimal control action of a vehicle. In this paper, a vehicle was treated as an agent to select its maneuver, that is, acceleration, cruise speed, and deceleration, according to dynamic conditions while approaching a signalized intersection equipped with I2V communication. An improved cellular automation model was utilized as the simulation platform. Three parameters, including the distance between the vehicle and the intersection, signal status, and instant vehicle speeds, were selected to characterize real-time traffic state. The total CO_2 emitted by the vehicle on the approach to the intersection serves as a measure of reward policy that informs the vehicle how good its operation was. The Q-learning algorithm was utilized to optimize vehicle driving behaviors for eco-driving. Vehicle exhaust emissions and traffic performance (travel time, stop duration, and stop rate) were evaluated in two cases: (1) an isolated intersection, and (2) a medium-scale realistic network. Simulation results showed that the eco-driving behavior obtained by RL can not only reduce emissions but also optimize traffic performance.

Vehicle emissions have become a major and costly problem in many countries with the growth of the population and motor vehicles and, more importantly, the increase of vehicle kilometers traveled. In 2014, the United States greenhouse gas emissions totaled 6,870 million metric tons of carbon dioxide equivalents, among which about 26% was from transportation (*1*). In China, a total of 45.437 million tons of vehicle emissions were produced in 2014, including 6.278 million tons of nitrogen oxide (NO_x), 4.244 million tons of hydrocarbons (HC), and 34.373 million tons of carbon monoxide (CO) (*2*). This huge amount of emissions has created significant environmental stress. Owing to this, emission reduction strategies have been promoted to relieve environmental pressures from the transportation sector.

"Eco-driving" is one such a strategy, which aims to reduce vehicles' fuel consumption and emissions through proper guidance to the detailed operation of traffic flow. Eco-driving means to maintain a steady speed, avoid hard acceleration and deceleration, anticipate well the traffic flow ahead, minimize the idling period, and so on. In this way, a vehicle's trajectory can become much smoother with higher fuel consumption efficiency and thereby reduced vehicle emissions (*3*).

[1]Department of Traffic and Transportation, College of Engineering, Zhejiang Normal University, Jinhua, Zhejiang, China
[2]Innovative Transportation Research Institute, Texas Southern University, Houston, TX
[3]Yangtze River Scholar and Adjunct Professor, Xuchang University and Beijing Jiaotong University
[4]Texas Southern University, Houston, TX
[5]College of Engineering, Zhejiang Normal University, Jinhua, Zhejiang, China

Corresponding Author:
Address correspondence to Fengxiang Qiao: fengxiang.qiao@tsu.edu

The benefits of eco-driving can be evaluated through: (1) real-world implementation, (2) traffic simulation, and (3) a combination of both. In real-world implementations, eco-driving can improve fuel efficiency from 4% to 16% (*4–6*), whereas estimated eco-driving benefits exhibit higher variability in the traffic simulation studies than in real-world tests, ranging from 8% to about 27% in fuel savings and CO_2 reduction (*7–10*). Via infrastructure-to-vehicle (I2V) communications, vehicles exchange information with roadside traffic control infrastructure, such as signal control devices, to be aware of real-time traffic control information, such as signal status ahead. In this way, drivers could gain sufficient time to prevent stopping by maneuvering their vehicles, typically by adjusting approaching speed. Thus, the I2V application could greatly improve drivers' eco-driving behaviors. In this paper, a simulation method is employed to evaluate the benefits of eco-driving behaviors in and around an urban road network.

Eco-driving within a road network could be formulated as a "trajectory optimization problem". Several techniques have been employed to optimize decision-making by minimizing fuel consumption. For example, dynamic programming has been utilized to optimize the passing sequence in helping a vehicle to optimize its speed profiles, in order to reduce both travel delay and fuel consumption (*11*). Milesich et al. applied a hierarchy of neural networks to create a vehicle model and optimization cycle that looks for parameters which influence vehicle maneuvers along a given path (*12*). Jiang et al. proposed a novel parallelizable algorithm to solve the coordination problem at traffic intersections under a given precedence order using a tailored variant of the augmented Lagrangian-based alternating direction inexact Newton method (*13*). However, these techniques use a complicated solving process and cannot be applied to different traffic conditions.

Unlike the techniques mentioned above, reinforcement learning (RL) has shown great potential for self-learning to take actions in an environment, so as to maximize some notions of cumulative reward for an optimal mapping between different traffic conditions and the corresponding optimal control action (*14*). In the field of traffic management and control, RL has been used in adaptive signal control (*15*), freeway ramp metering (*16*), and route planning (*17*). In RL, the agent (vehicle, ramp, traffic signal, and so on) interacts with the traffic environment in a closed-loop system to obtain the optimal mapping between the environmental traffic state and the corresponding optimal control action, offering an optimal operation law. The agent iteratively receives a feedback reward for actions taken, and adjusts the policy until it converges to the optimal operation policy. Applying RL to study eco-driving is associated with challenges. The state needs to be described reasonably to avoid "state space explosion", and an environment model reflecting the realistic traffic world needs to be built to enable vehicles to learn in the simulation platform. Once the acceleration pattern of vehicles can be correctly represented, the environmental performance can be correspondingly estimated.

In this paper, a RL-based evaluation method and a micro-simulation model are identified to investigate the impacts of eco-driving on vehicle emissions and traffic performance; these are expected to provide useful information to improve existing eco-driving strategies under an I2V environment. RL elements are described for building an eco-driving learning and evaluation platform. The applicability of using RL and microscopic traffic simulator is then explored through the example of a road. After that, the operation policy learned in the road is used in a medium-scale network to evaluate the effects of eco-driving.

Reinforcement Learning for Eco-Driving Platform

A RL problem is defined once an environmental model, states, actions, and rewards are clearly defined. To this end, these basic constructs are described in the context of a problem domain.

Environmental Model Configuration

As shown in Figure 1, a discrete event environment of RL is constructed as a signalized four-leg intersection equipped with I2V. Each approach has two lanes. The upstream lane is divided into L_{up} cells, while the downstream lane is divided into L_{down} cells. The length of each cell is 3.5 m, and each vehicle occupies two cells. Vehicles drive on the right-hand side of the road.

One discrete second refers to one time step in the discrete simulation environment. Let t indicate the current moment, and x_t, v_t, and a_t denote the position, speed, and acceleration of the vehicle at the tth time step, respectively. Each vehicle has a maximum speed $v_{max} = 5$ cell/s (i.e., 17.5 m/s), where, speed v_t is: $v_t = 0, 1, 2, \cdots, v_{max}$.

The distance between the vehicle and the downstream intersection stop line is denoted as λ_t, which can be calculated by Equation 1.

$$\lambda_t = L_{up} - x_t \tag{1}$$

A signal control mode is applied to this intersection. In Figure 2, T_{cycle} and T_{green} represent the signal cycle and

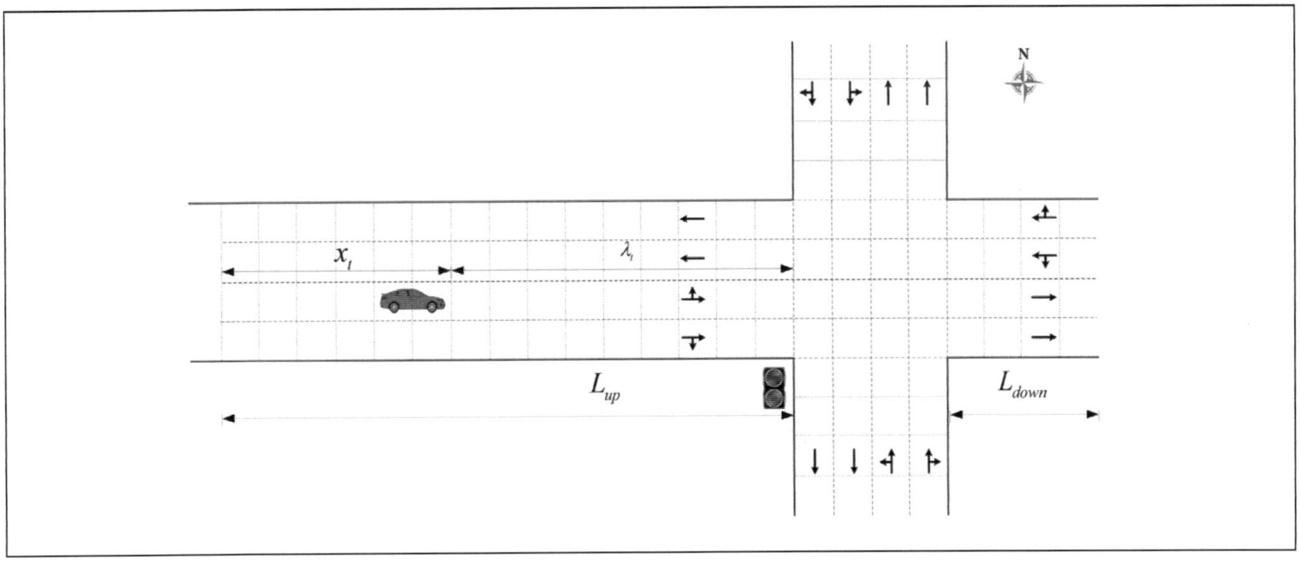

Figure 1. The sketch of the one-intersection four-approach scheme.

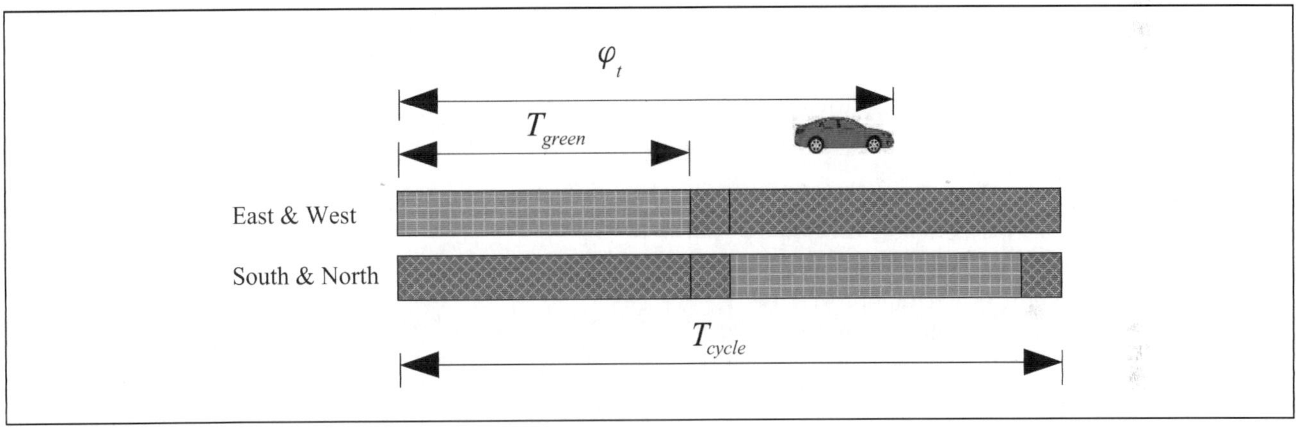

Figure 2. Signal control cycle.

the green time at the signalized intersection, respectively, while φ_t is the time from the start of the green light for the vehicle at the tth time step. An approaching vehicle receives real-time information on the signal status ahead via the installed I2V communication.

Vehicle States and Action Set

The state S_t of a vehicle is represented by a three-dimensional feature vector with the elements of the location x_t, the signal status φ_t and the speed v_t. The action of the vehicle at the tth time step is represented by $A_t \in \psi$, and the value of A_t could be -1, 0, and 1 for acceleration, cruise speed, and deceleration, respectively. At any given moment, one vehicle individually selects an action among the available action set ψ, depending on the state of the vehicle $S_t(x_t\,,\ \varphi_t\,, v_t)$.

States Update

At each simulation time step, the speed and position of each vehicle on the upstream are updated in parallel, according to the designed update rules shown in Equations 2–7.

$$v_t' = \min\left(\max\left(v_t + A_t',\ 0\right),\ v_{\max}\right) \tag{2}$$

where

$A_t' \in [-1,\ 0,\ 1]$ = the desired action;

v_t' = vehicle speed at the tth time step.

$$v_{t+1} = \begin{cases} \min\left(v_t',\ \lambda_t\right) & x_t \leq L_{\text{up}}\ \&\ \varphi_t > T_{\text{green}} \\ \min\left(v_t',\ L - x_t\right) & \text{otherwise} \end{cases} \tag{3}$$

where

v_{t+1} = the real speed that the vehicle will adopt at the tth time step.

$$a_t = v_{t+1} - v_t \tag{4}$$

where

a_t = vehicle acceleration rate.

$$A_t = \begin{cases} 1 & a_t > 0 \\ 0 & a_t = 0 \\ -1 & a_t < 0 \end{cases} \tag{5}$$

$$\varphi_{t+1} = \begin{cases} \varphi_t + 1 & \varphi_t < T_{\text{cycle}} \\ 1 & \varphi_t = T_{\text{cycle}} \end{cases} \tag{6}$$

$$S_t(x_t, \varphi_t, v_t) \rightarrow \begin{cases} S_{t+1}(x_t + v_t, \varphi_{t+1}, v_{t+1}) & x_t + v_t \leq L_{\text{up}} \\ S_{t+1}(L_{\text{up}} + 1, \varphi_{t+1}, v_{t+1}) & x_t + v_t > L_{\text{up}} \end{cases} \tag{7}$$

When the vehicle passes the stop line into an intersection, it will speed up to leave the intersection toward the downstream road by a desired speed $v_{\max}$.

Q-Learning Algorithm

The most common RL algorithm is Q-learning (*18*). The Q-learning agent learns an optimal mapping between the environmental state S and the corresponding optimal action A based on accumulating rewards $R(S, A)$. Each state–action pair (S, A) has a value called Q-factor that represents the expected long-run cumulative reward for the state-action pair (S, A). In each iteration, that is, t, an agent observes current states S and chooses to execute an action A that belongs to the available set of actions ψ; then, the Q-factor is updated according to the immediate reward $R(S, A)$ and the Q-factor of the next time step $t + 1$.

At an initial time, a vehicle is placed at the starting cell of the simulation platform with a random speed and then departure from this cell until passing through a stop line. In other words, the objective state is $S(L_{up} + 1, \varphi_t)$. To set this state as an objective, a reward value to each action is taken into account. The actions that lead immediately to the objective have an instant reward, depending on the total emissions. Other actions not directly connected to the target state have zero reward (go ahead) or negative reward (stop).

Imagine that the vehicle can learn through experiences. The vehicle can pass from one cell to another, but without the knowledge of the environment, and is not aware of which sequence of actions could lead to the lowest emission.

A matrix, "Q", $(L_{\text{up}} + 1)$ by T_{cycle}, is added to represent the memory of what the vehicle has learned through experiences. The rows of the matrix Q represent the position of vehicle x_t, and the columns is the state of the signal φ_t. Each element of the matrix includes a sub-matrix of 3 by 6, which means 3 actions and 6 speeds. The row, column, and the column in the sub-matrix represent the current state of the vehicle, and the row of the sub-matrix is the possible actions, leading to the next state.

A vehicle starts with knowing nothing and the matrix Q is initialized to zero. The vehicle will explore from state to state until it reaches the objective. Each exploration consists of the vehicle from the initial state to the objective state, and is an episode. Each time the vehicle arrives at the objective state, the program proceeds to the next episode. At each episode, the vehicle occurring at the beginning of the platform will be distributed at a random initial speed v_0 and a random signal status φ_0. They form the initial state $S_1(1, \varphi_1, v_1)$ and carry out following steps while the objective state has not been reached.

- Select one among all possible actions for the current state.
- Using this possible action, consider going to the next state.
- Get maximum Q value for this next state based on all possible actions.
- Calculating using Equation 8, and set the next sate as the current state.

$$Q_t(S_t, A_t) \leftarrow (1 - \alpha)Q_t(S_t, A_t) + \alpha[R(S_t, A_t) + \gamma \max_{A_{t+1} \in \psi} Q_{t+1}(S_{t+1}, A_{t+1})] \tag{8}$$

where

α = a learning rate;

$\gamma \in (0, 1]$ = a discount rate.

If γ is close to zero, the agent will tend to consider only immediate rewards. If γ is close to one, the vehicle will consider future rewards with greater weight, willing to delay the reward.

The algorithm above is used by the vehicle to learn from experience. The purpose of training is to enhance the "brain" of the vehicle, represented by matrix Q, with more training results in a more optimized matrix Q. To use the matrix Q, the vehicle simply traces the sequence of states, from the initial state to the objective state. The algorithm finds the actions with the highest Q-factors recorded in a matrix Q for a current state:

$$A_{t+1} \in \arg \max_{A_t \in \psi}[Q(S_t, A_t)] \tag{9}$$

Simulation Setup and Parameters

All simulations were executed in the Matlab environment. The length of cell is 3.5 m, and the maximum speed $v_{\max}$ is 5 cell/s (63.0 km/h), while each vehicle has 6 discrete speeds from 0 to 5 cell/s, indicating 0, 12.6, 25.2, 37.8, 50.4, and 63.0 km/h, respectively. The number of cells of an approach before the intersection L_{up} is 40 (i.e.,

140 m), and the number of cells of a downstream road after an intersection L_{down} is 20 (i.e., 70 m). Thus, the total length of a platform is 64 cells (i.e., 224 m). The signal cycle T_{cycle} is 80 s, while the green time T_{green} is 37 s. Thus, the size of the matrix Q is 41 × 80, and the size of the sub-matrix is 3 × 6. The learning rate α and the discount rate γ are set to 0.5. The negative reward (punishment) R_{stop} is 500 when the vehicle stops; otherwise, the R_{stop} is 0.

As CO_2 predominates over other components in exhaust emissions by a fuel combusted vehicle in weight (*19*), for purposes of illustration, the reward is specified based on CO_2 emissions. The reward R_{final} for the total CO_2 is subject to the maximum and minimum CO_2 exhaust emissions on the platform. Regarding this, a pre-training process was conducted to allow a vehicle to drive through the 224 m-long platform randomly 10,000 times, and its total CO_2 exhaust emissions for each run on the platform from the start cell to the last one were measured. It was found that the total CO_2 exhaust emissions range from 55 to 288 g and few CO_2 emission records are higher than 180 g or lower than 80 g. Therefore, 80 and 180 g were selected as critical values, based on which the total emission pattern is divided into seven equal parts as listed in Equation 10.

$$R_{final} = \begin{cases} 0 & CO_2 \geqslant 180 \\ 500 & CO_2 \in [160, 180) \\ 1000 & CO_2 \in [140, 160) \\ 1500 & CO_2 \in [120, 140) \\ 2000 & CO_2 \in [100, 120) \\ 2500 & CO_2 \in [80, 100) \\ 3000 & CO_2 < 80 \end{cases} \quad (10)$$

where

CO_2 = the total CO_2 emissions (g) produced by a vehicle that drives through a 224 m-long platform.

Once a vehicle completes its motion on the platform, the reward would be immediately applied to the vehicle, based on its total CO_2 emissions on the platform. The total CO_2 emissions are the sum of second-by-second CO_2 emission rate (g/s), which is determined by vehicles' real-time operational modes. The operational modes in the Motor Vehicle Emission Simulator (MOVES) were classified into 23 bins, according to driving speeds, acceleration rates, and vehicle-specific power (*VSP*). The *VSP* is the instantaneous tractive power per unit vehicle mass, which is determined by aerodynamic drag, acceleration, rolling resistance, grade, and the mass of the vehicle (*20*). The *VSP* for a light-duty vehicle is calculated using Equation 11.

$$VSP = v \times [1.1 \times a + 9.81 \times grade(\%) + 0.132] + 0.000302 \times v^3 \quad (11)$$

where

v = vehicle speed (m/s);

a = the acceleration rate (m/s^2);

grade (%) = vehicle vertical rise divided by slope length. It is assumed as 0% here.

In this study, the vehicle is assumed as a 5-year-old gasoline light-duty passenger car, the exhaust emission rate of which is estimated based on the corresponding average emission rate for the 23 operational bins in the MOVES model (*21*). The exhaust emission is mainly composed of four components, including CO_2, CO, HC, and NO_x. Whereas the CO_2 is a greenhouse gas, the other components are traffic-related air pollutants that pose a threat to public health.

In the MOVES model, the average emission rate of each exhaust component varies with vehicles' real-time operational mode. More specifically, the lowest emission rates are observed in Bin 0 and 1 for braking/deceleration, and an idle status, respectively. Whereas Bin 0 is defined as either the deceleration a equal to or smaller than negative 0.89 m/s^2 (−2 mi/h/s) or continuously smaller than negative 0.45 m/s^2 (−1 mi/h/s) for 3 s, Bin 1 refers to the speed v between negative and positive 1.61 km/h ([−1, 1) mph). The other bins can be divided into three speed zones: lower speed zone between 1.61 and 40.23 km/h ([1, 25)mph), medium speed zone between 40.23 and 80.47 km/h ([25, 50)mph), and high-speed zone equal to or greater than 80.47 km/h (50 mph). Within each speed zone, the average emission rate raises gradually along with incremental *VSP*. The higher emission rate is observed in the higher *VSP* in each speed zone. The total emissions of each component are the sum of second-by-second emission rates.

Training Process

At each episode, the total values of the matrix Q will change after the vehicle explores a random sequence of actions. The change value depends on the R_{stop}, R_{final} and the matrix Q of the last time step.

Figure 3 illustrates the total values' change in the matrix Q. One can observe that the Q value decreases first and then increases with the rise in episodes. The decrease in Q value is because, at the beginning, the matrix Q is "zero", and once the vehicle stops it receives a negative reward ($R_{stop} = -500$). Along with the increasing training times, the vehicle receives more and more rewards from $R_{final} \in [0, 3000]$, with the increase in the Q value. After enough episodes, the Q value will tend to a relatively converged value (4.3 × 10^6 in this case). The change process and the converged value depend on the setting of the rewards.

Table 1. Performance of Different Driving Behavior at the Intersection

Scenario \ MOE	Random-driving	Normal-driving	Eco-driving	Improvements Normal-driving vs. Random (%)	Improvements Eco-driving vs. Random (%)	Improvements Eco-driving vs. Normal-driving (%)
CO_2 (g)	123.81	90.00	86.06	27.31	30.49	4.38
CO (g)	1.89	1.69	1.43	10.58	24.34	15.39
HC (mg)	11.97	10.73	8.87	10.36	25.90	17.34
NO_X (mg)	44.31	36.85	31.24	16.84	29.50	15.22
Total stop duration (s)	13.16	11.46	4.30	12.92	67.36	62.48
Travel time (s)	45.08	25.86	25.19	42.64	44.12	2.60
Rate of passing the intersection without stop (%)	47.23	46.30	84.45	1.97	78.80	82.4

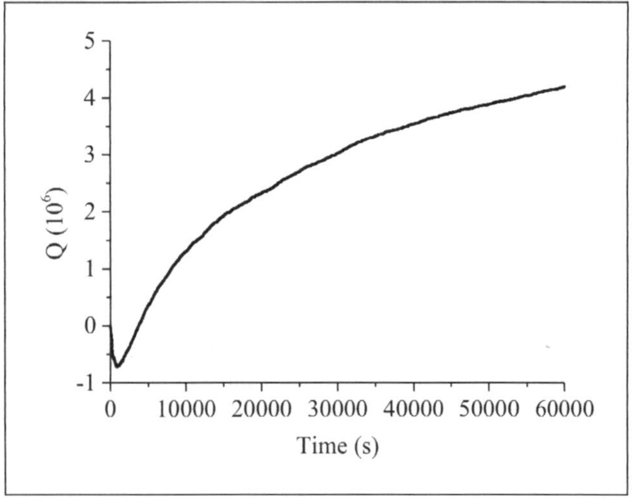

Figure 3. The Q values with the increase in episode.

Results and Discussion

In the Case of Isolated Intersection

Evaluation results are reported for three scenarios: (1) Random-driving, (2) Normal-driving, and (3) Eco-driving behavior. Random-driving means at each time step, the desired action A'_t is distributed as $+1$, 0, and -1, randomly. Normal-driving means the vehicle's movements are represented using the Nagel–Schreckenberg rules (*22*), that is, at each time step, A'_t always be $+1$ corresponding to the desire of the vehicle to travel at the speed limit ($v_{\max}$), which is subject to bang-bang control (*23*). Eco-driving here means at each time step, the vehicle selects the desired action A'_t according to the highest value in the column of the sub-matrix from the matrix Q. Taking into account the vehicle position, speed, and signal lights, the real action A_t is calculated. The vehicle selects action A_t to update its speed and position.

The measure of effectiveness (MOE) of each driving behavior is evaluated based on the measures of effectiveness, shown in Table 1. It shows a very promising overall performance of Eco-driving. After training, the vehicle exhaust emissions (CO_2, CO, HC, and NO_x) reduce by 30.49%, 24.34%, 25.90%, and 29.50%, respectively, if Eco-driving is in effect. The stop duration reduces from 13.16 to 4.30 s, and the average travel time passing the platform reduces from 45.08 to 25.19 s, meaning an improvement of 44.12%. The probability of passing the intersection without stop declines from 47.23% to 84.45% with an improvement by 78.80%, obviously. This means that the vehicle is more likely to reach the stop line at green light status.

It is clear that, comparing with the Normal-driving, the Eco-driving reduces all types of exhaust emission components. CO, HC, and NO_x emissions reduce by 15.39%, 17.34%, and 15.22%, respectively, whereas CO_2 emission reduction is relatively smaller with 4.38%. Meanwhile, the stop duration is 11.46 s on Normal-driving behavior, and 4.30 s on Eco-driving, where 62.48% of stop duration is prevented from unnecessary emissions.

It is interesting to find out that the travel time for Eco-driving is less than that of Normal-driving. This means that Eco-driving reduces emissions without an expense on travel time in this case. For example, Figure 4 illustrates the trajectories of the vehicle at $\varphi_0 = 65$s. When the vehicle sets off, the remaining red light is 15 s. Here the Eco-driving means optimal trajectory, and the Normal-driving close to real-world trajectory. For a Normal-driving scenario, a vehicle drives at the maximum speed 5 cell/s (63.0 km/h), and decelerates rapidly before an intersection, for the red light by an estimated arrival. The vehicle will wait for 8 s to restart. However, for an Eco-driving scenario, the vehicle slows down and drives at a lower speed to avoid the red light when arriving at the intersection. In this example, the travel time for the Eco-driving scenario is 21 s, whereas it is 22 s for the Normal-driving scenario.

Figure 5 illustrates the frequency of four emission components (CO_2, CO, HC, and NO_x) in different

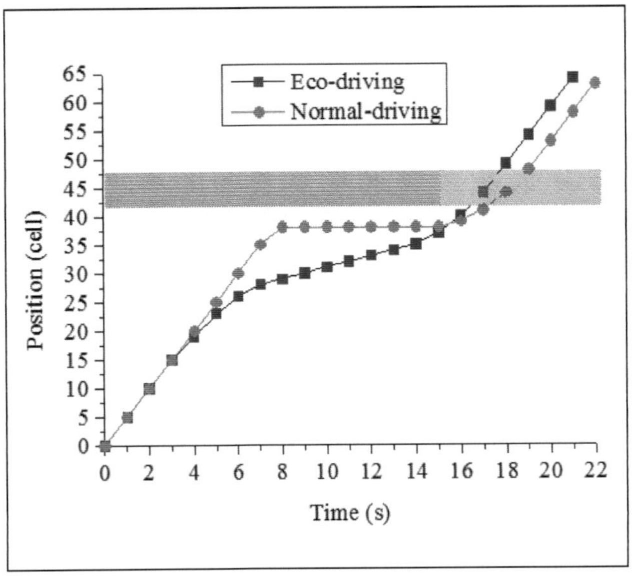

Figure 4. Comparison of Eco-driving trajectory and Normal-driving trajectory.

driving scenarios (Random-driving, Normal-driving, and Eco-driving). The distributions of all emissions are consistent with the findings shown in Table 1.

Figure 6 displays the frequency distributions of travel time. All of them have the minimum travel time about 12 s. The maximum frequency distribution is 0.03 when the travel time is 15 s under the Random-driving scenario. The frequency distribution of travel time under scenarios Normal-driving and Eco-driving is almost the same, whereas both the maximum travel time are 58 s and the maximum frequency distribution are 0.35 at the travel time of 13 s.

In the Case of Road Networks

The successfully operated eco-driving algorithm and the well-trained matrix Q were applied to a medium-scale traffic network on a simulated replica for further validation of the RL application-based eco-driving strategy.

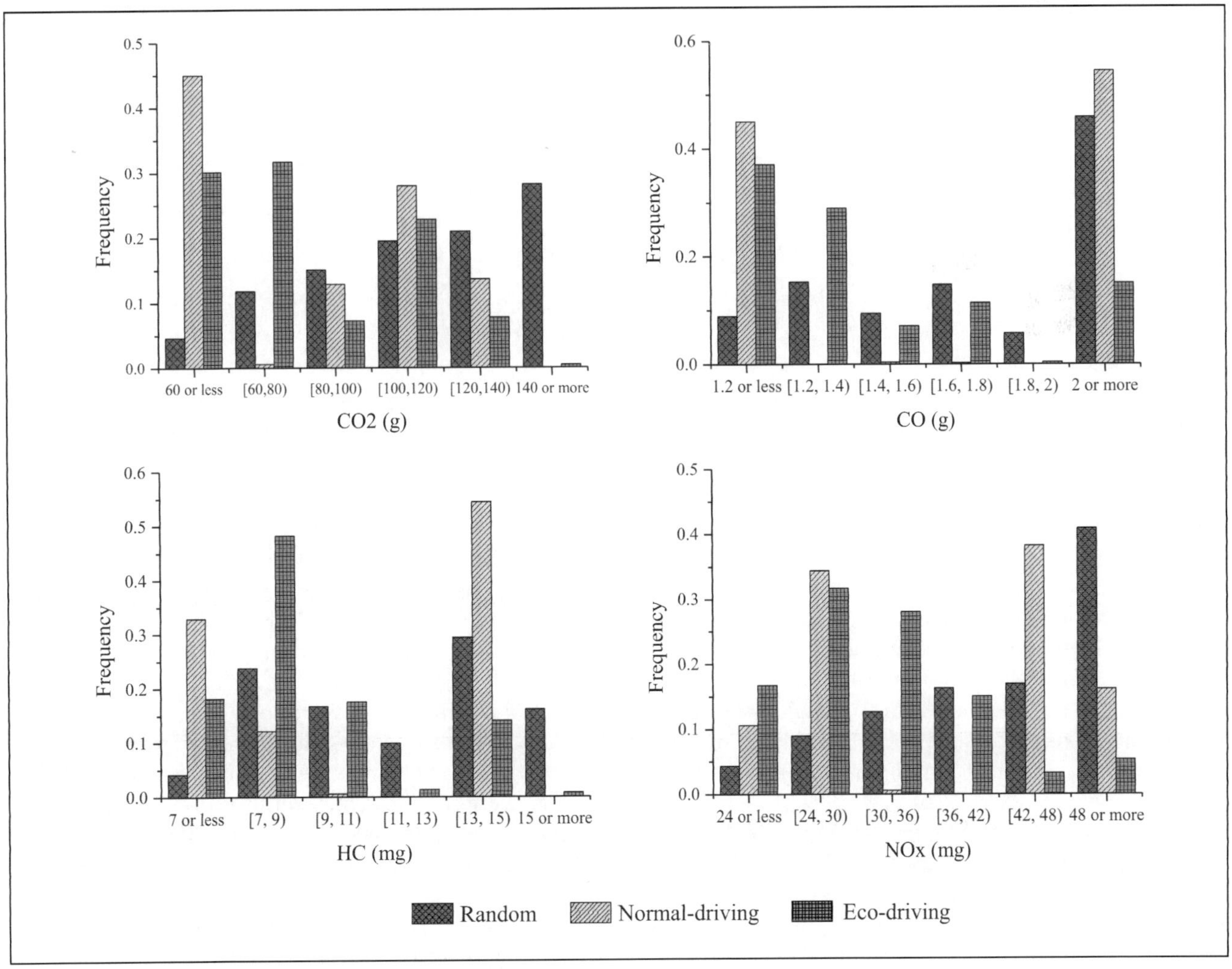

Figure 5. Frequency distributions of the average emission.

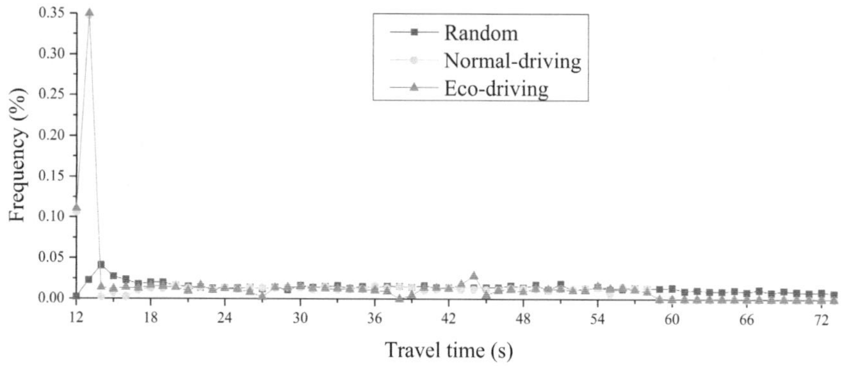

Figure 6. Frequency distributions of the travel time.

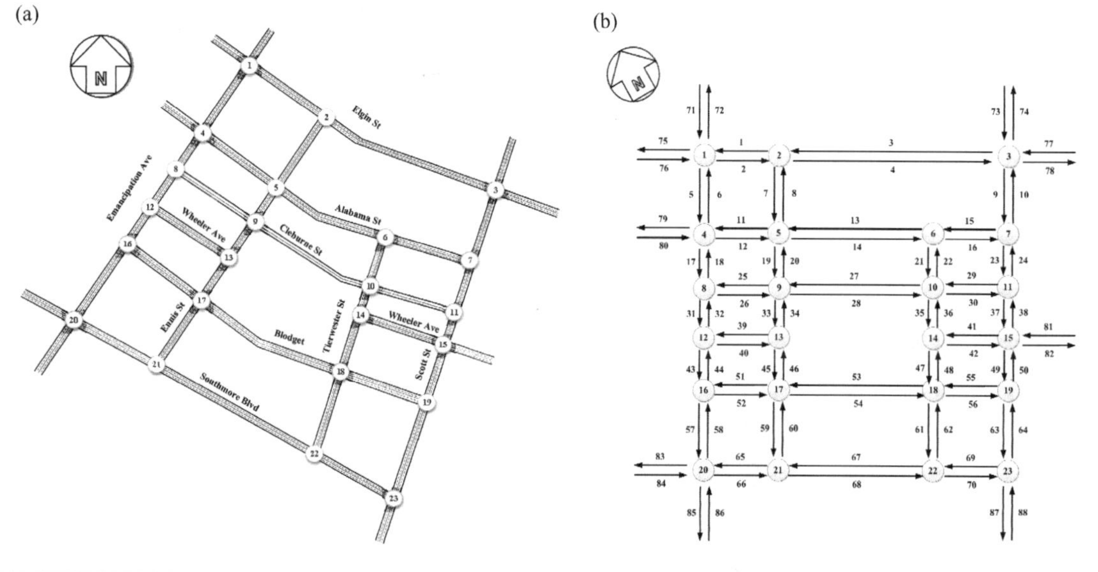

Figure 7. Road network around Texas Southern University and simulation; (*a*) selected road network around Texas Southern University (TSU) for case study, (*b*) the coded topological TSU network for simulation.

The Eco-driving strategy was tested on a simulated network around the Texas Southern University (TSU) in Houston, Texas, United States. The network in this study is bounded to the South by the Southmore Boulevard, to the West by the Emancipation Avenue, to the East by the Scott Street, and to the North by Elgin Street. This network includes 88 framed roads and 23 signal-controlled intersections. The analysis period considered in this application is the morning peak hour, with around 5,000 vehicular trips.

A stochastic cellular automaton model for urban two-way road network systems, described elsewhere (*24*, *25*), is utilized to demonstrate the medium-scale application of an eco-driving strategy on a simulated replica of the road network around the network in Figure 7. Figure 7*a* is the original network around TSU, and its topological representation is in Figure 7*b* including 88 links (roads). Vehicles enter the network from links (roads) 71, 73, 76, 77, 80, 81, 84, 86, and 88, and leave the network from links 72, 74, 75, 78, 79, 82, 83, 85, and 87. The vehicles can turn left, right, or go through at all intersections if the geometries permit. In case of Normal-driving, vehicles' movements are represented using the Nagel–Schreckenberg rules, that is, $A_t' = 1$, on all roads. In the case of eco-driving, if the distance between the vehicle and the downstream intersection λ_t less than 140 m, the vehicle's movements are indicated using the eco-driving rule, that is, A_t' depends on the state S_t and the Q-factors are learned from the forward platform of the road discussed in the previous section. Emissions and traffic parameters can be calculated through the simulator.

Table 2. Performance of Different Driving Behaviors in the Road Network

Scenario \ MOE	Normal-driving	Eco-driving	Improvements Eco-driving vs. Normal-driving
Avg. CO_2 (g)	2.27×10^3	2.08×10^3	8.37%
Avg. CO (mg)	43.20×10^3	37.9×10^3	12.27%
Avg. HC (mg)	273.21	238.24	12.80%
Avg. NO_X (mg)	1.14×10^3	1.02×10^3	10.53%
Avg. stop duration (s)	65.55	46.92	28.42%
Avg. travel time (s)	368.27	379.46	−3.03%
Avg. number of stops	2.27	2.10	7.50%

The performance of each driving behavior is evaluated based on the measures of effectiveness in Table 2: average exhaust emissions (CO, HC, and NO_x) per vehicle, average stop duration, travel time, and number of stops per vehicle.

The analysis of the results shown in Table 2 leads to the following findings. Compared with Normal-driving mode, the Eco-driving strategy can improve most of the measures of effectiveness, except for travel time. The reduction in emissions is around 10%. The stop duration and the number of stops decrease by 28.42% and 7.5%, respectively. The travel time reduces by 3.03%. This is because the Eco-driving scenario may slow down the vehicles from relatively higher speed in advance, thereby affecting the speed of the said vehicle as well as its following vehicles. For the same reason, the improvements of all measures are less than those in the first simulation case, where there was only one vehicle traveling in that signalized intersection (Table 1).

Conclusions and Future Work

In this paper, an approach was proposed to evaluate the total exhaust vehicle emissions at a signalized intersection equipped with an I2V technology. RL and micro-simulation models were identified to simulate eco-driving scenarios, which were applied to a single intersection road and a medium-scale network as examples to demonstrate the applicability of this approach. The evaluation results revealed positive impacts on exhaust emission reductions. In other words, eco-driving would improve not only traffic performance, but also air quality. More specific research findings are summarized as follows.

First, through the introduction of a reasonable penalty mechanism, the RL technology could help vehicles to build up an optimal mapping between traffic operations (acceleration, cruise speed, and deceleration) and environmental conditions (location, speed, and signal) to practice an eco-driving strategy.

Second, the eco-driving strategy was able to significantly reduce the total emissions in both a single intersection road and a network, therefore contributing to environmental sustainability. Importantly, the reduction in traffic-related air pollutants that pose a threat to public health, namely CO, HC, and NO_x, was more significant than in the greenhouse gas of CO_2 because of the reduced frequency of start and stop actions.

Finally, the eco-driving strategy allows vehicles to determine their acceleration according to the real-time distance to an intersection, speed, and signal state, thereby reducing the probability of encountering a red light. Therefore, the Eco-driving strategy could minimize vehicles' exhaust emissions by increasing the probability of passing a signalized intersection without a stop action.

Further study will lie in the improvement of simulation accuracy by increasing the degree of the speed discretization for even more significant enhancement of the eco-driving strategy in wider applications. Also, the RL technology could be extended to other driving scenarios such as routing choice, adaptive signal control, and lane change. The RL and micro-simulation models could be improved to simulate the interaction among vehicles on roads. More parameters could be taken into account within the optimization scheme, such as the lane where the vehicle is located, the distance between the vehicle and its following and leading vehicles on the same lane, and distance to vehicles on adjacent lanes.

Acknowledgments

This work was jointly supported by the Zhejiang Provincial Natural Science Foundation of China (LY18E080021), and the Project of Science and Technology of the Department of Transportation of Hubei Province, China (2017-538-3-2, 2014-721-3-13).

Author Contributions

The authors confirm contribution to the paper as follows: study conception and design: Junqing Shi, Fengxiang Qiao, Lei Yu; data collection: Qing Li; analysis and interpretation of results: Qing Li, Junqing Shi, Yongju Hu; draft manuscript preparation: Lei Yu, Fengxiang Qiao, Junqing Shi, Qing Li. All authors reviewed the results and approved the final version of the manuscript.

References

1. United States Environmental Protection Agency. *U.S. Greenhouse Gas Inventory Report: 1990-2014*. United States Environmental Protection Agency, Washington, D.C., 2015.
2. Ministry of Environmental Protection of the People's Republic of China. *China Vehicle Emission Control Annual Report: 2015*. Ministry of Environmental Protection of the People's Republic of China, Beijing, China, 2015.
3. Qian, G., and E. Chung. Evaluating Effects of Eco-driving at Traffic Intersections Based on Traffic Micro-simulation. *Proc., Australian Transport Research Forum 2011*, Adelaide, Australia, 2011.
4. Mensing, F., E. Bideaux, R. Trigui, J. Ribet, and B. Jeanneret. Eco-driving: An Economic or Ecologic Driving Style? *Transportation Research Part C: Emerging Technologies*, Vol. 38, 2014, pp. 110–121. https://doi.org/10.1016/j.trc.2013.10.013
5. Zarkadoula, M., G. Zoidis, and E. Tritopoulou. Training Urban Bus Drivers to Promote Smart Driving: A Note on a Greek Eco-driving Pilot Program. *Transportation Research Part D: Transport and Environment*, Vol. 12, No. 6, 2007, pp. 449–451. https://doi.org/10.1016/j.trd.2007.05.002
6. Ho, S. H., Y. D. Wong, and V. W. C. Chang. What Can Eco-driving Do for Sustainable Road Transport? Perspectives from a City (Singapore) Eco-driving Programme. *Sustainable Cities and Society*, Vol. 14, 2015, pp. 82–88. https://doi.org/10.1016/j.scs.2014.08.002
7. Xu, Y., H. Li, H. Liu, M. O. Rodgers, and R. L. Guensler. Eco-driving for Transit: An Effective Strategy to Conserve Fuel and Emissions. *Applied Energy*, Vol. 194, 2017, pp. 784–797. https://doi.org/10.1016/j.apenergy.2016.09.101
8. Li, J., M. Dridi, and A. El-Moudni. A Cooperative Traffic Control of Vehicle-intersection (CTCVI) for the Reduction of Traffic Delays and Fuel Consumption. *Sensors*, Vol. 16, No. 12, 2016, pp. 2175–2175. https://doi.org/10.3390/s16122175
9. Barth, M., and K. Boriboonsomsin. Energy and Emissions Impacts of a Freeway-Based Dynamic Eco-driving System. *Transportation Research Part D: Transport and Environment*, Vol. 14, No. 6, 2009, pp. 400–410. https://doi.org/10.1016/j.trd.2009.01.004
10. Schall, D. L., and A. Mohnen. Incentivizing Energy-Efficient Behavior at Work: An Empirical Investigation Using a Natural Field Experiment on Eco-driving. *Applied Energy*, Vol. 185, 2017, pp. 1757–1768. https://doi.org/10.1016/j.apenergy.2015.10.163
11. Li, J., M. Dridi, and A. El-Moudni. A Cooperative Traffic Control of Vehicle-Intersection (CTCVI) for the Reduction of Traffic Delays and Fuel Consumption. *Sensors*, Vol. 16, No. 12, 2016, pp. 2175–2175. https://doi.org/10.3390/s16122175
12. Milesich, T., J. Bucha, L. Gulan, and J. Danko. The Possibility of Applying Neural Networks to Influence Vehicle Energy Consumption by Eco Driving. *Proc.*, International Conference Mechatronics, Springer, Cham., 2017, pp. 372–379. https://doi.org/10.1007/978-3-319-65960-2_46
13. Jiang, Y., M. Zanon, R. Hult, and B. Houska. Distributed Algorithm for Optimal Vehicle Coordination at Traffic Intersections. *IFAC World Congress*, Vol. 50, No. 1, 2017, pp. 12082–12087. https://doi.org/10.1016/j.ifacol.2017.08.1511
14. Sutton, R. S., and A. G. Barto. *Reinforcement Learning: An introduction*. MIT Press, Cambridge, MA, 1998.
15. Abdulhai, B., R. Pringle, and G. J. Karakoulas. Reinforcement Learning for True Adaptive Traffic Signal Control. *Journal of Transportation Engineering*, Vol. 129, No. 3, 2003, pp. 278–285. https://doi.org/10.1061/(ASCE)0733-947X(2003)129:3(278)
16. Walraven, E., M. T. J. Spaan, and B. Bakker. Traffic Flow Optimization: A Reinforcement Learning Approach. *Engineering Applications of Artificial Intelligence*, Vol. 52, 2016, pp. 203–212. https://doi.org/10.1016/j.engappai.2016.01.001
17. Zolfpour-Arokhlo, M., A. Selamat, S. Z. M. Hashim, and H. Afkhami. Modeling of Route Planning System Based on Q Value-Based Dynamic Programming with Multi-Agent Reinforcement Learning algorithms. *Engineering Applications of Artificial Intelligence*, Vol. 29, 2014, pp. 163–177. https://doi.org/10.1016/j.engappai.2014.01.001
18. Watkins, C. J., and P. Dayan. Q-Learning. *Machine Learning*, Vol. 8, No. 3–4, 1992, pp. 279–292. https://doi.org/10.1007/BF00992698
19. United States Environmental Protection Agency. *Average Annual Emissions and Fuel Consumption for Passenger Cars and Light Trucks*. Transportation and Air Quality. EPA420-F-08-024, October 2008.
20. Jimenez-Palacios, J. L. *Understanding and Quantifying Motor Vehicle Emissions with Vehicle Specific Power and TILDAS Remote Sensing*. Massachusetts Institute of Technology, Cambridge, MA, 1998.
21. Frey, H. C., and B. Liu. *Development and Evaluation of a Simplified Version of MOVES for Coupling with a Traffic Simulation Model*. Transportation Research Board, Washington, D.C., 2013.
22. Nagel, K., and M. Schreckenberg. A Cellular Automaton Model for Freeway Traffic. *Journal De Physique I*, Vol. 2, No. 12, 1992, pp. 2221–2229. https://doi.org/10.1051/jp1:1992277
23. Wang, J., and H. A. Rakha. Fuel Consumption Model for Conventional Diesel Buses. *Applied Energy*, Vol. 170, 2016, pp. 394–402. https://doi.org/10.1016/j.apenergy.2016.02.124
24. Shi, J. Q., Y. J. Hu, S. L. Li, X. H. Zhang, and C. Y. Mao. Simulation and Analysis of Road Construction Traffic Flow in Urban Road Networks. *Advances in Mechanical Engineering*, Vol. 7, No. 11, 2015, pp. 1–6. https://doi.org/10.1177/1687814015618176
25. Shi, J. Q., L. Cheng, J. C. Long, and Y. L. Liu. A New Cellular Automaton Model for Urban Two-Way Road Networks. *Computational Intelligence & Neuroscience*, Vol. 2014, pp. 685047. https://doi.org/10.1155/2014/685047

The Standing Committee on Transportation and Air Quality (ADC20) peer-reviewed this paper (18-04322).

TRR
JOURNAL OF THE TRANSPORTATION RESEARCH BOARD

Article

Transportation Research Record
2018, Vol. 2672(25) 99–110
© National Academy of Sciences:
Transportation Research Board 2018
Article reuse guidelines:
sagepub.com/journals-permissions
DOI: 10.1177/0361198118796966
journals.sagepub.com/home/trr
SAGE

Investigating the Impact of Meteorological Conditions on Near-Road Pollutant Dispersion between Daytime and Nighttime Periods

Mohammad Hashem Askariyeh[1], Suriya Vallamsundar[2], and Reza Farzaneh[3]

Abstract

In urban areas in Texas and the rest of the United States, roadway work zone and construction activities are often conducted at night to reduce disruption to traffic and to prevent congestion caused by lane closures during peak hours. The reduced traffic delays due to nighttime construction have the potential to reduce traffic emissions. However, the air quality impacts associated with moving these activities from daytime to nighttime have not been studied in detail. Air quality impacts depend on two major factors: the traffic emissions and meteorological conditions. While the impact of traffic emissions between time periods has been studied in the literature, there is limited knowledge on understanding the impact of meteorological conditions on the dispersion of mobile source pollutants. This study specifically addresses this gap by evaluating the impact of the meteorological conditions on pollutant concentrations under different input settings related to the region, land use, distance from roadways, and averaging periods. The assessment of the impact of metrological conditions indicated that for the same amount of emissions, the nighttime period could result in higher pollutant concentration levels. However, given that traffic congestion and overall traffic volumes are generally substantially lower in the nighttime period, the findings do not imply that nighttime construction activities result in worse air quality in terms of pollutant concentrations. Thus, the relative difference in pollutant concentrations obtained from shifting construction activities from daytime to nighttime periods should be assessed based on a combination of meteorological and traffic conditions.

Nighttime construction is being used increasingly by state Departments of Transportation (DOTs) and other highway agencies to conduct highway maintenance and reconstruction projects mainly to reduce the impacts on congestion and mobility. Nighttime construction has both favorable and unfavorable effects for highway agencies and the public. Favorable aspects relate to reduction in congestion and delays as nighttime periods see traffic at its lowest levels, less air pollution as it is believed that lower tailpipe emissions lead to better air quality, lower traffic also related to lower fuel consumption and energy conservation, less inconvenience to the traveling public caused by lane closures at work zones, and lower economic impact to the surrounding businesses due to the work being performed in nighttime periods. Negative issues associated with nighttime projects include safety issues for workers, premium worker wages, material costs, added traffic control costs, reduced visibility, noise disturbance to the surrounding communities, and reduced availability of materials and equipment parts. While some of these positive and negative effects are well known and have been studied in detail, others have not been investigated sufficiently. For example, most studies have focused on examining the safety of nighttime work activities and concluded that the crash rate on sections of the roadway near work activity was higher than normal nighttime crash rates, and that crash rates were higher when lane closures were required compared with those when no lane closures were required (*1–4*). Studies have also examined the noise and vibration effects from

[1]Environment and Air Quality Division, Texas A&M Transportation Institute, College Station, TX
[2]Environment and Air Quality Division, Texas A&M Transportation Institute, Dallas, TX
[3]Environment and Air Quality Division, Texas A&M Transportation Institute, Austin, TX

Corresponding Author:
Address correspondence to Suriya Vallamsundar: s-vallamsundar@tti.tamu .edu

construction equipment and traffic congestion and found these issues to be detrimental especially if the construction activities are located closer to hospitals or schools (*5*).

From an air quality perspective, when construction activities are shifted to nighttime, congestion will be lower compared with the same project undertaken during the daytime. In turn, this is thought to reduce fuel consumption and vehicle emissions associated with nighttime versus daytime projects. There are two major factors driving the pollutant concentrations in the atmosphere, namely: (i) emissions released by the traffic and construction activities; and (ii) meteorological conditions. While emissions have a direct impact on pollutant concentrations, with higher concentrations being associated with higher emissions, the relationship between meteorology and pollutant concentrations is much more complicated. Meteorological conditions having an impact on how pollutant dispersion relates to characteristics of the lower and upper layers of the atmosphere and land-use conditions. Studies have shown the pollutant concentration levels to be significantly higher during pre-sunrise hours than during the daytime. The higher concentration levels are associated with nocturnal surface temperature inversion, low wind speeds, and high relative humidity (*6*). Ginzburg et al. (*7*) investigated the relationship between concentrations and meteorological parameters and found: (a) an inverse relationship between concentrations and temperatures; (b) concentrations decreasing with increases in wind speed; (c) no significant correlation between wind direction and concentrations; and (d) higher atmospheric buoyancy causing lower concentrations. In terms of seasonal variation, studies (*8*) found concentration levels to have a dominant peak in fall and winter seasons. Zhao et al. (*9*) found concentration variation to be greater in the summer season when temperature variations are also greater. Although meteorological conditions were found to have a dominant impact on pollutant concentrations, studies investigating the air quality effects of shifting construction activities between time periods examine the effects only from the traffic emissions point of view without considering the impact of meteorology (*10*). This highlights a need to better understand how the impact of meteorological conditions affects the expected benefits of shifting construction activities from daytime to nighttime periods.

The overall goal of this study was to evaluate the impact of meteorological conditions on pollutant dispersion emitted by shifting daytime construction activities to nighttime. A series of sensitivity analyses was performed to evaluate the impact of meteorological and land use factors (for the same set of emission rates) on pollutant concentrations between daytime and nighttime periods. Sensitivity analysis is performed using the Environmental Protection Agency's (EPA) approved AERMOD air dispersion model. This study is part of a larger research project conducted to investigate the emissions and air quality impacts of nighttime construction through case studies and to develop a decision-support framework. The study described in this paper aims at gaining a better understanding of air quality impacts specifically related to meteorological conditions that can be taken into account by public agencies in making informed decisions about whether to undertake nighttime construction. The rest of the paper is organized as follows. The following section presents the methodology used to assess the relative differences in pollutant concentrations between time periods. Results and findings obtained are then presented, followed by discussion and conclusions.

Methodology

This section outlines the methodology used to assess the relative difference in pollutant concentrations between daytime and nighttime periods. Two major factors that affect the dispersion of pollutants into the atmosphere are emission rates (the rate at which the emissions are released from an emission source) and meteorological conditions (wind speed, wind direction, atmospheric stability, and surface roughness etc.). The modeling approach consisted of performing a series of sensitivity analyses that varied key input parameters (for the same set of emission rates) to represent differences in nighttime and daytime conditions that may influence pollutant dispersion. Figure 1 illustrates the conceptual overview approach used for this study. The approach consists of the following major elements:

1. Set up case study.
2. Assess key input parameters for sensitivity analysis.
3. Perform air dispersion modeling and obtaining results.

In the three sections below, we first describe the set-up of the case study, followed by a description of the key input parameters assessed for sensitivity analysis, and finish with a description of the processing of the model runs to obtain results.

Case-Study Set-up

Sensitivity analysis was performed on a simplified representation of a roadway segment to isolate the impact of complicated roadway geometry, and emission rates. Figure 2 displays the source and receptor set-up of a hypothetical roadway line source with generic assumptions made for emission rates and source parameters.

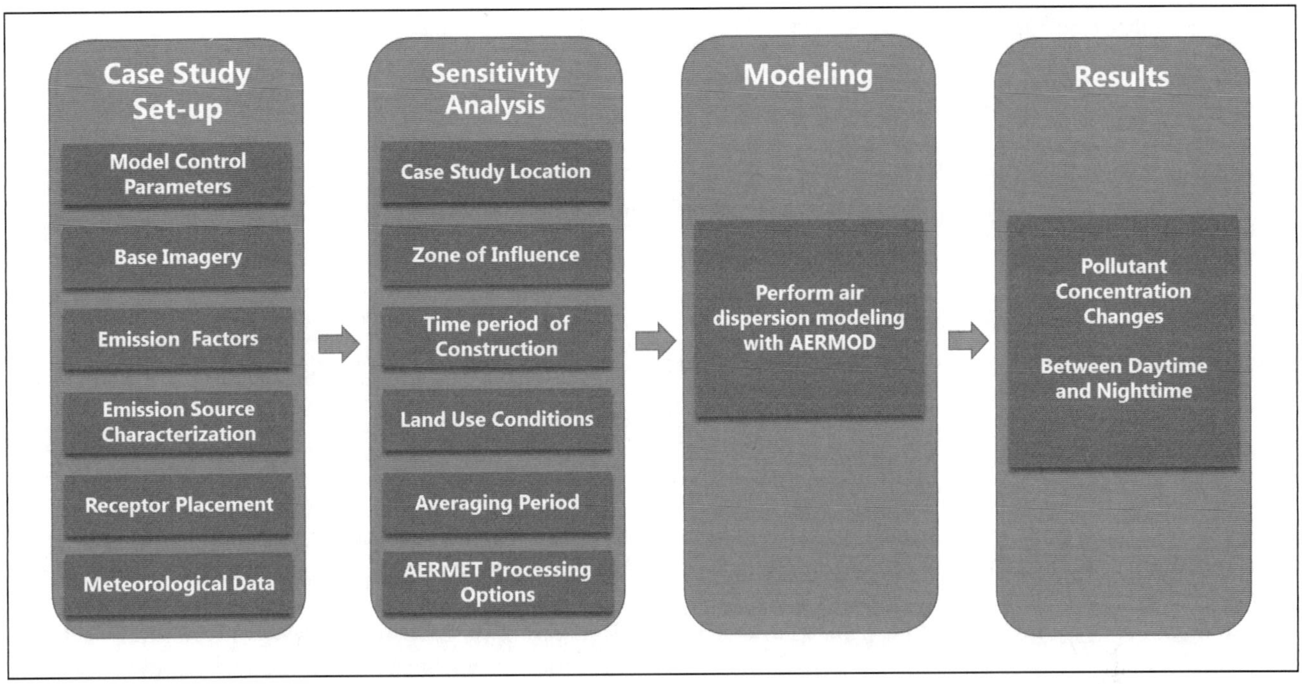

Figure 1. Conceptual methodology.

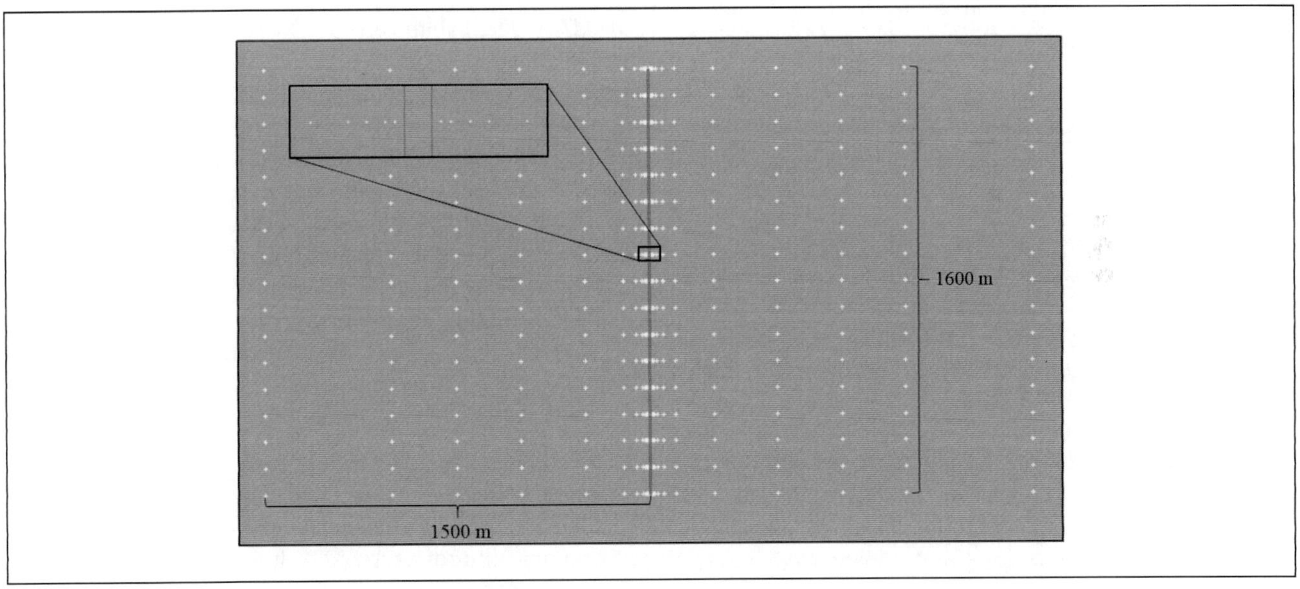

Figure 2. Source and receptor characterization.

The emission source consisted of a two-lane, at-grade highway segment extending to a length of 1,600 m (1 mile) and a width of 7.3 m (24 Ft). Inputs of highway configuration data (except for emission rates) were defined according to the EPA project-level hot-spot analysis. The highway segment was characterized using AERMOD area source with a source elevation of 0 m, a release height of 1.3 m, and an initial vertical dispersion parameter of 1.2 m. A unit emission rate was used for all modeling runs. Receptors were placed along the roadway from start of the roadway at 0 m and end of the roadway at 1,600 m at a spacing of 100 m. At each of these locations along the roadway, receptors were placed at 2.5 m, 5 m, 10 m, 15 m, 25 m, 50 m, 100 m, 250 m, 500 m, 750 m, 1000 m, and 1500 m from the edge of the road. AERMOD, being a Gaussian-based dispersion model is capable of predicting the dispersion patterns of any primary non-reactive pollutant with no chemical

Table 1. Key Input Parameters Assessed for Sensitivity Analysis

Parameters evaluated	Values
1. Case study location	Dallas Houston El Paso
2. Zone of influence	Near-road (0–250 m) from roadway edge Far-road (>250 m)
3. Time period of construction	TxDOT Specification Handbook Practical time period information from contractors
4. Land use	Rural Urban
5. Atmospheric stability	Stable, Unstable and Neutral stability classes are classified based on Monin-Obukhov length
6. AERMET	Regulatory default option Beta option (LOWWIND, ADJ_U*)

transformation. Accordingly, the dispersion patterns predicted by AERMOD apply to any primary non-reactive pollutant (such as carbon-monoxide (CO), primary particulate matter ($PM_{2.5}$ and PM_{10}), and primary nitrogen oxides (NO_x)).

Key Input Parameters for Sensitivity Analysis

The impact of meteorological conditions on pollutant concentrations between daytime and nighttime periods were assessed for different input parameter settings as listed in Table 1.

Description about key input parameters and their parameter values considered for sensitivity analysis is provided below.

Case Study Location. As it is difficult to generalize the meteorological conditions between different case study sites, sensitivity analyses were performed based on local parameters for three major areas in Texas, namely Dallas, Houston, and El Paso. Raw meteorological (surface and upper air) data specific to each case study location was processed using AERMINUTE, AERMET, and AERSURFACE preprocessors to produce data in a compatible format for AERMOD. The prevailing wind rose diagrams for daytime and nighttime periods (on an annual averaging basis) are shown in Figure 3. The predominant wind direction was the same for both daytime and nighttime periods for Dallas and Houston. However, for El Paso, the predominant wind direction changed drastically from blowing from the south-west (daytime period) to the north-east (nighttime period).

Time Period of Daytime and Nighttime Construction. Designation of time periods during which construction activities are performed during daytime and nighttime periods were obtained from two sources, namely: (a) the TxDOT Standard Specification Handbook; and (b) input obtained from construction contractors performing the activities. According to the TxDOT Standard Specification Book (*11*): "Nighttime work is defined as work performed from 30 minutes after sunset to 30 minutes before sunrise." Table 2 lists the designations of daytime and nighttime periods according to the seasonal sunrise and sunset times in Texas, based on the construction contractors' information.

Land Use. The urban/rural land use representativeness of a case study site is found to have an impact on the dispersed pollutant concentrations. Urban areas are generally hotter than nearby rural areas, especially at night, mainly because of heat retention by urban materials. Because of this heat retention, the vertical motion of the air is increased through convection, thereby leading to increased dispersion of pollutants (*12*). This phenomenon is referred to as the urban heat island effect. The purpose of evaluating this key parameter was to assess the impact of land use on the relative difference in concentration estimates between daytime and nighttime periods, given the same traffic characteristics and site configuration.

Atmospheric Stability. Among the different meteorological parameters, atmospheric stability which is responsible for mixing and dilution is found to have a significant impact on pollutant concentration between different time periods. Therefore, the variation of atmospheric stability was studied in detail. To evaluate the variation of pollutant concentration with atmospheric stability, pollutant concentrations were estimated at an hourly averaging period for four seasons for El Paso.

AERMET Meteorological Data Processing. The EPA noted, in 2007, issues with high concentrations due to the treatment of light winds in AERMOD. AERMOD exaggerates the nighttime concentration estimates due to the way it handles low winds (low wind speeds < 1 m/s) (*13*).

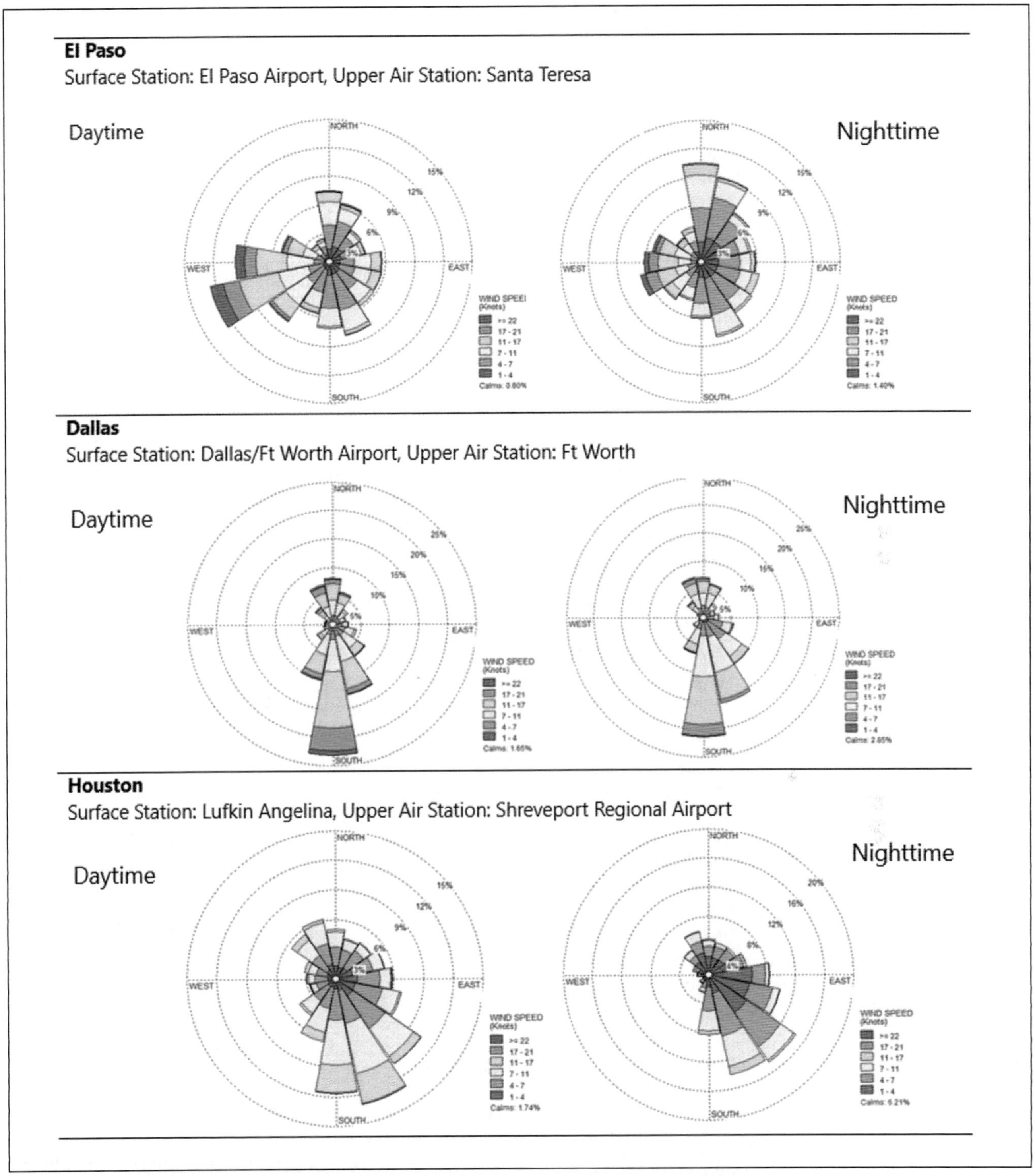

Figure 3. Wind rose diagrams for case study regions.

These low wind conditions typically occur during nighttime periods. To overcome the issue of over prediction, the EPA, in 2012, developed non-default BETA options for meteorological data processing in AERMET to improve AERMOD performance under low wind conditions (*13*). This included the LOWWIND BETA options on the MODELOPT keyword in AERMOD, and the ADJ_U* option included in stage 3 of the AERMET meteorological processor. The LOWWIND option increases the minimum value of sigma-v, and

Table 2. Daytime and Nighttime Construction Time Period

(A) Construction time period according to TxDOT specification handbook

Season	Sunrise/sunset	Daytime construction period	Nighttime construction period
Winter (Dec., Jan., Feb.)	Sunrise 7:30 a.m., Sunset 5:30 p.m.	7 a.m.–6 p.m.	6 p.m.–7 a.m.
Spring (Mar., Apr., May)	Sunrise 7 a.m., Sunset 8 p.m.	6:30 a.m.–8:30 p.m.	8:30 p.m.–6:30 a.m.
Summer (June, July, Aug.)	Sunrise 6:30 a.m., Sunset 8:30 p.m.	6 a.m.–9 p.m.	9 p.m.–6 a.m.
Fall (Sep., Oct., Nov.)	Sunrise 7:30 a.m., Sunset 7 p.m.	7 a.m.–7:30 p.m.	7:30 p.m.–7 a.m.

(B) Construction time period according to construction contractors

Season	Practical daytime construction period	Practical nighttime construction period
Winter (Dec., Jan., Feb.)	6 a.m.–9 p.m.	9 p.m.–6 a.m.
Spring (Mar., Apr., May)	6 a.m.–9 p.m.	9 p.m.–6 a.m.
Summer (June, July, Aug.)	6 a.m.–9 p.m.	9 p.m.–6 a.m.
Fall (Sep., Oct., Nov.)	6 a.m.–9 p.m.	9 p.m.–6 a.m.

replicates the centerline concentration accounting for horizontal meander, but utilizes an effective sigma-y and eliminates upwind dispersion. The ADJ_U* option adjusts the surface friction velocity (U*) to improve the performance of the model in dealing with low wind speed and stable conditions (*14*). These non-default options are not approved for regulatory purposes but can be used for research purposes with the approval of appropriate reviewing authority. Meteorological data was processed using these non-default options to assess their impact on pollutant concentrations between different time periods.

Performing Air Dispersion Modeling and Obtaining Results

Among the different air dispersion models, the EPA approves AERMOD for a wide range of regulatory applications including roadways and off-road networks (i.e., construction sites) in all types of terrain. AERMOD is a steady-state Gaussian plume model and uses an advanced method to characterize stability compared with its processor models. AERMOD uses a continuous function called Monin-Obukhov length to characterize atmospheric stability. The two regulatory components of AERMOD include the meteorological preprocessor (AERMET) and the terrain data preprocessor (AERMAP). AERMET processes the meteorological data from the National Weather Station (NWS) and onsite data. AERMET produces output files containing the surface scalar parameters and the vertical profile of meteorological data. AERMAP preprocesses complex terrain data and generates receptor grids, using USGS digital elevation data (*15*). The AERMOD model requires emission factors to be specified for all hours of the day and does not allow the user to run the model for specific time periods. To overcome this issue, the AERMOD model was set to run for the daytime and nighttime time periods as follows:

- Daytime model runs: emission rates were assigned a value of 1 for the daytime period hours and were given a value of 0 for nighttime period hours
- Nighttime model runs: emission rates were assigned a value of 0 for the daytime period hours and were given a value of 1 for nighttime period hours

Whenever a given hour is specified a value of 0, AERMOD does not compute concentration estimates for that hour. For example, in a daytime model run, AERMOD concentration estimates were based only on emissions occurring during daytime periods. In a nighttime model run, AERMOD concentration estimates were based only on nighttime period emissions. The resulting pollutant concentrations obtained were normalized by the corresponding time periods.

Results

To evaluate and compare the effects of different time periods, time period designations, land use, areas, and zones of influence from the roadway edge on pollutant concentrations, different combinations of these factors were defined to model pollutant dispersion using AERMOD. Figure 4 shows the variation in normalized concentration predictions as a function of time period designation, zone of influence or distance from the roadway, and land use. Results shown in Figure 4 are based on the default regulatory option in AERMOD.

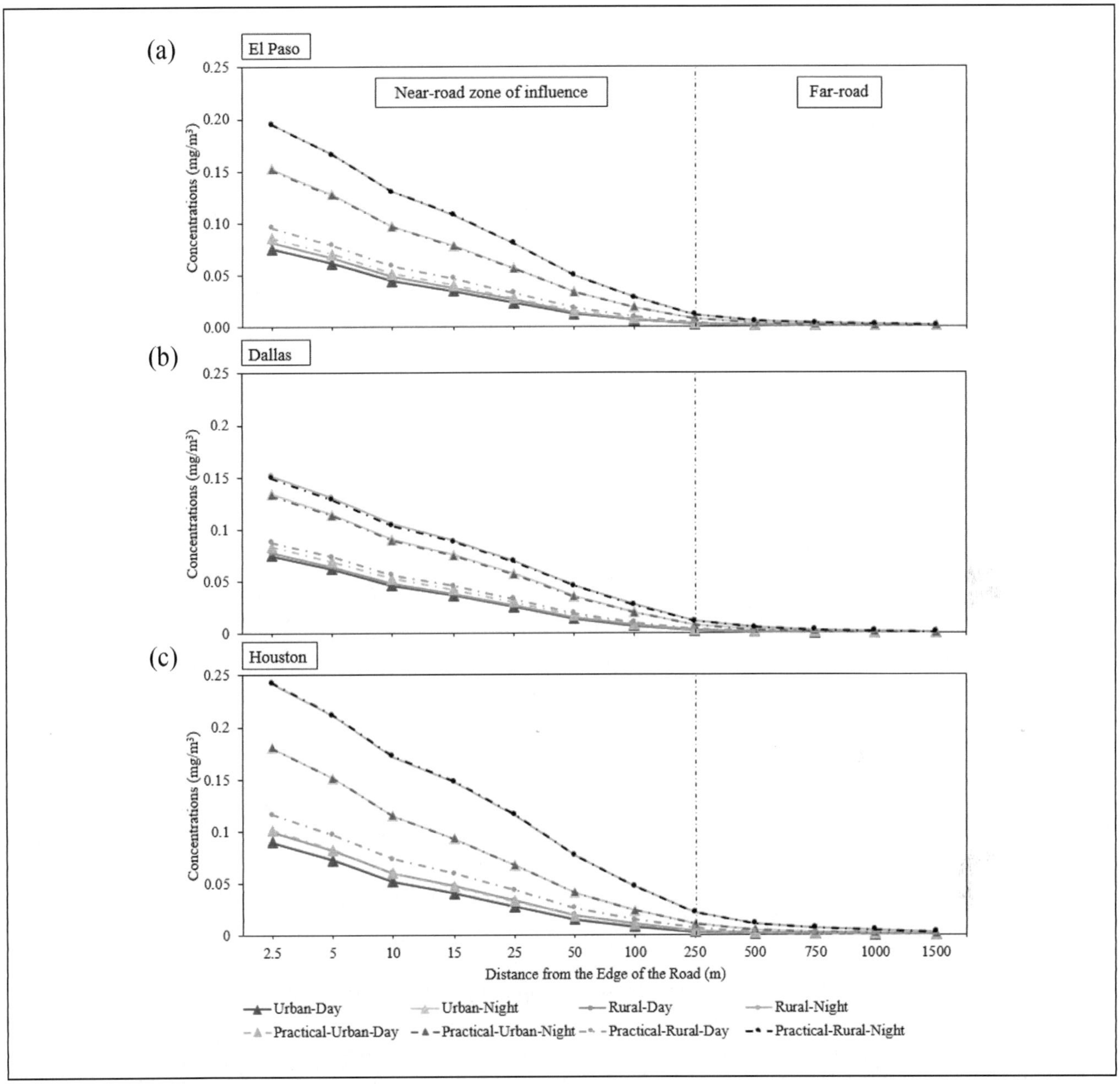

Figure 4. Normalized concentrations with distance from roadway edge, land use, and time period designation for (*a*) El Paso; (*b*) Dallas; and (*c*) Houston.

Effect of Distance from Roadways

As shown in Figure 4, concentration estimates were found to be the highest near the roadway link and then gradually decreased with distance from roadway links. The rate of decline in concentration estimates was found to be steep for a distance of 250 m and the concentration estimates flattened for distances beyond 250 m. These findings were found to be consistent with the literature (*16–20*). Karner et al. (*20*) analyzed 41 roadside monitoring studies between 1978 and 2008 and concluded that almost all pollutants decay to background levels at a distance of 115 to 570 m from the edge of the road and the decay rate varies from one pollutant to another. Venkatram et al. (*21*) analyzed data from three near-road pollution measurements and AERMOD dispersion model and found the concentration of an inert pollutant to decay rapidly to less than one-fifth of its initial strength from roadway edge.

Effect of Time Period

For all the case study areas (El Paso, Dallas, and Houston) concentration estimates were found to be higher for nighttime periods compared with daytime and

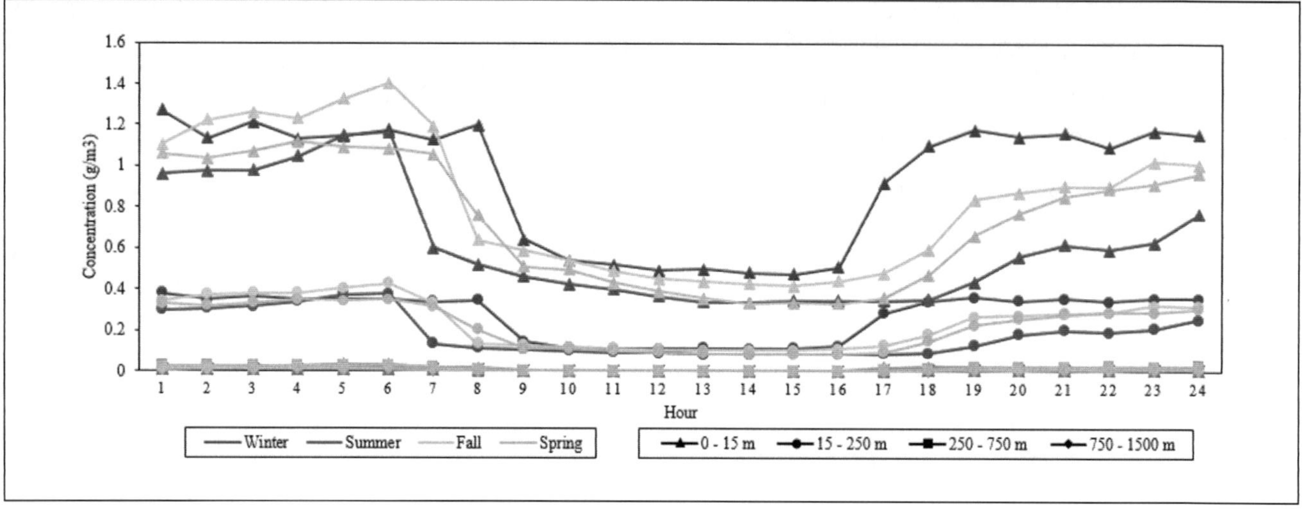

Figure 5. Pollutant concentrations at an hourly averaging period varied by season for El Paso.

rural land use compared with urban land use conditions. The reason for high concentration estimates during nighttime compared with daytime periods was because of the stable atmospheric conditions during nighttime periods. Sunlight during the daytime helps in the mixing/dispersion of pollutants in the atmosphere. When the sunlight strikes the Earth's surface during the day, it heats up more quickly and the heat is transferred to the air immediately above the ground, causing the warm air to rise and mix with the cooler air above. When the sun sets during the evening and nighttime periods, the Earth's surface cools down much faster than air, resulting in cooler (heavier) air near the ground and warmer (lighter) air staying on top. This fairly stable atmospheric condition during nighttime periods leads to much reduced mixing and dispersion of pollutants and thereby higher concentrations. Thus, the low atmospheric transport and dispersion characteristics were the reason for higher concentration estimates (*22*) during nighttime periods.

Effect of Land Use

Lower concentration estimates observed in an urban land-use setting compared with a rural setting was because of the urban heat island effect. In this regard, buildings, roads, and structures in urban areas absorb more radiation and energy compared with almost flat terrain conditions in rural areas. Because of this heat retention, the vertical motion of the air was increased through convection, yielding better mixing, stronger vertical air flux, and eventually better dilution resulting in increased dispersion of pollutants (*22*).

Effect of Modeling Options in AERMOD

In addition to the default regulatory option, AERMOD was run with the BETA option (meteorological data processed with LOWWIND option in AERMOD and use of ADJ_U* in AERMOD). Concentration estimates obtained using BETA option were compared with the regulatory option. The regulatory option was found to predict higher concentration estimates compared with BETA option. However, the comparison exhibited a minor difference in concentration levels ranging between 2% and 4%. A potential reason for the low difference could be that the BETA options are more impactful for studies involving tall stacks (*23*).

Effect of Atmospheric Stability

To investigate the variation of pollutant concentration with atmospheric stability, pollutant concentrations were estimated at an hourly averaging period for El Paso. Hourly concentrations are further classified by seasons and are shown in Figure 5. The higher concentration values in Figure 5 are because of the lower averaging period of an hour compared to Figure 4, which is based on an annual averaging period.

Atmospheric stability affects the dispersion of vehicle emissions downwind of the highway and is governed by heat and momentum forces in the environment. AERMOD provides a continuous measure of atmospheric stability based on an energy balance in the planetary boundary layer. The energy balance is represented by the sensible heat flux, which depends on net radiation and surface characteristics such as available surface moisture. Atmospheric stability in AERMOD is

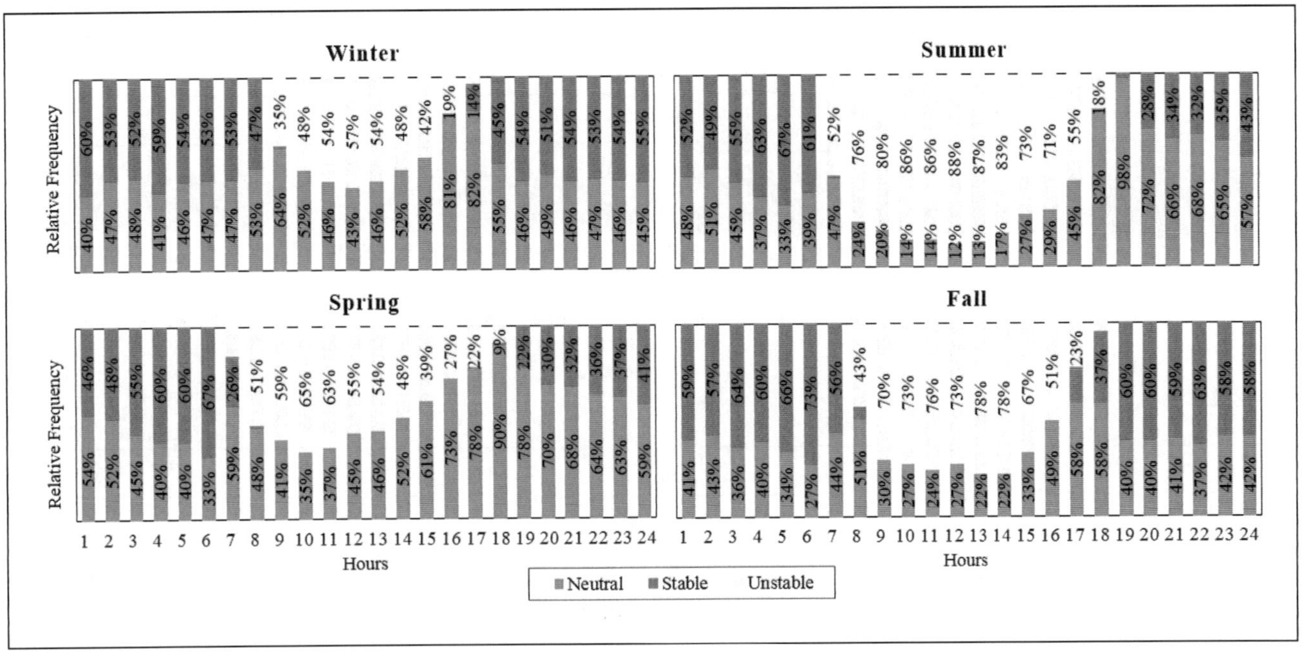

Figure 6. Distribution of atmospheric stability conditions for El Paso.

represented as a function of the Monin-Obukhov length (L[m]) (*24*). The atmospheric stability is obtained from the surface and upper air data processed by AERMET. Unstable atmospheric conditions refer to convective conditions when the atmosphere is not stable leading to increased dispersion and lower pollutant concentration estimates. Stable atmospheric conditions, typically observed during nighttime periods, have low atmospheric transport and dispersion leading to higher pollutant concentration estimates and neutral conditions are in the middle between stable and unstable conditions.

The hourly distribution of atmospheric stability conditions, based on AERMET provided Monin-Obukhov length for all seasons specific to El Paso, is shown in Figure 6. In this regard, Monin-Obukhov length ranging from $0-10^5$ m is classified as Unstable, followed by 0–10^5 m classified as Stable and values greater than 10^5 m classified as Neutral stability class (*25*). Comparing Figure 5 with Figure 6, the variation of concentration estimates closely followed the variation in atmospheric stability conditions. Concentration levels were lower between hours 8 to 17, coinciding with the unstable atmospheric conditions. Concentration levels were higher between hours 1 to 8 and 17 to 24 due to the presence of predominantly stable conditions and limited neutral conditions. The magnitude of concentration levels was higher during winter and fall seasons compared with summer and spring seasons. This was because of the higher prevalence of stable atmospheric conditions (less sunlight time compared with summer and spring) leading to reduced levels of mixing and pollutant dispersion during fall and winter seasons. Considering seasonal atmospheric conditions, concentration estimates were found to be higher for winter and fall nighttime periods, followed by summer and spring nighttime periods, winter and fall daytime periods, and summer and spring daytime periods.

The relative difference (expressed as a percentage) in average concentration levels between nighttime and daytime periods is summarized in Table 3. The relative difference expresses how much the concentration estimates are higher in nighttime periods compared with daytime periods for the same traffic activity, and source characterization. The relative difference was found to increase with distance from roadway edge because of the very low concentration values obtained at distances greater than 250 m. A higher relative difference was observed for rural land-use conditions with time periods defined by the TxDOT specification handbook. The difference in concentration estimates between case study regions (i.e., El Paso, Dallas, and Houston) was due to their different meteorological conditions.

Combining all estimates from Table 3, on average for any urban area, annual average pollutant concentrations dispersed during nighttime conditions were higher by 100%–120% compared with daytime periods. This relative difference increased to 150%–200% for rural land use conditions. This finding was found to be consistent with other near-road studies examining the difference between daytime and nighttime air pollutant impact. Ultrafine particle (UFP) concentrations at night were reported by Zhu et al. (*6*), who conducted measurements upwind (300 m) and downwind (500 m) of a freeway from

Table 3. Relative Difference in Average Pollutant Concentrations between Nighttime and Daytime Periods

Distance from roadway	0–15 m	15–250 m	250–750 m	750–1500 m
		El Paso		
Urban – S	110%	176%	269%	267%
Rural – S	151%	254%	370%	367%
Urban – P	81%	119%	166%	164%
Rural – P	109%	167%	226%	228%
		Dallas		
Urban – S	88%	172%	326%	366%
Rural – S	109%	237%	496%	589%
Urban – P	67%	117%	189%	197%
Rural – P	80%	154%	275%	296%
		Houston		
Urban – S	106%	174%	302%	333%
Rural – S	159%	300%	495%	536%
Urban – P	80%	121%	183%	191%
Rural – P	118%	194%	274%	281%

Note: S = TxDOT specification handbook; P = Construction contractors' information. *Relative difference is expressed as (Nighttime Concentration - Daytime Concentration)/Daytime Concentration. Relative difference* in average concentration levels (expressed as a percent)

10:30 p.m. to 04:00 a.m. Although traffic volumes were much lower at night (about 25% of peak) particle number concentrations were about 80% of the daytime peak concentrations along a major freeway in Los Angeles. Hu et al. (*26*) measured air pollutant concentrations along the I-10 freeway in west Los Angeles, one–two hours before sunrise in the winter and summer months using an electric vehicle mobile platform equipped with fast-response instruments. Although traffic volumes during the pre-sunrise hours were lower than during the day, the UFP concentrations were significantly higher in the pre-sunrise period due to strong atmospheric stability, low wind speeds, low temperatures and high humidity values. In conclusion, the study found that the combination of sufficient traffic flow with meteorological conditions during pre-sunrise hours resulted in elevated concentrations of UFP, nitrogen-oxides, and polycyclic aromatic hydrocarbons during pre-sunrise hours.

Conclusion

Key findings from the series of sensitivity analyses performed to evaluate the impact of meteorological conditions on pollutant concentrations between daytime and nighttime periods include the following:

- Peaking effects in pollutant concentrations were observed near-road and concentrations declined with distance from roadway edge. The rate of decline was dependent on meteorological conditions and varied by season.
- Concentrations were higher in rural areas when compared with urban land use conditions due to the retention of heat by urban materials that increased the vertical motion of air, leading to increased pollutant dispersion in urban conditions.
- Concentrations were higher during nighttime compared with daytime (when emissions levels from the source are held equal) because of the stable atmospheric conditions, lower mixing heights, and lower wind speeds leading to higher concentrations of pollutants at the near-ground level during nighttime periods.
- Considering all combinations of input parameters, concentration estimates were found to be higher for the rural-night period, followed by urban-night, rural-night and rural-day.
- The higher relative difference in concentrations between daytime and nighttime periods were observed in far-road areas (i.e., farther than 250 m from the roadway) because of extremely low values when the concentrations fall back to background levels.
- On average, for any urban area, annual average pollutant concentrations dispersed during nighttime conditions were higher by 100%–120% compared with daytime periods. This relative difference increased to 150%–200% for rural land use conditions.

In summary, from a pollutant dispersion perspective, the assessment of the impact of meteorological conditions indicated that, for the same amount of emissions (i.e., if traffic volumes were held equal between daytime and nighttime), higher pollutant concentration levels are expected overnight. However, given that traffic congestion and overall traffic volumes are generally substantially lower during the nighttime period, the findings do not imply a net increase in pollutant concentrations in the region due to nighttime construction. Thus, the relative difference in pollutant concentrations obtained from shifting construction activities to nighttime from daytime periods should be assessed based on a combination of meteorological and traffic conditions.

Authors acknowledge the limitations of the results, as they were based on hypothetical case study settings evaluating the impact of meteorological conditions with no change in traffic activities. However, a review of the literature shows the conclusions and findings of this study, with regards to the difference in dispersion patterns between daytime and nighttime periods, and land use, were consistent with other studies. For future work, the authors recommend a methodology based on actual traffic data and near-road ambient air monitoring observations collected for daytime and nighttime to validate the findings obtained from the sensitivity analysis.

Acknowledgments

This study conducted as part of a research project titled "*Investigate the Air Quality Benefits of Nighttime Construction in Non-attainment Counties*," is funded by the Texas Department of Transportation (TxDOT). The authors would like to thank members of the TxDOT Project Committee for their guidance and support in conducting this research.

Author Contributions

The authors confirm contribution to the paper as follows: study conception and design: S. Vallamsundar and R. Farzaneh; data collection: M. Askariyeh; analysis and interpretation of results: M. Askariyeh and S. Vallamsundar; draft manuscript preparation: S. Vallamsundar and M. Askariyeh. All authors reviewed the results and approved the final version of the manuscript.

References

1. Hitchins, J., L. Morawska, R. Wolff, and D. Gilbert. Concentrations of Submicrometre Particles from Vehicle Emissions Near a Major Road. *Atmospheric Environment*, Vol. 34, No. 1, 2000, pp. 51–59.
2. Shepard, F. D., and B. H. Cottrell Jr. Benefits and Safety Impact of Night Work-Zone Activities. *Transportation Research Record: Journal of the Transportation Research Board*, 1986. 1086: 31–36.
3. Sullivan, E. C. *Accident Rates During Nighttime Construction*. Report No. UCB-ITS-RR-89-11. Institute of Transportation Studies, University of California-Berkeley, Berkeley, CA, 1989.
4. Hall, J. W., and V. M. Lorenz. Characteristics of Construction-Zone Accidents. *Transportation Research Record: Journal of the Transportation Research Board*, 1989. 1230: 20–27.
5. Colbert, D. *Productivity and Safety Implications of Night-Time Construction Operations*. Independent Research Study Report. Purdue University, West Lafayette, IN, 2003.
6. Zhu, Y., T. Kuhn, P. Mayo, and W. Hinds. Comparison of Daytime and Nighttime Concentration Profiles and Size Distributions of Ultrafine Particles near a Major Highway. *Environmental Science & Technology*, Vol. 40, No. 8, 2006, pp. 2531–2536.
7. Ginzburg, H., X. Liu, M. Baker, R. Shreeve, R. K. M. Jayanty, D. Campbell, and B. Zielinska. Monitoring Study of the Near-Road PM2.5 Concentrations in Maryland. *Journal of the Air & Waste Management Association*, Vol. 65, No. 9, 2015, pp. 1062–1071.
8. Li, L., J. Qian, C.-Q. Ou, Y.-X. Zhou, C. Guo, and Y. Guo. Spatial and Temporal Analysis of Air Pollution Index and its Timescale-Dependent Relationship with Meteorological Factors in Guangzhou, China, 2001–2011. *Environmental Pollution*, Vol. 190, 2014, pp. 75–81.
9. Zhao, S., Y. Yu, D. Yin, J. He, N. Liu, J. Qu, and J. Xiao. Annual and Diurnal Variations of Gaseous and Particulate Pollutants in 31 Provincial Capital Cities Based on In–Situ Air Quality Monitoring Data from China National Environmental Monitoring Center. *Environment International*, Vol. 86, 2016, pp. 92–106.
10. Baldauf, R., E. Thoma, M. Hays, R. Shores, J. Kinsey, B. Gullett, S. Kimbrough, V. Isakov, T. Long, R. Snow, A. Khlystov, J. Weinstein, F.-L. Chen, R. Seila, D. Olson, I. Gilmour, S.-H. Cho, N. Watkins, P. Rowley, and J. Bang. Traffic and Meteorological Impacts on Near-Road Air Quality: Summary of Methods and Trends from the Raleigh Near-Road Study. *Journal of the Air & Waste Management Association*, Vol. 58, 2008, pp. 865–878.
11. The Texas Department of Transportation. *Standard Specifications for Construction and Maintenance of Highways, Streets, and Bridges*. Texas Department of Transportation, Austin, TX, 2014. ftp://ftp.dot.state.tx.us/pub/txdot-info/des/spec-book-1114.pdf.
12. U.S. EPA. *Heat Island Effect*. 2016. https://www.epa.gov/heat-islands.
13. U.S. EPA. *Regional, State and Local Modelers Workshop*. Virginia Beach, VA, 2007.
14. Qian, W., and A. Venkatram. Performance of Steady-State Dispersion Models Under Low Wind-Speed Conditions. *Boundary-Layer Meteorology*, Vol. 138, No. 3, 2011, pp. 475–491.
15. U.S. EPA. *Addendum User's Guide for the AMS/EPA Regulatory Model - AERMOD*. 2015, pp. 1–189. https://www3.epa.gov/ttn/scram/dispersion_prefrec.htm#aermod.
16. Lawson, S. J., I. E. Galbally, J. C. Powell, M. D. Keywood, S. B. Molloy, M. Cheng, and P. W. Selleck. The Effect of Proximity to Major Roads on Indoor Air Quality in

Typical Australian Dwellings. *Atmospheric Environment*, Vol. 45, No. 13, 2011, pp. 2252–2259.
17. Barros, N., T. Fontes, M. P. Silva, and M. C. Manso. How Wide should be the Adjacent Area to an Urban Motorway to Prevent Potential Health Impacts from Traffic Emissions? *Transportation Research Part A: Policy and Practice*, Vol. 50, 2013, pp. 113–128.
18. Cohen, J., R. Cook, C. R. Bailey, and E. Carr. Relationship between Motor Vehicle Emissions of Hazardous Pollutants, Roadway Proximity, and Ambient Concentrations in Portland, Oregon. *Environmental Modelling & Software*, Vol. 20, No. 1, 2005, pp. 7–12.
19. Janssen, N., P. van Vliet, F. Aarts, H. Harssema, and B. Brunekreef. Assessment of Exposure to Traffic Related Air Pollution of Children Attending Schools Near Motorways. *Atmospheric Environment*, Vol. 35, No. 22, 2001, pp. 3875–3884.
20. Karner, A., D. S. Eisinger, and D. A. Niemeier. Near-Roadway Air Quality- synthesizing the Findings from Real-World Data. *Environmental Science & Technology*, Vol. 44, No. 14, 2010, pp. 5334–5344.
21. Venkatram, A., M. Snyder, V. Isakov, and S. Kimbrough. Impact of Wind Direction on Near-Road Pollutant Concentrations. *Atmospheric Environment*, Vol. 80, 2013, pp. 248–258.
22. Turner, D. B. *Workbook of Atmospheric Dispersion Estimates: An Introduction to Dispersion Modeling*, 2nd ed. Lewis Publishers, Boca Raton, FL, 1994.
23. Federal Register. *Appendix W to Part 51—Guideline on Air Quality Models*, 2005, pp. 68229–68261. http://www.epa.gov/scram001/guidance/guide/appw_05.pdf.
24. Golder, D. Relations among Stability Parameters in the Surface Layer. *Boundary-Layer Meteorology*, Vol. 3, No. 1, 1972, pp. 47–58.
25. Seinfeld, J. H., and S. N. Pandis. *Atmospheric Chemistry and Physics: From Air Pollution to Climate Change*, 2nd ed. John Wiley & Sons, Hoboken, NJ, 2006.
26. Hu, S., S. Fruin, K. Kozawa, S. Mara, S. E. Paulson, and A. M. Winer. A Wide Area of Air Pollutant Impact Downwind of a Freeway during Pre-Sunrise Hours. *Atmospheric Environment*, Vol. 43, 2009, pp. 2541–2541.

The Standing Committee on Transportation and Air Quality (ADC20) peer-reviewed this paper (18-04561).

Article

TRR
JOURNAL OF THE TRANSPORTATION RESEARCH BOARD

Transportation Research Record
2018, Vol. 2672(25) 111–122

DOI: 10.1177/0361198118797208
journals.sagepub.com/home/trr
SAGE

Integrating Engine Start, Soak, Evaporative, and Truck Hoteling Emissions into MOVES-Matrix

Xiaodan Xu[1], Haobing Liu[1], Hanyan "Ann" Li[1], Michael O. Rodgers[1], and Randall Guensler[1]

Abstract

The MOVES (MOtor Vehicle Emissions Simulator) model was developed by the U.S. Environmental Protection Agency (USEPA) to estimate emissions from mobile sources and is required to be used for regional air quality planning and conformity analysis in all states except California. However, the MOVES interface is complicated, and assessing emissions from dynamic large-scale transportation networks can be difficult. To aid in these analyses, the MOVES-Matrix modeling tool was developed as an alternative to the direct application of the MOVES model. MOVES-Matrix employs a massive multidimensional array of MOVES outputs created by running MOVES with every allowable combination of input variables. Once this output array has been generated, subsequent energy and emissions analyses can be conducted quickly and dynamically. Until recently, MOVES-Matrix has only been used to analyze running exhaust. In this study, MOVES-Matrix has been used expanded to include emissions from engine starts, truck hoteling, evaporative sources, and brake/tire wear as well as running exhaust. A case study is conducted for the metropolitan Atlanta, GA to verify the feasibility of using this expanded version of MOVES-Matrix and to ensure that the approach obtains the exact same results as applying MOVES directly. The travel activity inputs come from regional travel data generated by the Atlanta Regional Commission's Travel Demand Model. The emission results from MOVES-Matrix were compared with MOVES output to verify the equivalence of this approach.

Motor vehicle emissions are generated by a variety of processes including running exhaust, engine starts, fuel evaporation, and truck hoteling. The magnitude of these vehicle emissions depends on how, and under what conditions, the vehicles are operating or parking (*1*), and thus these emission-producing vehicle activities and processes are further partitioned by operating mode in estimating emission inventories. The U.S. Environmental Protection Agency (USEPA) developed the MOVES (MOtor Vehicle Emissions Simulator) model to estimate emissions from on-road vehicles in the United States (*2*), and MOVES is the approved regulatory emission model that must be applied to current emissions inventory development and transportation conformity analysis (unless a project-level conformity screening tool has been approved for use in the region).

In MOVES terminology, pollutants are emitted by "sources" via one or more "processes." MOVES uses a "modal-based" emission modeling approach in which, the mobile source emission inventories are estimated by quantifying specific emission-producing vehicle activities (e.g., starts or on-road operations) and coupling these activities with activity-specific emission rates (*3*). On-road gram/mile or gram/second emission rates are applied to the on-road vehicle activity (miles or seconds of operation). Similarly, the MOVES model also treats some individual processes, such as start exhaust and truck hoteling exhaust, as discrete emission-producing activities, or individual "puffs" of emissions (*2*). The total emission inventory is the sum of emissions from individual activities for the entire vehicle population. With local or regional activity inputs prepared for different emission process, users can assess the potential environmental impacts of transportation design and operation strategies using MOVES (*3*).

In previous studies, MOVES has been widely applied in assessing emissions generated from various processes and for a variety of purposes (for example: *4–6*). Analysis

[1]School of Civil and Environmental Engineering, Georgia Institute of Technology, Atlanta, GA

Corresponding Author:
Address correspondence to Xiaodan Xu: xxu312@gatech.edu

of a particular transportation scenario typically requires development of a new set of MOVES inputs. This requirement makes MOVES difficult to implement for assessing large-scale transportation networks across a variety of operating scenarios. Also, existing MOVES pre-processors or post-processors are typically designed for county-level rather than link-level emissions, limiting the ability to map spatial emission distributions, perform hot-spot assessments, conduct microscale pollutant dispersion modeling, and undertake other small spatial scale analyses.

In developing MOVES-Matrix, the goal of the research team was to develop a fast and efficient modeling support tool that could be used across any type and scale of analysis and obtain the exact same emissions outputs as using MOVES directly (*7*). This has been achieved by running the MOVES model across all combinations of input variables, storing the outputs in a massive data array, and then pulling the required individual emission rates into the analytical work. MOVES-Matrix is essentially a multidimensional array of pre-run MOVES emission rates, and can generate any composite emission rate for use in modeling more than 200 times faster than running the MOVES directly. Fuel consumption and emission rates from MOVES-Matrix can be used directly in a variety of applications without ever needing to re-run the MOVES model. In previous studies, the team developed linkages between MOVES-Matrix and travel demand models (*8*), Vissim simulations (*9*), CALINE4 and AERMOD dispersion models (*10*), and instrumented vehicle data (*11*).

This paper describes the integration of engine start, soak, evaporative, and truck hoteling emissions into MOVES-Matrix and presents an application of MOVES-Matrix in developing a regional inventory for Atlanta, GA including all MOVES emission processes. For this case study, vehicle activity estimates were produced by a regional travel demand model (TDM) that provides a reasonable representation of regional traffic conditions. For the emission rates, MOVES-Matrix was upgraded to integrate start exhaust, truck hoteling, and evaporative emissions based on MOVES 2014a (the current regulatory version). The resulting emission inventories are generated at high spatial (transportation link-level) resolution, which is equivalent to running a MOVES "project-level" assessment for the entire region, and were demonstrated to be exactly the same as obtained when MOVES is applied directly.

Methodology

The emission inventory analysis in MOVES can be described as applying applicable emission rates to specific vehicle activities across an entire vehicle population ("stype" in Equation 1) (*2*) for the given set of environmental conditions, fuel specifications, and so forth. The relationship among the three elements can be illustrated using the following equation:

$$Emissions = \sum_{Stype=i}^{n} (activity\ by\ OpMode\ fraction \times emission\ rate\ by\ OpMode\ bin) \quad (1)$$

For all processes, the vehicle population is defined by the MOVES source type population and vehicle age distribution, and these distributions may differ spatially and temporally across transportation network links. Most users prepare the fleet input based on locally derived data, typically regional vehicle registration data, or some combination of local data with MOVES default distribution (*12*).

In MOVES project-level analysis, the vehicle activities and emission rates are differentiated by various emission processes. The MOVES project-level processes and activities covered by this study are listed in Table 1. In MOVES, the relevant vehicle activities consist of vehicle miles traveled (VMT), source hours, hoteling hours, and number of engine starts. The instantaneous operating condition for certain activities is represented by operating mode, where activity falls onto one of a series of operating mode bins (*OpMode* Bins) in MOVES. Refueling emissions depend on fuel/energy consumption, from running exhaust, engine start, and hoteling; hence, refueling emissions are a byproduct of other emitting processes. Because refueling occurs at gasoline service stations, these emissions can (and should) be linked to stationary source activities. The tank vapor venting process (diurnal evaporation) is not included in this case study because it applies the MOVES project-level approach. Diurnal evaporation is estimated using in the MOVES county-level model, and requires 24-hour profiles for vehicle operations and ambient temperatures, whereas MOVES project-level model only allows single-hour scenario input.

The approach taken to prepare MOVES-Matrix for various emissions processes was essentially the same as previously performed to generate exhaust emission rates (*7*). The MOVES project-level model was run iteratively across all possible combination of temperature and humidity, for specific inspection and maintenance (IM) program and fuel specifications, to generate the matrix of emission rates per source type, model year, unit of activity, and operating mode bin. The emission rates in MOVES-Matrix for each process are in the form of

Table 1. MOVES Project-Level Processes and Activity (*13*)

Process ID	Process name	Specific activity	Operating mode bin (*Opmode*)	*Opmode* bin ID
1	Running exhaust	VMT	Vehicle-specific power (VSP)/ Scaled tractive power (STP) bin	0–40, 300
2	Start exhaust	Number of engine start	Soak time bin	100–108
9	Brake wear (PM only)	VMT	VSP/STP bin, brake wear stopped	0–40, 501
10	Tire wear (PM only)	VMT	Average speed bin	400–416
11	Evap permeation	Source hour	Hot/cold soak and operating	150–151, 300
13	Evap fuel leaks	Source hour	Hot/cold soak and operating	150–151, 300
15	Crankcase running exhaust	VMT	VSP/STP bin	0–40, 300
16	Crankcase start exhaust	Number of engine start	Soak time bin	100–108
17	Crankcase extended idle exhaust	Hoteling hours	Extended idling	200
18	Refueling displacement vapor loss	Energy consumption	See process 1, 2, 90, 91	
19	Refueling spillage loss	Energy consumption	See process 1, 2, 90, 91	
90	Extended idle exhaust	Hoteling hours	Extended idling	200
91	Auxiliary power exhaust	Hoteling hours	hoteling diesel auxiliary	201

Note: The ID numbers in the Table 1 are defined by the USEPA for MOVES; details can be found at: (https://www.epa.gov/moves/moves-algorithms).

Table 2. Emission Process Category in this Study

No.	Category	Process	Activity
1	Running emissions	Running exhaust, crankcase running exhaust, brake wear, tire wear	VMT
2	Start emissions	Start exhaust, crankcase start exhaust	Number of engine starts
3	Hoteling emissions	Extended idle exhaust, auxiliary power exhaust, and crankcase extended idle exhaust	Hoteling hours
4	Evaporative emissions	Fuel permeation, fuel leaks	Source hours
5	Refueling emissions	Refueling displacement vapor loss, refueling spillage loss	Combination of 1, 2, 3

grams of emissions or joules of energy consumption per unit of activity, by operating mode.

The major task in this study was to develop the iteration methodology used to prepare emission rates for each emission process, and to develop the output matrices of emission rates for the various processes. The methodology for each emitting process was different, so each was explored in more detail in the following sections. To simplify the analysis, some of the emitting processes were combined when they shared the same activity inputs. The five categories of emitting processes are listed in the Table 2.

Running Emissions Modeling Regime

In MOVES, running operation refers to operation of internal-combustion engines after the engine and emission control systems have stabilized at operating temperature (*13*). The running exhaust is estimated based on vehicle VMT or vehicle hours traveled (VHT) with respect to different speeds. The VMT is further partitioned by speed and vehicle-specific power (VSP) for light-duty vehicles or scaled tractive power (STP) for heavy-duty vehicles. In MOVES, running emissions are contributed from running exhaust and crankcase running exhaust. In addition, particulate matter from brake wear and tire wear emissions can be included in the running emission package, because they essentially share the same inputs as running exhaust. To account for running emissions, the following inputs are prepared for vehicle activities:

- **Link data**: The link-level VMT, average speed, and road grade are required inputs.
- **VMT fractions**: The VMT fractions by source type and model year should be prepared to properly partition total VMT by vehicle population.
- **Operating mode fractions**: Users need to employ at least one of the following three on-road operating conditions to generate *OpMode* fractions: average speed, operating mode distribution, or second-by-second speed profile.

MOVES-Matrix is prepared for all four running process, in emission-per-hour by individual source type, model year and operating mode bin (0–40, 300, 400–416, and 501). For each process, VHT is calculated from VMT and average speed by link. Then, emissions are estimated by multiplying VHT by source type, model year, and *OpMode* Bin with corresponding emission rates and then summing emissions by link.

Start Emissions Modeling Regime

Vehicle emission rates are elevated for the first few minutes after an engine is started (*14*). The vehicle activity data needed for estimating engine start emissions include the number, location, and time of engine starts. Start exhaust and crankcase start exhaust can be considered as start emissions as they both use the number of starts as their major activity. To properly account for start emissions, following inputs should be prepared for estimating start activities.

- **Number of engine starts by zone:** Engine starts occur at the origin location of trip origin–destination pairs. In this case, the number of starts can be prepared for pre-defined traffic analysis zones (TAZs) in a TDM, neighborhood zones, or for an entire region.
- **Start fractions by vehicle population:** The start fractions by source type and model year are prepared to partition the number of starts by vehicle population.
- **Soak time distribution:** Users prepare the soak time distribution by time of days and vehicle type. The parking durations at different times of day can vary significantly as a function or trip chaining and previous activities.

MOVES-Matrix is prepared for the two start processes (start and crankcase) in grams-per-start by individual source type, model year, and soak time bin. For each process, emissions are estimated by multiplying number of starts by source type, model year, and *OpMode* Bin with corresponding emission rates for each zone.

Hoteling Emissions Modeling Regime

In MOVES, "hoteling" is defined as any long period of time during which truck drivers sit in their vehicles during mandated driver break periods (*2*). Hoteling emissions are only available for long-haul combination trucks (MOVES source type 62). Extended idling, auxiliary power exhaust, and crankcase idling exhaust are considered as they occur during total hoteling hours. To properly account for truck hoteling emissions, the following inputs should be prepared for estimating hoteling activities:

- **Hoteling hours:** Hoteling hours are required for calculating hoteling emissions.
- **Truck population:** The long-haul combination truck population by model year should be prepared to properly represent fleet composition.
- **Extended idle/auxiliary power hour fraction:** The fraction of extended idling or idling with auxiliary power units should be defined to partition hoteling hours for individual process.

MOVES-Matrix is prepared for three hoteling processes, in grams of emission per hoteling hour by model year for source type 62. For each process, the hoteling emissions are estimated by multiplying hoteling hours per model year, by hour fraction of process, by corresponding emission rates.

Evaporative Emissions Modeling Regime

A significant portion of unburned fuel evaporates from vehicles throughout the day. Evaporative processes occur during vehicle refueling, while parked, and while driving, and contribute to a large portion of gaseous hydrocarbon emissions from gasoline vehicles (*15*). Evaporative processes are different from exhaust emissions because they do not directly involve combustion. Although refueling emissions are included in MOVES evaporative emissions, refueling losses employs different vehicle activity inputs and should be modeled as a stationary source activity. To properly account for evaporative emissions, the following inputs should be prepared for estimating evaporative activities.

- **Source hours operating and parked:** Source hours operating (SHO) and source hours parked (SHP) should be prepared by zone. SHO can use the on-road VHT in running emissions, and SHP can share the same zonal information as engine start.
- **Vehicle population:** Vehicle operation and parking should be tracked by source type so that source hours of operating and source hours of parking can applied respectively.
- **Soak time distribution:** Users should prepare the fractions of hot soak and cold soak to split the service hours of parking.

MOVES-Matrix is prepared for two evaporative processes (vehicle operating and parked), in emissions per source hour, by source type, model year, for hot/cold soak and operating. For each process, emissions are estimated by multiplying source hours per model year, by

Table 3. Summary of Model Inputs for Different Emission Process

No.	Category	Population inputs	Vehicle activity inputs	Emission rates
1	Running emissions	Link volume; VMT fraction by source type and model year	Link VMT, average speed; Average grade; STP/VSP bin fraction	Emission rates per hour by source type, model year and VSP/STP bin
2	Start emissions	Vehicle population by source type and model year	Number of engine start by zone; Soak time distribution	Emission rates per start by source type, model year and soak time bin
3	Hoteling emissions	Long-haul combination truck population by model year	Hoteling hours; Hour fraction for individual process	Emission rates per hoteling hour by model year
4	Evaporative emissions	VMT fraction by source type and model year; Vehicle population by source type and model year	SHO and SHP; Soak time distribution	Emission rates per source hour by source type, model year and soak time bin/ operating
5	Refueling emissions	See 1, 2 and 3	See 1, 2 and 3	See 1, 2 and 3

hour fraction of process, by corresponding emission rates.

Refueling Emissions Modeling Regime

Refueling emissions refers to the displaced fuel vapors and fuel spillage when liquid fuel is added to the tank at a gas station (*15*). Refueling emissions are estimated from the total volume of fuel dispensed (gallons); therefore, refueling losses are a function of energy consumption. As energy consumption of all vehicles are estimated from running, engine start, and hoteling process, the refueling emissions are allocated to fuel consumed in on-road driving activity, engine start activity, and truck hoteling activity. The activities for refueling emissions include VMT, number of engine starts, and truck hoteling hours.

The model inputs required for each category of emissions are summarized in Table 3. Although refueling emissions are calculated in this case study, they will not be used in comparative analysis later in the paper because refueling emissions are better represented in a stationary source emissions inventory, given that the emissions occur at the refueling location.

An Atlanta Case Study

The Atlanta, GA metropolitan region was used as a case study to estimate emissions from all processes. The Atlanta Regional Commission's (ARC's) TDM was used to populate most of the model inputs (*16*). The ARC is the metropolitan planning organization for the metropolitan Atlanta, GA area. The emission results were analyzed for its speciation profiles, temporal distributions, and spatial distributions. Finally, a verification of MOVES-Matrix was performed to demonstrate that the emission results generated from MOVES-Matrix were the same as using MOVES directly.

Model Input

The ARC's TDM estimates regional vehicle activity using their activity-based model for the 20-county non-attainment area (*16*). The model forecasts regional travel activity at 30-minute resolution, and predicts trips and link-level network travel. The model covers 5,981 TAZs and 74,500 roadway network links in the 20-county metropolitan area, with detailed land use and road characteristic information. In this study, the vehicle activities from ARC's TDM model were applied in estimating spatial and temporal emission distributions on the network through linkage to MOVES-Matrix, using methodology introduced above. The preparation of vehicle population, vehicle activity, and emission rate inputs will be introduced in the following sections.

Vehicle Population. The regional registration data were used to provide vehicle population by source type and model year. The MOVES default relative mileage accumulation rate was used to project vehicle population composition into VMT fraction by source type and model year. The vehicle distribution was simplified to be uniform for the entire region. However, in areas where high-resolution fleet composition data may be available, it is not a problem to assign local fleets to freeways and arterials, zones, or even individual links. The MOVES-Matrix approach provides this flexibility if fleet composition is tracked by the user.

Vehicle Activity. The ARC's TDM outputs were post-processed to obtain the vehicle activity inputs for emission modeling. If the data or fraction were not available through ARC's output, the MOVES default data or fraction were applied as a surrogate.

- **Link data**: Link-level VMT and average speed were readily available from TDM output, and directly applied in the modeling process.
- **Average grade**: Average grade was assumed to be 0 for all the links.
- **VSP/STP bin composition**: The MOVES default VSP/STP bin distribution was used.
- **Number of engine starts**: Total number of engine starts was trip production by TAZ in TDM model, and was derived by processing the trip output.
- **Soak time distribution**: The MOVES default soak time distribution was used.
- **Hoteling hours**: Hoteling hours were assumed to be a set portion of total heavy-duty truck VHT using the MOVES default hoteling hour fraction.
- **Hoteling process hour fraction**: The MOVES default fraction was used.
- **Source hours operating**: SHO was the same as link-level VHT derived from VMT and average speed.
- **Source hours parked**: SHP was derived based on number of vehicle parked by TAZ, which was calculated by the total vehicle population subtract vehicles in operation.

MOVES-Matrix. The regional default fuel and IM program were used to prepare MOVES-Matrix. The matrices were prepared for all 13 emission processes. Based on these modeling runs, the overall emissions composition and the estimated temporal and spatial distributions of hydrocarbon (HC) and particulate matter ($PM_{2.5}$) are presented below. The number of MOVES runs needed for developing MOVES-Matrix for Atlanta, with 21 calendar years × 3 fuel months × 111 temperature bins (0–110 F with 1 F interval) × 21 humidity bins (0–100% with 5% interval) is 1,909,089, with the number of runs for each process listed below:

- Running exhaust: 146,853
- Start exhaust: 1,174,824
- Evaporative emissions: 293,706
- Hoteling emissions: 293,706
- Other emissions can also be output by current runs (no additional runs are needed)

Currently, the research team has access to 202 computing nodes with 8,200 cores. The team can generate a full Matrix of emission rate outputs for a region (one fuel program and one I/M program) in one week on this system. In this study, MOVES-Matrix was used for a typical summer weekday in 2017.

Emissions Results

The following section presents the regional emission outputs, including the emission speciation, temporal distributions, and spatial distributions. Finally, the emission results were compared with the emission outputs from MOVES2014a to demonstrate that the MOVES-Matrix generates exactly the same results as MOVES.

Speciation Profile for Emissions. After estimating the emissions from different processes, the daily species profile for hydrocarbon (HC) and $PM_{2.5}$ can be determined by emission process (Figure 1). More than 70% of HC emissions were generated from non-combustion processes, such as start and evaporative process. The crankcase emissions and truck hoteling emissions were negligible for HC. However, most $PM_{2.5}$ emissions were still coming from on-network resources, including running, brake wear, and tire wear. PM start and crankcase emissions, and hoteling emissions were relatively low.

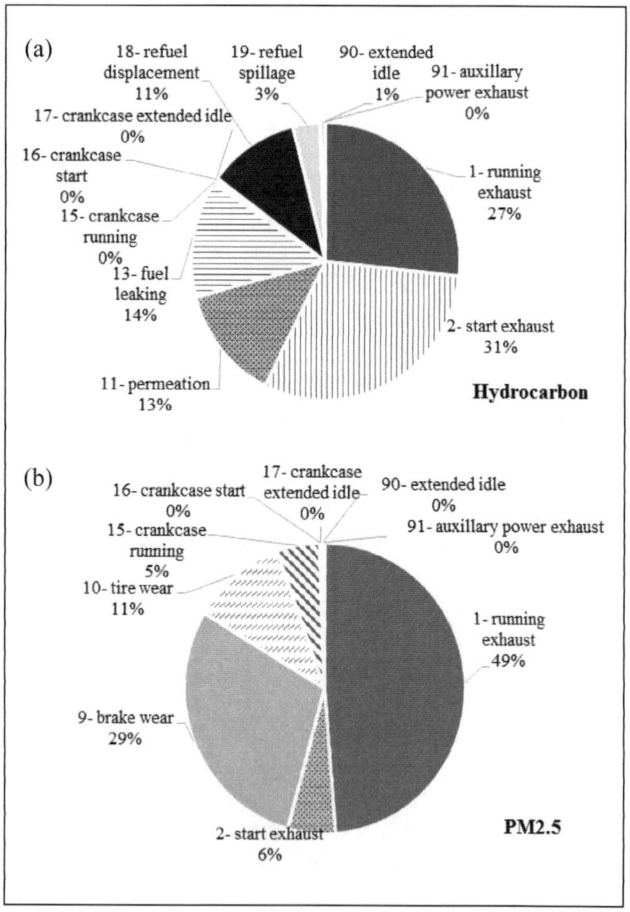

Figure 1. (*a*) The species profile of HC by process and (*b*) the species profile of $PM_{2.5}$ by process.

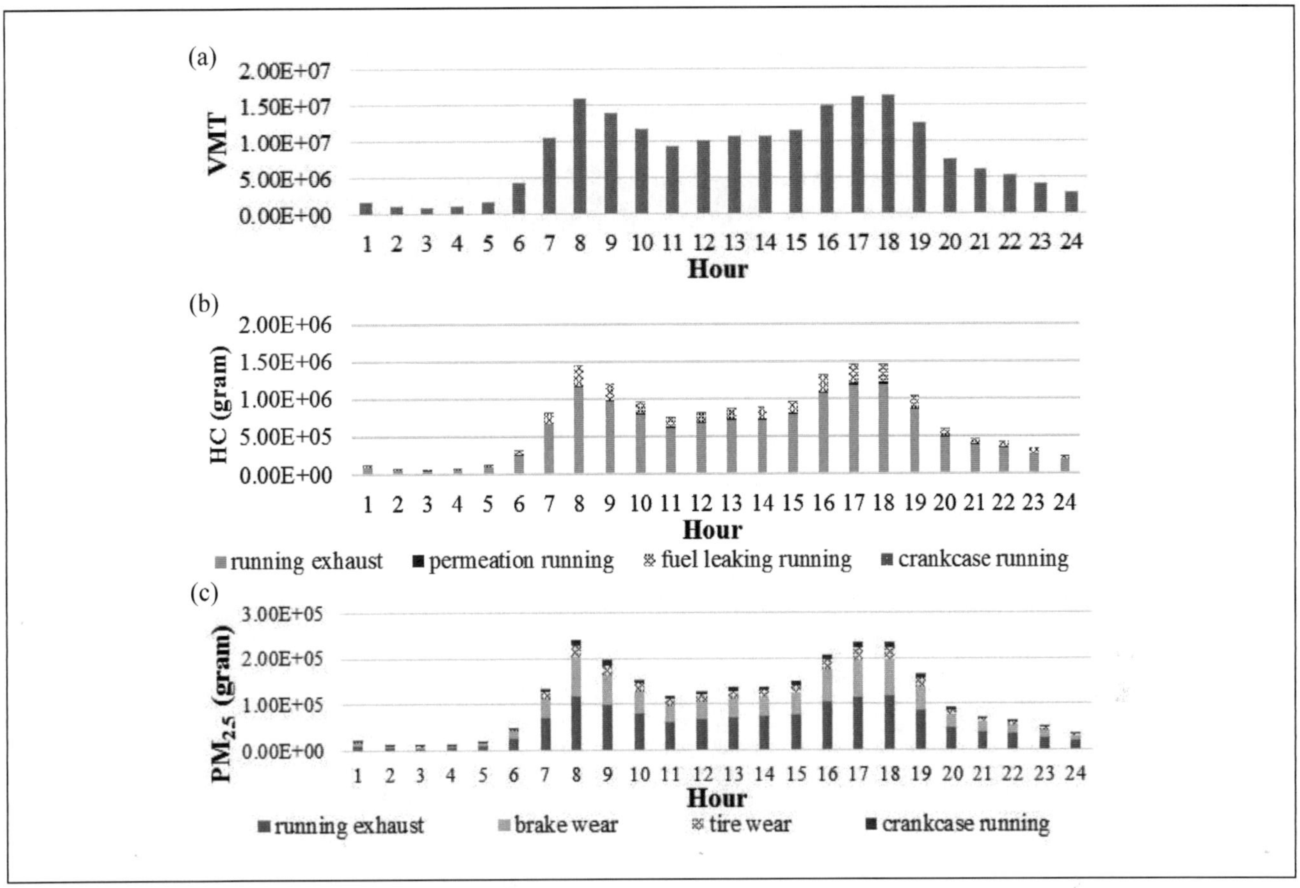

Figure 2. Temporal profile of (*a*) VMT; (*b*) HC; and (*c*) $PM_{2.5}$ from running resource.

Temporal Distribution of Activities and Emissions. The 24-hour activity and emission profiles were generated for each of the categories introduced above to assess the temporal pattern of activity and emissions. Hoteling emissions were not included because they represent such a small portion, and refueling emissions were not included as discussed earlier. The running results are shown in Figure 2.

Figure 2 indicated that emissions from the running portion generally followed the same pattern as VMT, with a major morning and afternoon peak. The HC emissions were dominated by running exhaust, whereas $PM_{2.5}$ emissions included a significant portion of brake wear and tire wear (mostly brake wear). The temporal profile of starts and start emissions are provided as Figure 3. Start emissions of HC and $PM_{2.5}$ still followed the same pattern as number of starts. However, the morning peak had significantly higher hourly emissions, given the longer soak times before start.

Finally, the emissions during parking were plotted as well as SHP and meteorology profiles. Figure 4 illustrates that evaporative emissions of HC from parking were more likely to be affected by both SHP and meteorology together. HC would significantly increase if the temperature reached a higher maximum in the afternoon. In general, the temporal distribution of HC and $PM_{2.5}$ from running and start followed the same pattern of activity, whereas the distribution of evaporative HC emissions is a function of activity and meteorology.

Spatial Distribution of Activities and Emissions. The spatial distribution of activities and emissions from 8:00–9:00 a.m. were generated for the entire 20-county area and are plotted for various vehicle activities in Figure 5. VMT was more heavily distributed around major transportation corridors and highways. Engine start emissions were dispersed in major trip generating zones, and parked vehicles were more concentrated in large residential areas and central business district. HC and $PM_{2.5}$ were also plotted for running and off-network (start and evaporative during parking), respectively, as shown in Figure 6.

Network HC and $PM_{2.5}$ closely followed the regional VMT pattern. The off-network HC was similar to the pattern for number of starts and parked vehicles. Off-network $PM_{2.5}$ emissions were very low. The spatial distribution of emissions showed that that HC derive from on-network and off-network activity, whereas $PM_{2.5}$ emissions were mostly generated through on-road operations.

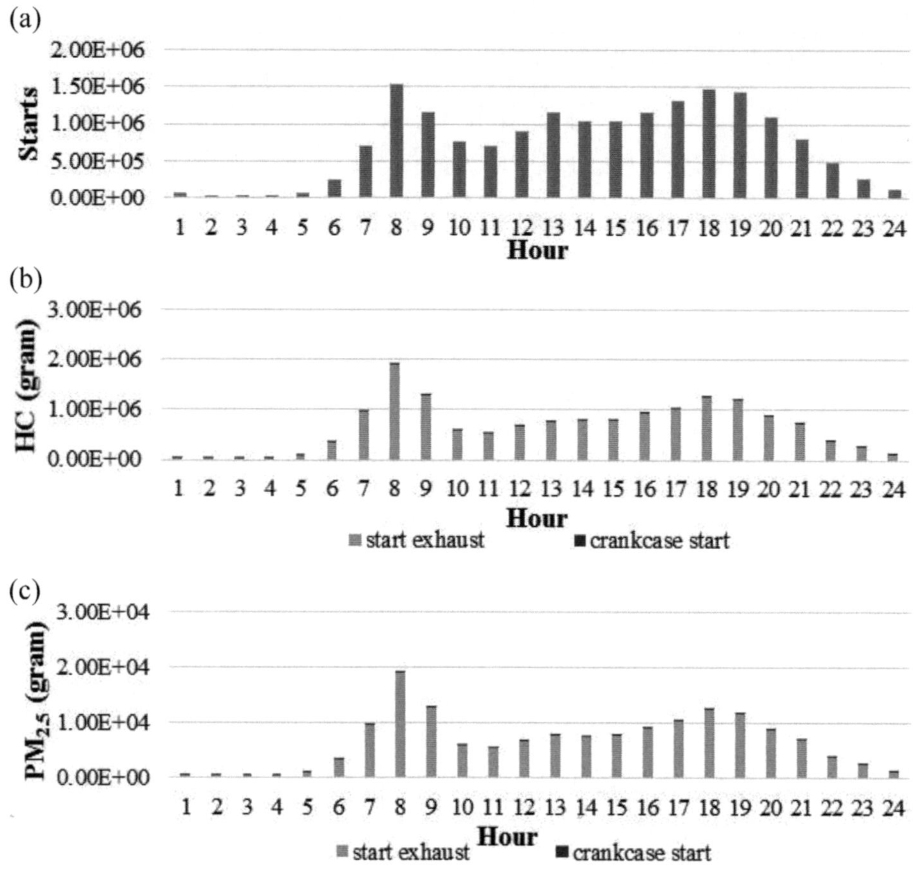

Figure 3. Temporal profile of (*a*) starts, (*b*) HC, and (*c*) $PM_{2.5}$ from start resource.

MOVES-Matrix Verification

Finally, the results from MOVES-Matrix were compared with results using MOVES 2014a directly to verify the emission analysis. Because each MOVES run required to generate a link-level emission rate takes a long time to code and process (74,000 links for 24 hours), the verification was conducted for a large subset of links and hours. For the on-network portion of emissions (including process IDs 1, 9, 10, 15, part of 11, 13, 18, 19), 200 links were randomly selected from the network during for first hour verification. The comparison between MOVES and MOVES-Matrix results is shown in Figure 7. The emissions from network resources estimated by MOVES-Matrix were exactly the same as MOVES results. For the off-network portion of emissions (including process ID 2, 16, 17, 90, 91, part of 11, 13, 18, 19), the first hour of data was used for verification. The comparison between MOVES and MOVES-Matrix results are shown in Figure 8.

As expected, because the emissions in the matrices were derived from individual MOVES runs, the emissions from MOVES runs and MOVES-Matrix runs are the same. Hence, the MOVES-Matrix process generates the exact same regional emission inventory estimations as MOVES. More verification will be conducted on the distributed computing server farm to verify MOVES-Matrix for all links, all hours, and all pollutants.

Summary

In this study, a MOVES-Matrix was updated to incorporate MOVES emission rates for engine starts, truck hoteling, and evaporative processes in MOVES project-level model. Hence, MOVES-Matrix can now be applied to both regional inventory development and full project-level modeling. The vehicle activity and vehicle population inputs required for each of the emission processes were identified and discussed. Emissions by process are generated by coupling vehicle activities with corresponding emission rates in MOVES-Matrix, and summed for the vehicle population. A case study of Atlanta, GA was conducted by applying this method to the 20-county non-attainment area. The modeling results

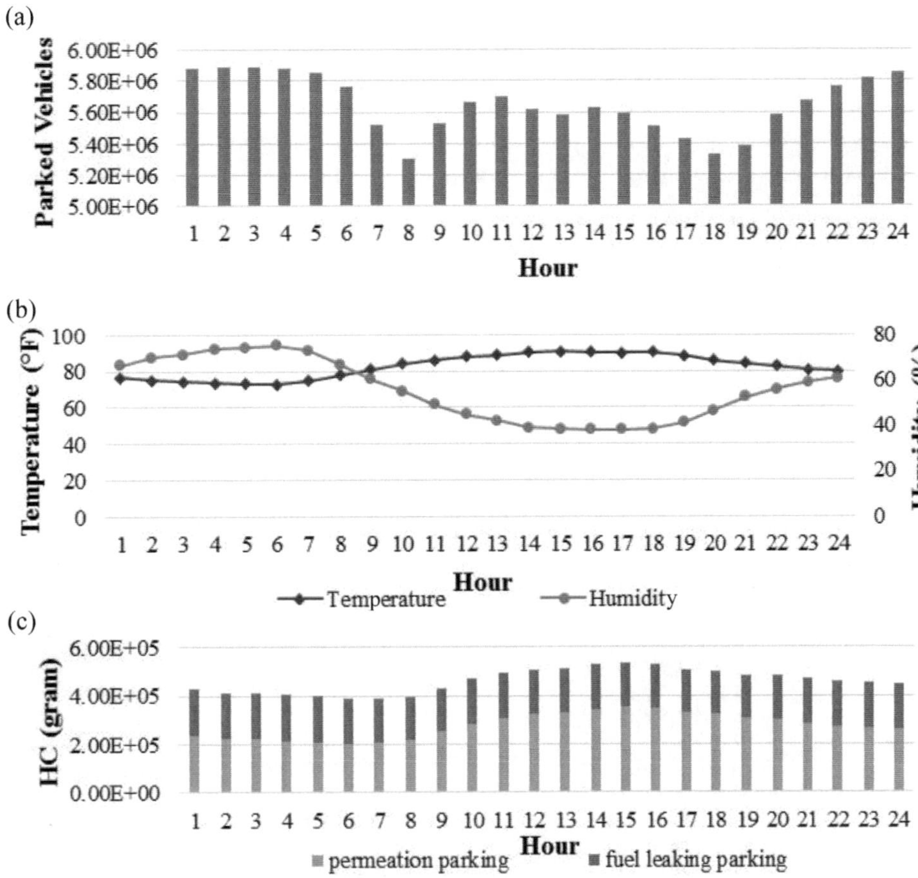

Figure 4. Temporal profile of (*a*) parked vehicle; (*b*) meteorology; and (c) HC from parking.

demonstrate that the emission profiles for selected pollutants can be populated for the entire region, with fleet and operations resolution at link level. Finally, a verification process was undertaken to demonstrate that the MOVES-Matrix generates the exact same inventory results as applying MOVES 2014a directly. MOVES-Matrix operates much more efficiently because MOVES has already been run for every applicable condition (improves modeling speed and flexibility). Users can specify information down to the link level if desired without ever having to re-run the MOVES model. By implementing MOVES-Matrix, planning and policy analysis can be performed rapidly and all scenarios can be automatically assessed via distributed computing. In addition, applying MOVES-Matrix would also benefit for conformity as it allows modelers to assess local fleet and roadway conditions in their analysis.

The research team believes that additional research efforts are warranted. First, the tank vapor venting (diurnal evaporation) emission rates are not included in this study, and should be brought from County-level MOVES runs if users need to use it for their analysis. The team is developing the approach for integrating diurnal evaporation through an on going project. Second, the team applied the MOVES default operating mode bin distributions and county-level fleet composition for emission estimation, which might be significantly improved when link-level data are employed. Additional verification efforts are also underway to demonstrate that MOVES-Matrix generates the same results as MOVES across a wide variety of scenarios. The team will continue to test and verify MOVES-Matrix and place the latest version online for public testing by the end of 2018.

Acknowledgments

This work was supported by the National Center for Sustainable Transportation, a National University Transportation Center sponsored by the U.S. Department of Transportation. The research team also acknowledges and appreciates the Atlanta Regional Commission for providing data and technical support for this project.

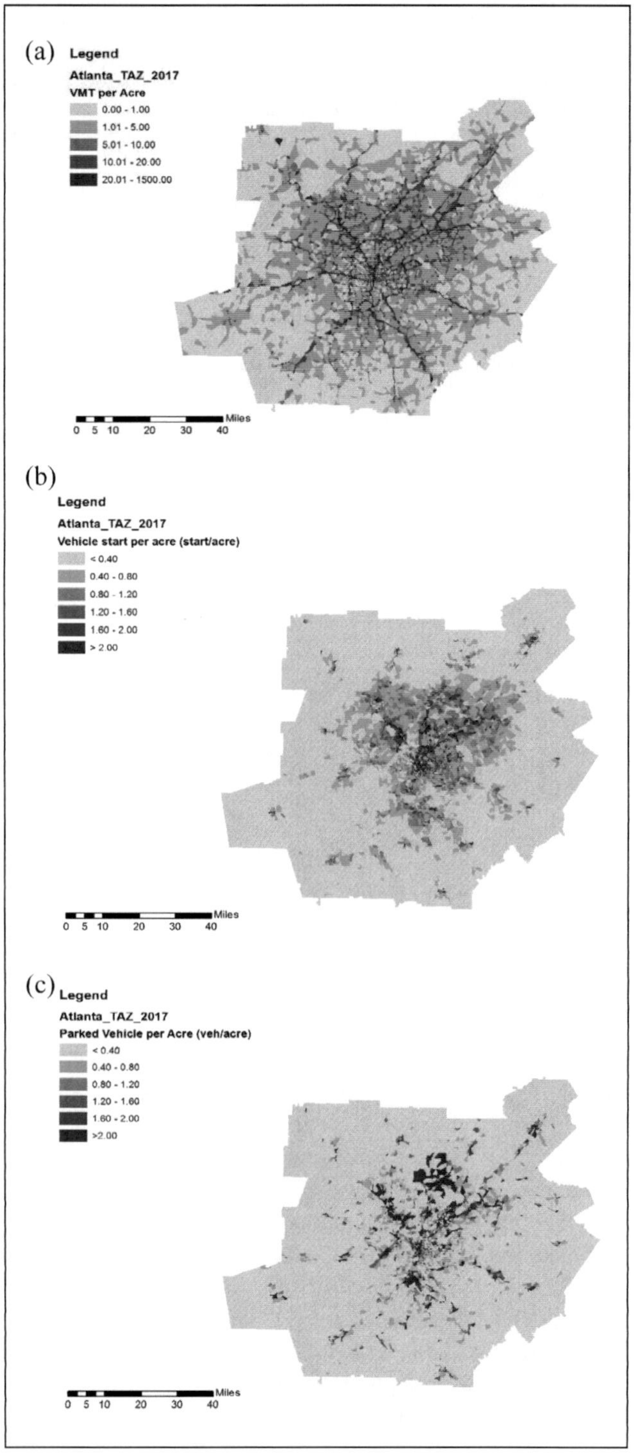

Figure 5. Spatial distribution of (*a*) VMT; (*b*) starts; and (*c*) parked vehicles.

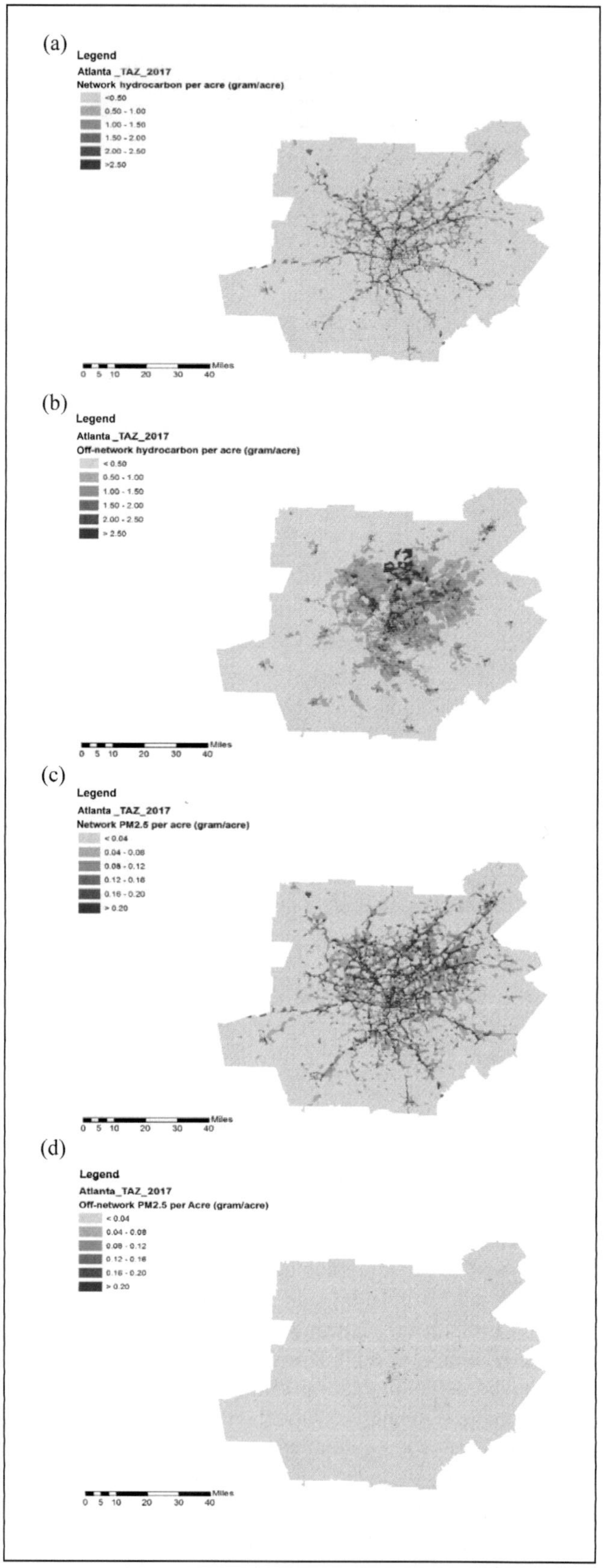

Figure 6. Spatial distribution of (*a*) network HC; (*b*) Off-network HC; (*c*) network $PM_{2.5}$; and (*d*) Off-network $PM_{2.5}$.

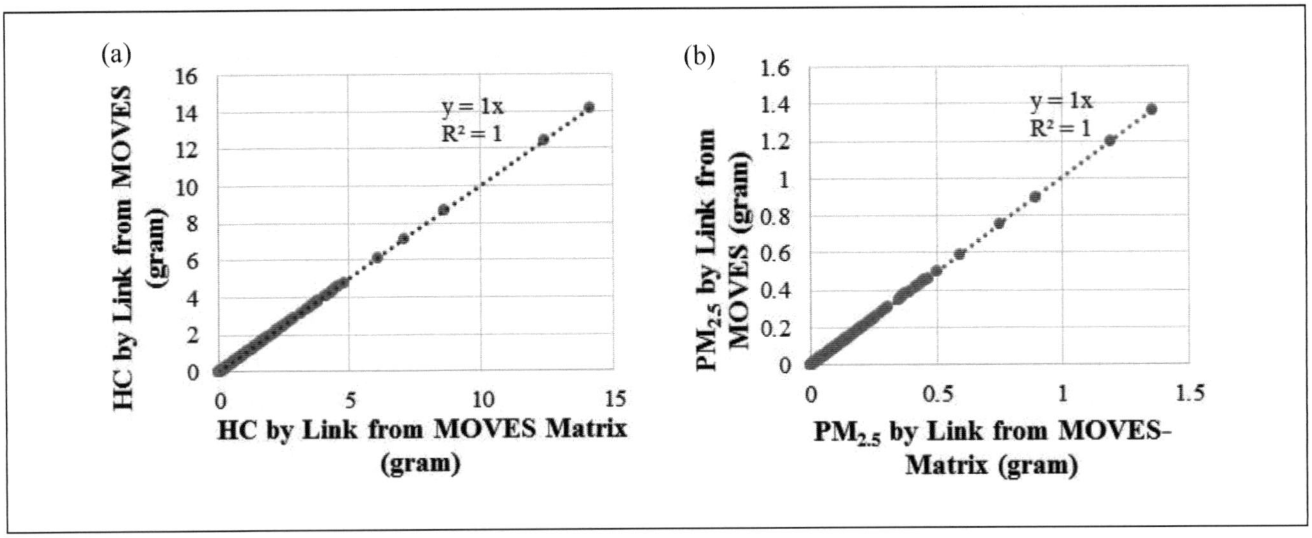

Figure 7. The comparison between MOVES and MOVES-Matrix results for (*a*) HC and (*b*) $PM_{2.5}$ from network resources.

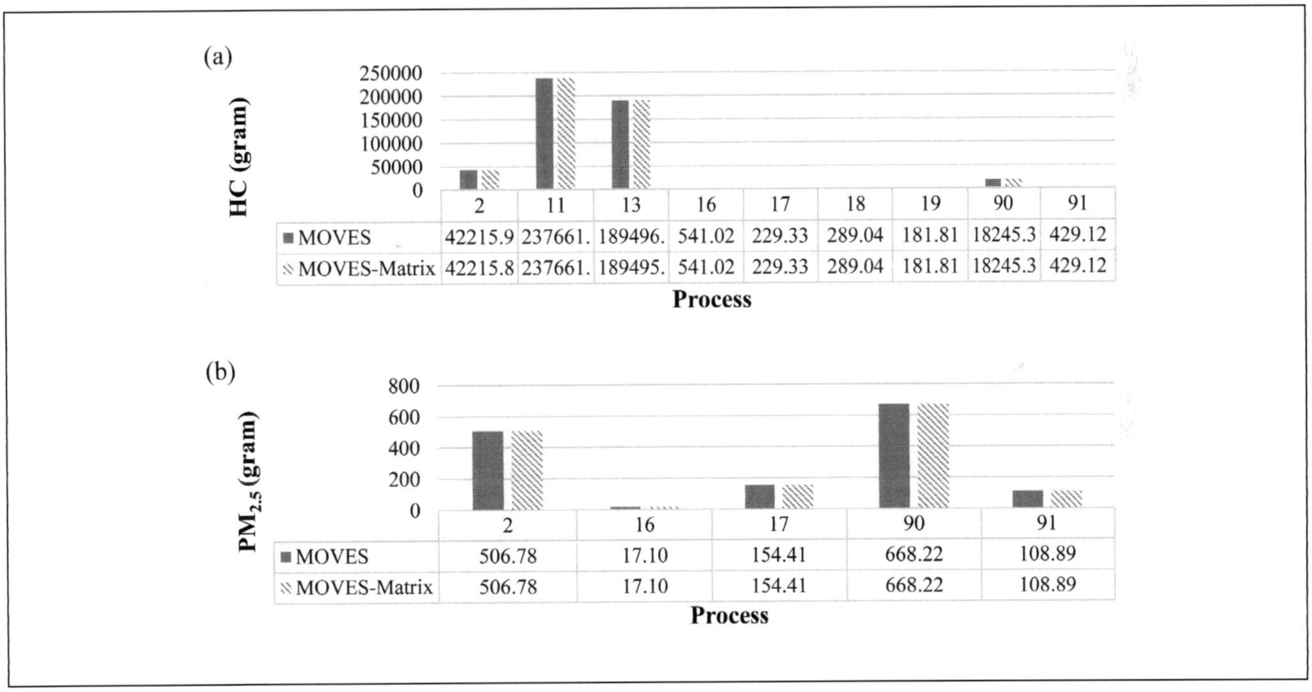

Figure 8. The comparison between MOVES and MOVES-Matrix results for (*a*) HC and (*b*) $PM_{2.5}$ from off-network resource.

Author Contributions

The authors confirm contribution to the paper as follows: study conception and design: Xiaodan Xu, Haobing Liu, Hanyan "Ann" Li, Michael O. Rodgers, Randall Guensler; data collection: Xiaodan Xu, Haobing Liu; analysis and interpretation of results: Xiaodan Xu; draft manuscript preparation: Xiaodan Xu, Haobing Liu, Hanyan "Ann" Li, Michael O. Rodgers, Randall Guensler.

All authors reviewed the results and approved the final version of the manuscript.

References

1. Guensler, R., D. Sperling, and P. Jovanis. *Uncertainty in the Emission Inventory for Heavy-Duty Diesel Powered Trucks*. Publication UCD-ITS-RR-91-02. Institute of Transportation Studies, University of California, Davis, Davis, CA, 1991.
2. USEPA, U.S. Environmental Protection Agency. *MOVES2014 and MOVES2014a Technical Guidance: Using MOVES to Prepare Emission Inventories for State Implementation Plans and Transportation Conformity.*

Publication EPA-420-B-15-093. U.S. Environmental Protection Agency, U.S. Government Printing Office, Washington, D.C., 2015.
3. USEPA, U.S. Environmental Protection Agency. *Population and Activity of On-Road Vehicles in MOVES2014.* Publication EPA-420-R-16-003a. U.S. Environmental Protection Agency, U.S. Government Printing Office, Washington, D.C., 2016.
4. Fujita, E., D. Campbell, B. Zielinska, J. Chow, C. Lindhjem, A. DenBleyker, G. Bishop, B. Schuchmann, D. Stedman, and D. Lawson. Comparison of the MOVES2010a, MOBILE6. 2, and EMFAC2007 mobile source emission models with on-road traffic tunnel and remote sensing measurements. *Journal of the Air & Waste Management Association*, Vol. 62, No. 10, 2012, pp. 1134–1149. DOI: 10.1080/10962247.2012.699016.
5. Hoekzema, A. *Modeling Truck Idling Emissions in Central Texas.* Capital Area Council of Governments, Austin, TX, 2015. http://www.capcog.org/documents/airquality/reports/2015/Modeling_Truck_Idling_Emissions_in_Central_Texas_-_2015-05-05.pdf. Accessed July 25, 2017.
6. Jobson, T., and Y. Huangfu. *Impact of Cold Climates on Vehicle Emissions: The Cold Start Air Toxics Pulse.* Publication CESTiCC Project 101409. Department of Civil and Environmental Engineering, Washington State University, Pullman, WA, 2016.
7. Guensler, R., H. Liu, X. Xu, Y. Xu, and M. Rodgers. MOVES-Matrix: Setup, Implementation, and Application. Presented at 95th Annual Meeting of the Transportation Research Board, Washington, D.C., 2016.
8. Xu, X., H. Liu, Y. Xu, M. Rodgers, and R. Guensler. Regional Emission Analysis using Travel Demand Models and MOVES-Matrix. Presented at 2016 Transportation Planning and Air Quality Conference, Minneapolis, MN, August 4–5, 2016a.
9. Xu, X., H. Liu, J. M. Anderson, Y. Xu, M. Hunter, M. Rodgers, and R. Guensler. Estimating Project-Level Vehicle Emissions with VISSIM and MOVES-Matrix. *Transportation Research Record: Journal of the Transportation Research Board*, 2016b. 2570: 107–117.
10. Liu, H., X. Xu, M. O. Rodgers, Y. A. Xu, and R. L. Guensler. MOVES-Matrix and distributed computing for microscale line source dispersion analysis. *Journal of the Air & Waste Management Association*, Vol. 67, No. 7, 2017, pp.763–775. DOI: http://dx.doi.org/10.1080/10962247.2017.1287788
11. Guensler, R., H. Liu, Y. Xu, A. Akanser, D. Kim, M. Hunter, and M. Rodgers. Energy Consumption and Emission Modeling of Individual Vehicles Using MOVES-Matrix. *Transportation Research Record: Journal of the Transportation Research Board*, 2017. 2627: 93–102.
12. Porter, C., D. Kall, D. Beagan, R. Margiotta, C. Systematics, I. Cambridge, J. Koupal, S. Fincher, and A. Stanard. *Input Guidelines for Motor Vehicle Emissions Simulator Model.* Publication Contractor's Report for NCHRP Project 25-38. Cambridge Systematics, Inc., Cambridge, Mass., Eastern Research Group, Inc., Austin, TX, 2014.
13. USEPA, U.S. Environmental Protection Agency Environmental Protection Agency. *Exhaust Emission Rates for Light-Duty On-road Vehicles in MOVES2014.* Publication EPA-420-R-15-005. U.S. Environmental Protection Agency, U.S. Government Printing Office, Washington, D.C., 2015.
14. Weilenmann, M., J. Favez, and R. Alvarez. Cold-Start Emissions of Modern Passenger Cars at Different Low Ambient Temperatures and their Evolution over Vehicle Legislation Categories. *Atmospheric Environment*, Vol. 43, No. 15, 2009, pp. 2419–2429. DOI:10.1016/j.atmosenv.2009.02.005.
15. USEPA, U.S. Environmental Protection Agency. *Evaporative Emissions from On-Road Vehicles in MOVES2014.* Publication EPA-420-R-14-014. U.S. Environmental Protection Agency, U.S. Government Printing Office, Washington, D.C., 2014.
16. Atlanta Regional Commission. *Activity-Based Travel Model Specifications: Coordinated Travel – Regional Activity Based Modeling Platform (CT-RAMP) for the Atlanta Region.* Atlanta Regional Commission, Atlanta, GA, 2012.

The Standing Committee on Transportation and Air Quality (ADC20) peer-reviewed this paper (18-04825).

The contents of this paper reflect the view of the authors who are responsible for the facts and accuracy of the data presented herein. The contents do not necessarily reflect the official view or policies of the Department of Transportation. This paper does not constitute a standard, specification, or regulation.

Article

TRR
JOURNAL OF THE TRANSPORTATION RESEARCH BOARD

Transportation Research Record
2018, Vol. 2672(25) 123–138

DOI: 10.1177/0361198118797456
journals.sagepub.com/home/trr
SAGE

Field Evaluation of Vehicle to Infrastructure Communication-Based Eco-Driving Guidance and Eco-Signal System

Byungjin Ko[1], Lian Cui[2], Saerona Choi[3], Byungkyu Brian Park[2], and Seunghan Ryu[2]

Abstract

In urban areas, the main factor causing excessive fuel consumption is stop-and-go, which is prevalent at urban intersections. To decrease fuel consumption at the intersections, various methods have been proposed and evaluated using simulation models and driving simulators. However, it is not known how reliable the driving simulator results would be when compared with those of the field test. In this paper, eco-driving guidance and eco-signal system based on vehicle-to-infrastructure communication are developed and tested in the field in a multi-vehicle environment. They were also tested using a driving simulator and the results were compared with those of the field test. This was intended to assess the reliability of driving simulator-based study. The field test results at an isolated intersection indicated that the eco-driving guidance could save the fuel consumption around 20%–40% and the eco-signal system could reduce the fuel consumption up to 45%. Although the absolute fuel consumptions between the driving simulator test and the field test did not match, the results from both tests confirmed better performance with eco-driving guidance and eco-signal system when compared with the base scenarios. This indicates that a driving simulator is an effective tool for assessing relative benefits of eco-driving guidance and eco-signal system.

Transportation is the second largest sector responsible for the greenhouse gas (GHG) emissions in the United States (*1*). Thus, many researchers have explored the benefits of various technologies, such as developing fuel-saving devices, improving engine technologies, and providing eco-driving guidance toward fuel consumption savings and greenhouse gas reductions. In the transportation field, eco-driving is known as an effective method to save fuel consumption at signalized intersections. Eco-driving technology provides optimal speed information, which allows a vehicle to reduce unnecessary fuel consumption, through vehicle-to-vehicle or vehicle-to-infrastructure (V2X) communication.

In general, there are several methods to assess the effectiveness of the eco-driving technology: simulation-based evaluation, driving simulator-based experiment, and field test. For most of the studies on eco-driving, simulation tools have been mainly used due to their convenience and the cost effectiveness. Microscopic traffic simulation tools (e.g., VISSIM and PARAMICS) have been used to analyze the benefits of eco-driving. The preliminary simulation results showed significant reductions by approximately 20% in fuel consumption, with eco-driving controls (*2–8*). The simulation-based evaluations usually have a strong assumption that the vehicles can perfectly follow optimal speeds suggested by the eco-driving control algorithms. However, in the real world, drivers might not able to exactly follow the recommendation speed because of human drivers' perception reaction time, compliance behavior and time needed to reach desired speed. Therefore, a microscopic simulation-based method might not be suitable for assessing eco-driving behaviors because it does not properly model actual drivers' behaviors.

[1]Department of Information and Communication Engineering, Daegu Gyeongbuk Institue of Science and Technology, Korea
[2]Department of Civil and Environmental Engineering, University of Virginia, Charlottesville, VA
[3]Transportation Safety Research and Development Institute, Korea Transportation Safety Authority, Korea

Corresponding Author:
Address correspondence to Lian Cui: lc9ud@virginia.edu

Therefore, field-test-based evaluation is considered as a more reasonable method than a simulation method. For example, The International Commission for Driver Testing (CIECA) eco-driving project (*9*) was introduced to improve the fuel economy of vehicles in Europe. In California, Stillwater and Kurani (*10*) reported that 40% of 98 participants drove more economically when they received fuel usage information (i.e., energy feedback). Tulusan et al. (*11*) implemented eco-driving using mobile technology in the field and showed fuel consumption reductions around 3%–8%. Barth and Boriboonsomsin (*12*) tested a dynamic eco-driving system in both the simulation and the field, which provides eco-speed information to the drivers. The dynamic eco-driving advice system led to fuel reductions of approximately 10%–20%, whereas the reduction rate of field-testing results was slightly lower than simulation results. Kurani et al. (*13*) installed devices providing energy feedback at the dashboard in vehicles and collected data from 118 drivers for two months in California. They found a significant reduction rate of 2.7% in fuel consumption with the energy feedback. Many previous studies mentioned that field tests should be considered because fuel usages are affected by many factors including road geometry (e.g., grade and road surface friction), environmental conditions (e.g., wind speed and direction), traffic conditions, and driver's behavior (*3–6*).

However, field tests have limitations such as expensive experimental cost and safety issues, which trigger that the driving simulator is frequently used to assess many transportation applications. For eco-driving studies, many research groups have been using driving simulators to evaluate various eco-driving strategies, which provide eco-driving information via variable message signs (VMS) or an in-vehicle device (*14–19*). The results showed that the fuel consumption was varied by the experiment scenarios and the reduction rates were around 10%–20% compared with without eco-driving. Another advantage of using a driving simulator is that many scenarios can be analyzed in less time and cost constraints compared with the field test. Those repetitive tests can reduce impacts of external factors so that it allows focus on capturing the main features of experiments.

In this study, an eco-driving guidance and an eco-signal system using V2X communication in a multi-vehicle environment were developed. There is no doubt that field tests are better than driving simulator-based tests in the aspects of reality and accuracy. But the effect of the driving simulator is also worthy of investigation. Therefore, the proposed system was evaluated through field tests. Then the driving simulator-based tests were followed, in which the driving environment was the same as the field, to compare the test results. The primary objectives of this study were to reduce fuel consumption using eco-driving guidance and eco-signal mechanisms and to verify the confidence of a driving simulator by comparing with field test results. For calculating recommendation speeds, the signal phase and timing (SPaT) information, locations, and speeds of a subject vehicle, which are transmitted through V2X communication, were used.

The main contributions of this study are as follows. First, an eco-driving guidance system is proposed so that it can provide speeds to minimize fuel consumption of the subject vehicle and prevent critical safety situations at the same time by considering maneuvering of the leading vehicle at the intersections. Second, an eco-signal mechanism is combined to the eco-guidance system to effectively decrease fuel consumption by considering both vehicle trajectories and traffic signal phases. Third, the proposed system is evaluated with both field tests and laboratory tests using a driving simulator, then the results are compared and analyzed in terms of fuel consumption which are collected by on-board diagnostics II (OBD II) data or estimated using the VT-Micro model. Fourth, we assessed the reliability of driving simulator and verified that a driving simulator is an effective tool for evaluating relative benefits of eco-driving guidance and eco-signal system.

Proposed System

To improve the fuel consumption at the signalized intersections in a multi-vehicle condition, the recommendation speed is calculated for the driver based on the SPaT information, current speed, and the remaining distance from the intersection. The basic settings are as follows: a leading vehicle's (V_L) trajectory information to ensure safe driving; the leading vehicle (V_L) equipped with an on-board unit (OBU), but not receiving an eco-driving guidance; and a following vehicle (V_F), which is a subject vehicle equipped with OBU, following an eco-driving guidance. In this section, the proposed system, including the eco-driving and the eco-signal mechanisms, is described in detail.

System Architecture

As shown in Figure 1*a*, a roadside unit (RSU) is installed at the intersection, and the V_L and the V_F are approaching the intersection. As mentioned above, only V_F follows the eco-driving guidance system of which the flow chart is represented in Figure 1*b*. Every vehicle broadcasts a basic safety message (BSM) including the vehicle's GPS (g_f is for V_F; g_l is for V_L) and speed information (v_f is the current speed of V_F; v_l is the current speed of V_L) in every 0.1 s using OBU. The RSU that is connected to

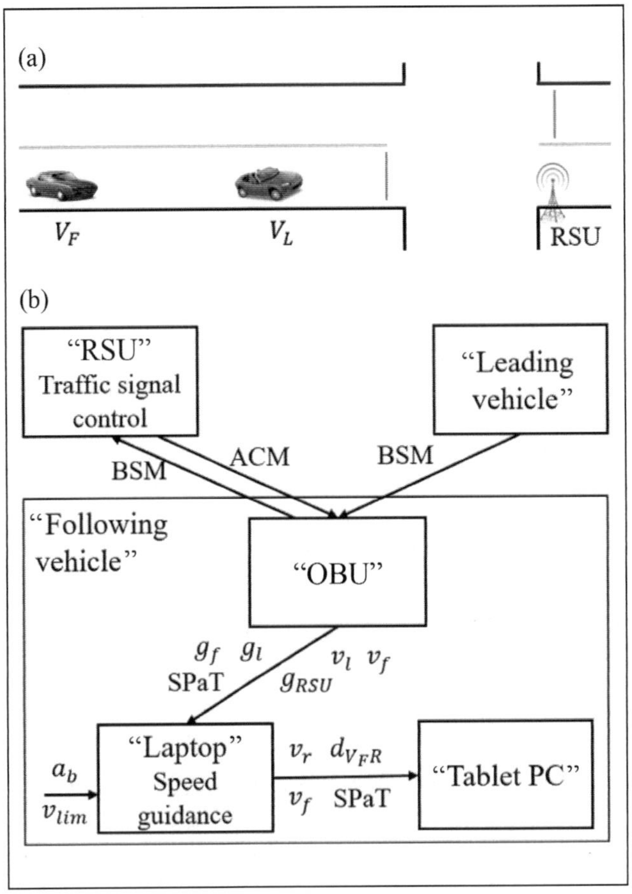

Figure 1. System architecture showing (*a*) road environment; (*b*) flow chart of proposed system.

traffic signal control extends the remaining green time based on the received BSM and the proposed eco-signal mechanism when V_F can only cross the intersection with additional green time. The RSU broadcasts an a-la-carte message (ACM) including SPaT and GPS (g_{RSU}) information. Note that the message formats of BSM and ACM can be referred in (*20*). The OBU of V_F transmits the received information from the RSU and V_L to the laptop using user datagram protocol (UDP) over the wired network. The laptop calculates the recommendation speed (v_r) based on the proposed eco-driving mechanism, the driver's average acceleration (a_b), and a speed limit (v_{lim}). Then, the recommendation speed and vehicle maneuver (e.g., acceleration, speed maintenance, and deceleration) are transmitted to the tablet PC via Bluetooth, along with the current speed, SPaT, and the remaining distance information.

Eco-Guidance and Eco-Signal Mechanisms

The vehicle trajectory planning developed in (*21*) is adopted to reduce fuel consumption at a signalized

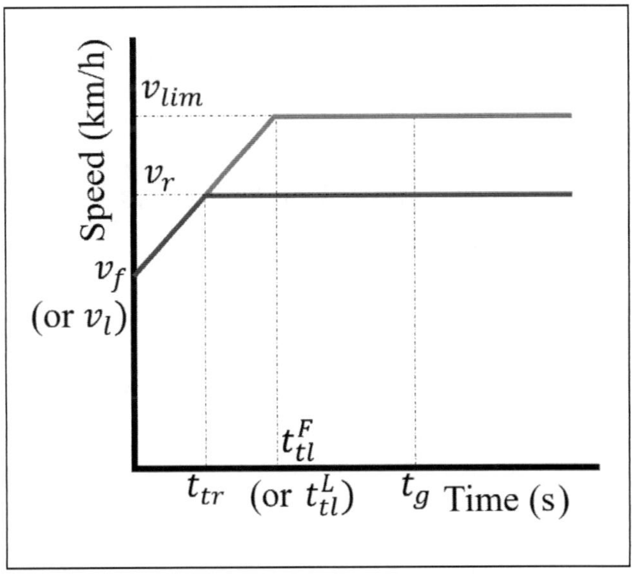

Figure 2. Acceleration profile.

intersection. The trajectory planning in (*21*) attempts to minimize the fuel consumption by minimizing the total tractive power demand and the idling time. Thus, the trajectory of vehicle was designed to accelerate to a target speed quickly and then keep the speed until passing the intersection. The acceleration profile was designed with a trigonometric increase, which can be referred to in (*21*). Since we could not be assured that the acceleration profile is effective to drivers who have different driving behaviors, we used constant acceleration model reflecting the driver's acceleration behavior. Moreover, to further reduce the fuel consumption, the eco-signal mechanism proposed in (*22*), which controls green splits towards the minimum waiting time of vehicles, is adopted.

Eco-Speed Guidance Mechanism. A recommendation speed is determined based on the information of SPaT, the vehicles' speed, and GPS of the vehicles and RSU. The proposed system calculates a recommendation speed differently according to the signal state. The trajectory information and braking distance V_L are also considered for the safety issue on the speed recommendation.

Green Time. During the green time, V_F first determines whether the vehicle can cross the intersection within the remaining green time by estimating the expected maximum moving distance of V_F and V_L (d_{eV_F} and d_{eV_L}). For d_{eV_F}, the constant acceleration model, whose gradient is a_b as shown in Figure 2, is used in order to reflect the driver's acceleration behavior. For d_{eV_L}, the acceleration of V_L (a_{V_L}) is used as the gradient instead of a_b due to the uncertainty of coming trajectory of V_L. Therefore, d_{eV_F}

and d_{eV_L} are defined as follows, which are derived by measuring the area of Figure 2 indicating that a vehicle speeds up to $v_{\lim}$ and keeps the speed during the remaining green time (t_g).

$$d_{eV_F} = \begin{cases} \left(2v_f + a_b t_{tl}^F\right)\frac{t_{tl}^F}{2} + v_{\lim}\left(t_g - t_{tl}^F\right), & t_{tl}^F < t_g \\ \left(2v_f + a_b t_g\right)\frac{t_g}{2}, & t_{tl}^F \geq t_g \end{cases} \quad (1)$$

$$d_{eV_L} = \begin{cases} \left(2v_l + a_{V_L} t_{tl}^L\right)\frac{t_{tl}^L}{2} + v_{\lim}\left(t_g - t_{tl}^L\right), & t_{tl}^L < t_g \\ \left(2v_l + a_{V_L} t_g\right)\frac{t_g}{2}, & t_{tl}^L \geq t_g \end{cases} \quad (2)$$

Note that t_{tl}^F and t_{tl}^L indicate the time to the speed limit from the current speeds of V_F and V_L respectively, which are computed with a_b and a_{V_L}.

When V_F Can Cross the Intersection within the Remaining Green Time. If d_{eV_F} is longer than the distance between the RSU and V_F ($d_{V_F R}$) and d_{eV_L} is longer than the distance between the RSU and V_L ($d_{V_L R}$), the V_F determines that it can cross the intersection within the remaining time. The recommendation speed depends on the driving manner of V_L. If the driver of V_L prefers to speed up more than the speed limit while crossing an intersection, V_F does not need to consider the trajectory of V_L because the recommendation speed may not be higher than the speed limit. However, if V_L approaches the intersection eco-friendly, which means a smooth speed trajectory and is close to the eco-driving behavior, V_F needs to consider the trajectory for the speed recommendation of V_L because a wrong recommendation speed may cause a car accident with V_L. Therefore, two types of recommendation speed are calculated for the safety issue during the green time. The first type is to compute the recommendation speed regardless of the trajectory of V_L. Namely, it is applicable to the case that V_L speeds up more than $v_{\lim}$. To this end, the proposed method in (*23*) is adopted for the speed recommendation. The second type is to compute the recommendation speed using additional information, such as braking distance of V_F and trajectory of V_L. It applies to the case that V_L approaches the intersection eco-friendly. The recommendation speed is calculated as follows.

Since V_L approaches the intersection eco-friendly in the second type, V_F needs to follow V_L with the braking distance to cross the intersection within the remaining green time safely. It is expected that V_F is the braking distance away from V_L when the signal changes to yellow light. Therefore, the expected moving distance of V_F during t_g satisfies Equation 3.

$$\left(v_f + v_r\right)\frac{t_{tr}}{2} + \left(t_g - t_{tr}\right)v_r = d_{eV_L} + d_{V_F V_L} - d_b \quad (3)$$

This is derived from Figure 2, where t_{tr} indicates the time to the recommendation speed from v_f; $d_{V_F V_L}$ means the distance between V_F and V_L; d_b is the braking distance of V_F; and d_b is represented by:

$$d_b = v_f t_{pr} + \frac{{v_f}^2}{2\mu g} \quad (4)$$

where t_{pr} means perception reaction time, and μ and g indicate the coefficient of friction between the road surface and the tires and the gravity of Earth respectively. t_{pr} and μ are set as 1.5 s and 0.7 respectively, that are common baseline values incorporating the ability of the vast majority of drivers under normal road conditions (*24*, *25*). In order to find v_r from Equation 3, it is transformed to Equation 5, which is a quadratic equation for t_{tr}, to find t_{tr} first. After finding t_{tr} using the quadratic formula, v_r can be computed by Equation 6:

$$\left(2v_f + a_b t_{tr}\right)\frac{t_{tr}}{2} + \left(a_b t_{tr} + v_f\right)\left(t_g - t_{tr}\right) = d_{eV_L} + d_{V_F V_L} - d_b \quad (5)$$

$$v_r = a_b t_{tr} + v_f \quad (6)$$

After finding the two types of v_r, the system selects the minimum v_r as a speed recommendation.

When V_F Can Cross the Intersection with Additional Green Time. If d_{eV_F} and d_{eV_L} are longer than $d_{V_F R}$ and $d_{V_L R}$, respectively only with the increased green time, the V_F decides to cross the intersection by expecting that the RSU will increase the remaining green time according to the eco-signal mechanism, which is described later in this section. Here V_F expects that t_g will be increased to $t_g + \beta_{\max}/2$ where $\beta_{\max}$. is the maximum additional time in the eco-signal mechanism. The methods to compute d_{eV_F}, d_{eV_L}, and v_r are the same as the above methods except that $t_g + \beta_{\max}/2$ is used instead of t_g in Equations 1–5.

When V_F Cannot Cross the Intersection Despite the Additional Green Time. If d_{eV_F} is shorter than $d_{V_F R}$, or d_{eV_L} is shorter than $d_{V_L R}$ despite the increased green time, V_F considers that it cannot cross the intersection within the remaining time. In order to decrease fuel consumption, the system guides the driver with a speed that can avoid full stop. Namely, the system leads the vehicle to cross the intersection at the next green time. The recommendation speed is computed by

$$v_r = \frac{d_{SV_F}}{t_g + y + r} \quad (7)$$

where d_{SV_F} indicates the distance between the stop bar and V_F; y and r mean yellow and red intervals, respectively.

Yellow and Red Time. When V_F approaches a stop bar during the yellow or red time, the proposed system leads the vehicle to cross the intersection without deceleration when the red time ends. Namely, as an ideal scenario, the system guides V_F to pass the stop bar when the signal changes from red to green. However, since V_L may pass the stop bar or wait for the green light at the stop bar, it is challenging to predict the recommendation speed due to the safety issue. In order to find the recommendation speed, it is assumed that V_L may stop at the stop bar before green time. When calculating a recommendation speed, the system leads V_F to have the same speed as V_L after t seconds and maintain the braking distance corresponding to the speed. t means a variable time after changing from red to green light. That is to say, the vehicle crosses the stop bar without deceleration during $t_y + r + t$ at the yellow time or $t_r + t$ at the red time. Here t_y and t_r indicate the remaining yellow time and the remaining red time respectively. The recommendation speed is computed with:

$$\frac{d_{SV_F} - n_{V_L}\gamma + 0.5a_b t^2 - d_b}{T + t} = a_{V_L} t \quad (8)$$

where T is represented by $T = t_y + r$ or $T = t_r$, a_{V_L} means the average value of V_L's acceleration, and n_{V_L} and γ represent the number of leading vehicles and the average length of vehicles including the safety distance of the stopped vehicle. After finding the t using the quadratic formula, v_r can be computed by:

$$v_r = \frac{d_{SV_F} - n_{V_L}\gamma + 0.5a_b t^2 - d_b}{T + t} \quad (9)$$

Note that v_r depends on the trajectory of V_L and it cannot be precisely predicted every time due to the condition or driver of V_L. Therefore, in every situation, v_r will be v_l when $d_{V_F V_L}$ is less than d_b to prevent cars colliding.

Eco-Signal Mechanism. For the eco-signal mechanism, the RSU needs to verify whether the eco-driving vehicle can cross the intersection with/without the additional time β_{max} seconds through Equation 1. If the vehicle can cross the intersection only with the additional time, the RSU extends the remaining green time as described in Equation 10. The optimal signal timing plan is selected among feasible signal timing plans by minimizing total delay of n vehicles. More detailed information regarding the eco-signal mechanism is described in (*22*).

$$\min_{p^v \in X, \psi \in Y} \sum_v D(p^v, \psi) \quad (10)$$

Subject to

$$\psi = \left\{ (t_g + \beta, y, r)_1, \ldots, (t_g + \beta, y, r)_n \right\}, 1 \leqslant n \quad (10.1)$$

$$p^{v-1} \leqslant p^v \leqslant p^{v+1} \quad (10.2)$$

$$0 \leqslant \beta \leqslant \beta_{max} \quad (10.3)$$

where

β = green extension time;

ψ = traffic signal timing plan;

p^v = arrival time at the stop bar of an individual vehicle v; and

D = total delay (sec).

Delay is assigned to a vehicle which cannot cross the stop bar within the remaining green time (i.e., when p^v – current time$>t_g$). The vehicle's delay is accumulated until passing the stop bar at the next green signal. The delay is described as the time difference between when the vehicle crosses the stop bar with the current speed by ignoring signal timing plan and when the vehicle crosses the stop bar at the next green signal. RSU can estimate the delay based on the signal timing plan and the vehicle's trajectory. Namely, when a vehicle cannot cross an intersection within the remaining green time, RSU calculates the vehicle delay when it crosses the stop bar. With regard to the delay of stopped vehicle waiting for green signal, RSU can update the delay with the signal timing plan if the remaining red time is increased. To extend the remaining green time, we select the β towards the minimum total delay of vehicles. Although the delay of vehicle waiting for a green signal would be increased by extending the red time (i.e., extension by β), the total delay of every vehicle in the intersection can be minimized by the eco-signal mechanism by adding a constraint that the green time is only extended if it minimizes the total intersection delay. Note that the optimal value β of the genetic algorithm mentioned in (*22*) is not exploited since only two vehicles are considered for this test. Moreover, the RSU is allowed to extend the remaining green time once to prevent severe delay of vehicles.

Evaluations and Results

The proposed system was evaluated in the field as well as with a driving simulator based on PreScan, an off-the-shelf program explicitly modeling vehicle dynamics and sensor errors, to verify the confidence of the simulation-based eco-driving and eco-signal studies. As to the performance metric, the fuel consumption was used to prove the effects on fuel savings. The length of test-bed was 110 m, and it was assumed that the test-bed consists of two intersections; V_F moved from the first intersection to the next intersection. The distance between the stop bar and RSU was 15 m. v_{lim} was 40 km/h (11.12 m/s). Green and red intervals were 20 s each and yellow interval was 3 s. β_{max} was 12 s. a_b was set as a driver's average acceleration value, and a_{V_L} was set as 1.7 m/s^2, which was the experimental value of the four drivers' average acceleration during 3 s after departure.

Test Scenarios

To evaluate the proposed system, four scenarios were designed depending on traffic signal and vehicles' location.

- Scenario 1: V_L and V_F departed when the remaining green time was 16 s and 14 s respectively; the distance between V_F and the RSU was 110 m; the distance between V_F and V_L was 10 m.
- Scenario 2: V_L stopped at the stop bar; V_F departed when the remaining red time was 10 s; the distance between V_F and the RSU was 110 m.
- Scenario 3: V_L and V_F departed when the remaining red time was 15 s and 13 s respectively; the distance between V_F and the RSU was 110 m; the distance between V_F and V_L was 10 m.
- Scenario 4: V_L and V_F departed when the remaining green time was 11 s and 10 s respectively; the distance between V_F and the RSU was 110 m; the distance between V_F and V_L was 10 m.

Through Scenario 1, it was proved that the proposed system prevented unnecessary acceleration for crossing the intersection during the remaining green time. In Scenarios 2 and 3, it was verified that the system guided the trajectory of V_F to approach the intersection without deceleration after accelerating to the recommendation speed during the remaining red time although V_L was ahead. Scenario 4 proved that fuel consumption could be further saved by the eco-signal mechanism.

Field Test. The devices were deployed on the road and in the vehicle as shown in Figure 3*a* and *b*. The recommendation speed, remaining distance, and SPaT information were displayed on the tablet PC as shown in Figure 3*c*. When the current speed exceeded the recommendation speed, the shading color of recommendation speed was changed to red to warn the driver as shown in Figure 3*c*. Two Hyundai Avante MD vehicles were used for the field test as V_F and V_L. Each scenario was tested for five pairs (with and without guidance) of trials. The fuel consumption, which is the amount of fuel injected into the engine, was measured through OBD II. OBD II data indicated in-vehicle sensor data, in real time and compared it with fuel consumption measured using the VT-Micro model. In the VT-Micro model, fuel consumption is calculated based on vehicle's speed and acceleration.

Driving Simulator Test. For the driving simulator test, the software PreScan and the hardware Logitech were used to implement the eco-driving and eco-signal system (*26*). To ensure a fair comparison, the network and the scenarios were built identical to those of field. The driving simulator environment was as shown in Figure 4, in which the driver can see the actual speed as well as the recommendation speed in different colors. For the leading vehicle, the trajectories in the field test (of the leading vehicle) were imported in order to make sure the driving

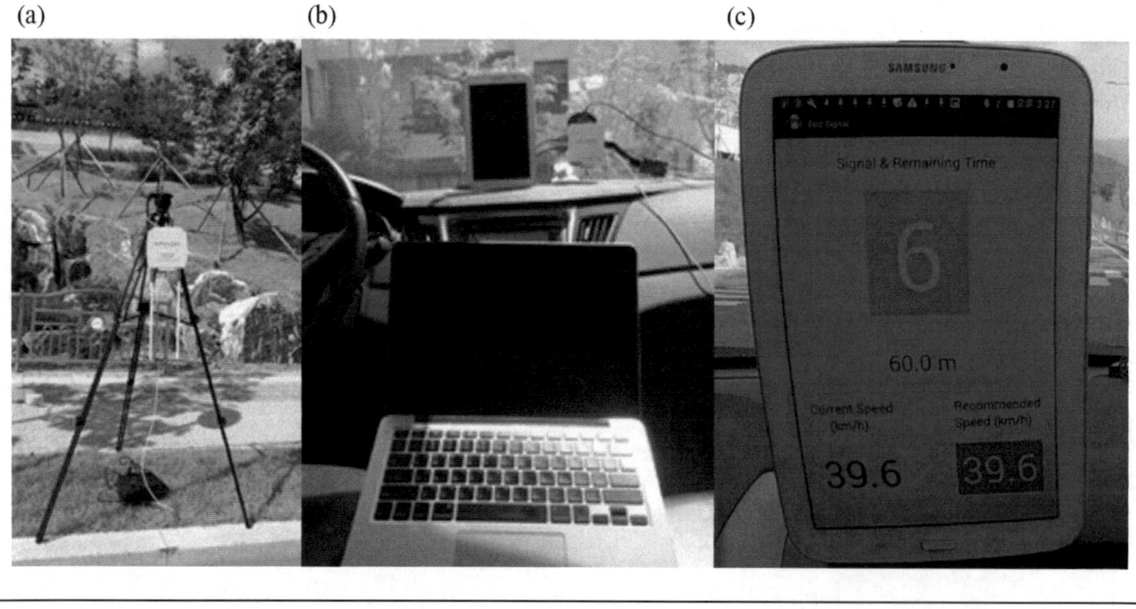

Figure 3. Devices for field test: (*a*) RSU; (*b*) laptop, tablet PC, and OBU; and (c) guidance display.

Figure 4. Driving simulator settings.

Table 1. Field Measurements using Real Data and the VT-Micro Model

Scenarios	Fuel consumption (ml, field) with	Fuel consumption (ml, field) without	Difference	Fuel consumption (ml, VT-Micro) with	Fuel consumption (ml, VT-Micro) without	Difference
Scenario 1 (Green)	24.7	31.7	−22.2%	22.0	24.7	−10.6%
Scenario 2 (Red stop)	22.0	29.5	−25.7%	21.8	26	−16.1%
Scenario 3 (Red)	21.5	35.7	−39.8%	21.9	31.2	−29.8%
Scenario 4 (Eco signal)	28.2	51.4	−45.0%	23.9	40.7	−41.9%

environment in the driving simulator was as similar as possible to the field. All five trajectories for each scenario were tested, and each trajectory was tested ten times by each driver (two drivers participated in the test).

Test Results

Based on the field test, the average fuel consumptions, which were measured as well as calculated with the VT-Micro model using the measured speed and acceleration data, are summarized in Table 1. The data for every trial is attached in Appendix A. Here, we are focusing on two points: first, to evaluate the effectiveness of the eco-driving guidance in the field; second, to observe how the fuel consumption calculation model VT-Micro distorts the results since the driving simulator results rely on the theoretical calculation of fuel consumption.

From Table 1, it can be seen that the eco-guidance could reduce the fuel consumption approximately 20%–40% in the normal signal scenarios (Scenarios 1–3). The trajectory comparisons in Figure 5 can show the changes made by the eco-driving guidance. Under the guidance, the driver had more smooth speeds, especially avoiding unnecessary decelerations. The benefit was more obvious in Scenarios 2 and 3 than Scenario 1 since the red signal was forcing the vehicles to stop. The eco-signal helped with fuel saving the most at 45.0%. Considering that the eco-signal enabled green-time to lengthen so that the vehicle could avoid a stop, the result is as expected.

For the theoretically estimated fuel consumption, the overall values were underestimated (i.e., the fuel consumption reduction from VT-Micro was around 10% lower than that of the field measured). The differences were also narrowed, compared with those measured in the field. That is because the model was not calibrated to any specific vehicle since limited field fuel consumption data were available. Besides, the gear level change is not considered for the calculation of fuel consumption in the VT-Micro model. In detail, a vehicle consumes more fuel after just achieving a higher gear level than the calculated amount by the VT-Micro model although the speed is not changed or increased only 1 km/h. Namely, the gear level changes could have caused differences between the measured amount and the calculated amount using the

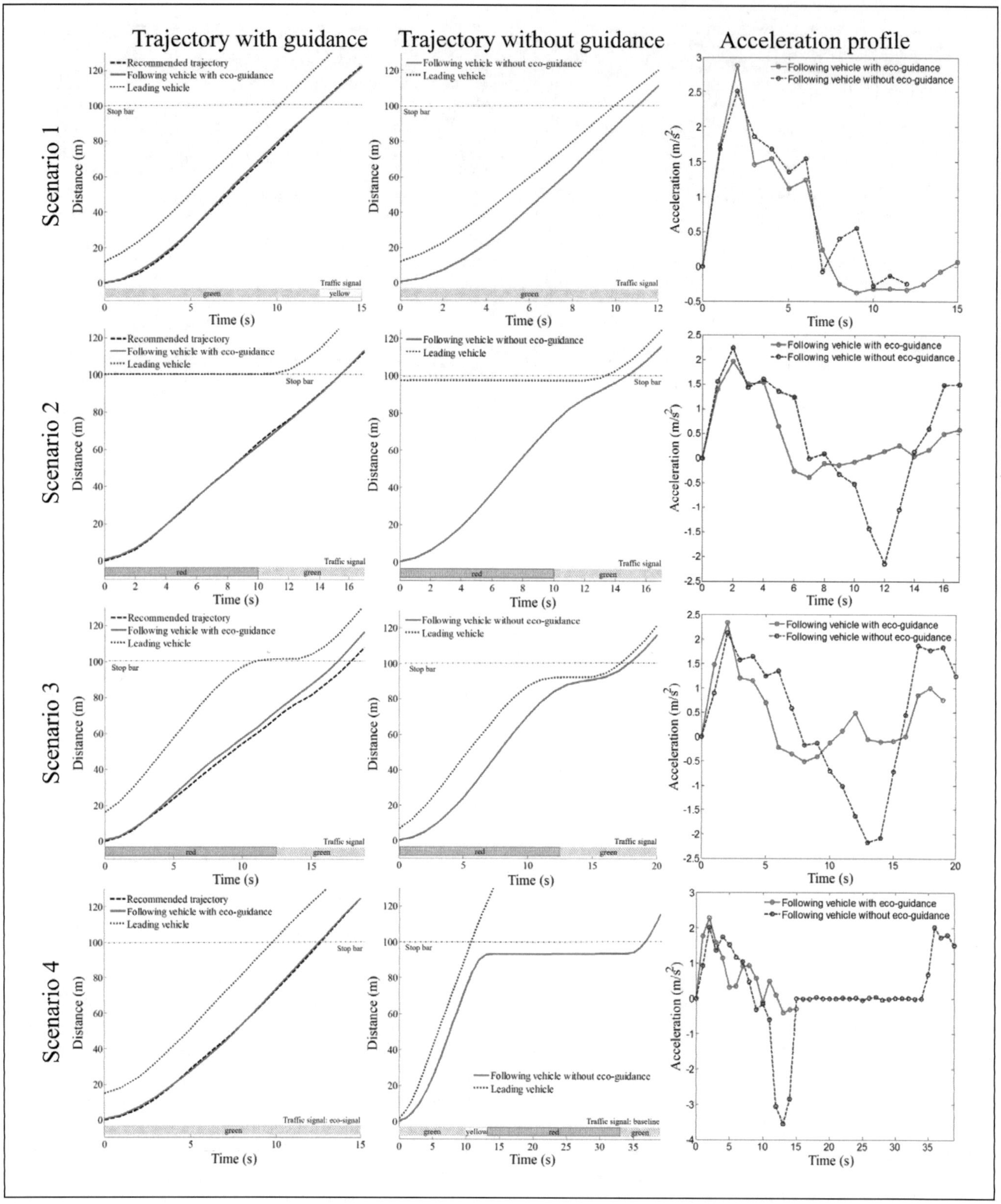

Figure 5. Comparing trajectories with and without guidance.
Note: s = seconds.

VT-Micro model. Considering that the maximum gear level of V_F was always higher in the tests without the guidance (i.e., up to level 4 in Scenarios 1 and 4 with the guidance; up to level 5 in Scenarios 1 and 4 without the guidance; up to level 3 in Scenarios 2 and 3 with the guidance; up to level 4 in Scenarios 2 and 3 without the guidance), the relative difference between the measured amount and the theoretically calculated value is bigger in the test without the guidance. Therefore, the fuel reduction from the VT-Micro is lower than the one based on the measured values in Table 1. Moreover, given that V_F kept the speed after acceleration without unnecessary deceleration and the gear level is not considered in the VT-Micro model, the calculated fuel consumption values were less sensitive than the measured values when following the guidance.

The driving simulator test results are represented in Table 2. The data for every trial is attached in Appendix A. The difference cells shaded with gray represent significant differences at 95% confidence level. The four histograms in each scenario indicate the fuel consumptions with and without the eco-driving guidance of Driver1 and Driver2, respectively.

On average, the fuel consumption reductions by the eco-driving guidance as well as the eco-signal were statistically significant. The range of the reduction rates was pretty wide: + 0.6%~−8.3% in Scenario 1; −16.1%~−30.8% in Scenario 2; −9.1%~−32.9% in Scenario 3; and −29.9%~−32.4% in Scenario 4. The fuel consumptions in the driving simulator test were always higher than the average in the field test. This can be due to the diminished feeling of the simulator toward the traffic noise and pavement friction, which led to the drivers driving more aggressively unexpectedly compared with the field. That is to say, when given the same recommendation speed, the driver in the driving simulator would have harder pressure on the throttle or brake in a short period to adjust the speed compared with in the field. Scenario 1 was not showing an obvious fuel consumption reduction, even though the differences were proved to be significant. The reduction rates were all smaller than that in the field test. For Scenarios 2 and 3, the fuel consumption reduction rates were closest to the field test results, where the reduction values in the field test were covered by the reduction rate range in the simulator test. Eco-signal case (Scenario 4) had the most fuel consumption reduction rate at around 30%, which was also lower than the field test.

From the results, it can be observed that the theoretical fuel consumption calculation gives a lower reduction rate compared with the measured result. On the basis of this, the driving simulator result made a further underestimation of the eco-driving guidance system. The main reason can be considered as the unexpected aggressive driving behaviors such as more accelerations and decelerations in the driving simulator due to the diminished feeling of the noise and shaking arisen from the pavement friction. This also caused the higher fuel consumptions in the simulator test compared with those in the field test.

Though the fuel consumption reductions had significant differences with eco-driving guidance and eco-signal vs. without them, Scenario 1 did not present obvious fuel savings compared with the other scenarios and the same scenario in the field test, and the fuel reduction rate varied between + 0.6%~−8.3%. Scenarios 2 and 3 had similar fuel consumption reduction rates in the simulator test, varying from −16.1%~−30.8% and −9.1%~−32.9%, respectively, compared with the field test (−16.1% and −29.8%, respectively). Finally, Scenario 4 in the simulator test had the biggest fuel saving at around 30% but was also lower than the field test result (41.9%).

Conclusions

In this paper, the effectiveness of eco-driving technology on fuel consumption was evaluated in the field as well as using a driving simulator. To implement the experiments, an eco-driving guidance and an eco-signal system using V2X communication in a multi-vehicle environment were proposed. By testing several pairs (with and without guidance) of trials for each scenario, the fuel consumption reduction rates were calculated to evaluate the benefit of the eco-driving system, and also to validate the confidence of the driving simulator through comparing with the field test result.

The field test results showed that the eco-driving guidance could save the fuel consumption around 20%–40% and the eco-signal system could achieve fuel consumption reduction up to 45%. The trajectory and the acceleration profile proved that the eco-driving can successfully help the participants drive more smoothly and avoid some unnecessary braking. Compared with the measured fuel consumption, the theoretically calculated fuel consumptions with the speed and acceleration data using the VT-Micro model were underestimated, and their fuel consumption reductions were around 10% lower than those of the field measured results. That is to say, the driving simulator results could vary depending on the fuel consumption estimation model as well as the vehicle used in the field, although it seems the relative benefits are valid.

For the driving simulator test, the average fuel consumption reductions by the eco-driving system were statistically significant. But the range of the reduction rates was pretty wide, and only Scenarios 2 and 3 had similar reduction rates to the field, in which the values in the field test were included in the reduction rate range in the simulator test. Compared with other scenarios, Scenario 1 was not showing an obvious fuel consumption reduction (+ 0.6%~−8.3%) though the differences were

Table 2. Average Fuel Consumption in the Driving Simulator Test

Scenario 1 (Green)	Driver 1 (ml)		Difference	Driver 2 (ml)		Difference
	With	Without		With	Without	
Field test	22.0	24.7	−10.6%	22.0	24.7	−10.6%
1	25.4	26.7	−4.7%	26.8	28.2	−4.9%
2	24.8	27.1	−8.3%	27.5	29.1	−5.4%
3	24.6	25.1	−1.9%	26.8	27.2	−1.3%
4	25.6	25.4	0.6%	26.2	27.2	−3.7%
5	25.8	25.6	0.6%	26.6	27.0	−1.6%
Average	25.2	26.0	−2.8%	26.8	27.7	−3.4%

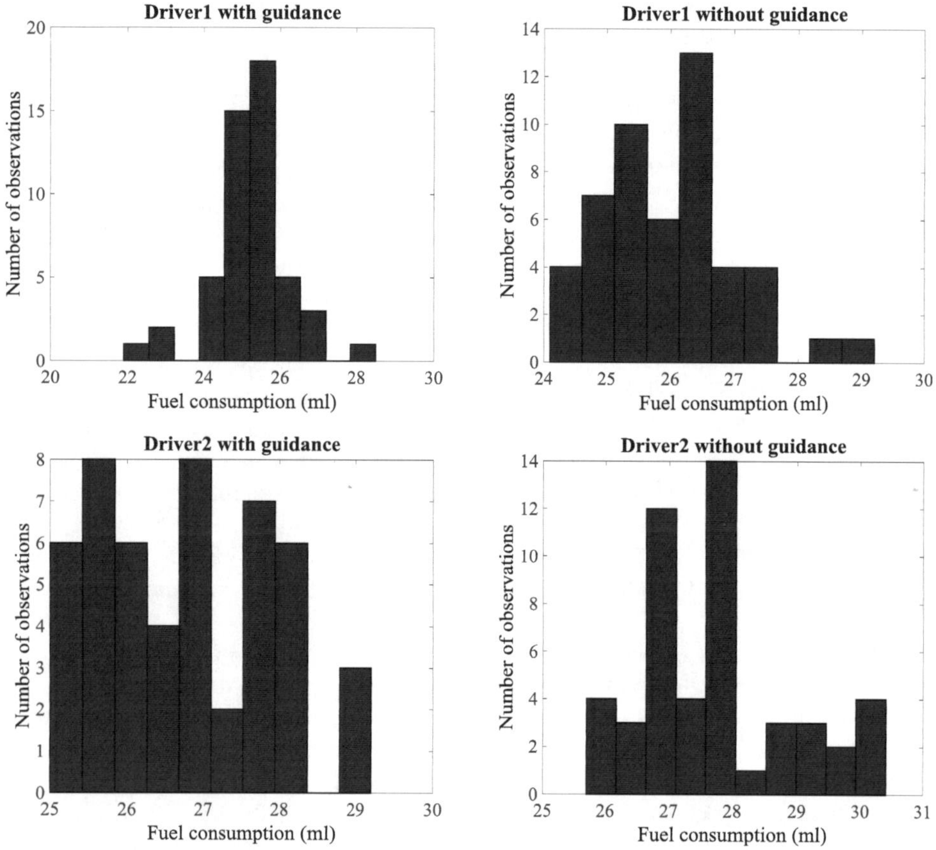

Scenario 2 (Red stop)	Driver1 (ml)		Difference	Driver2 (ml)		Difference
	With	Without		With	Without	
Field test	21.8	26	−16.1%	21.8	26	−16.1%
1	24.3	29.4	−17.6%	29.6	39.1	−24.3%
2	24.6	29.3	−16.1%	30.5	39.2	−22.2%
3	24.5	29.3	−16.5%	29.2	40.1	−27.2%
4	22.8	29.6	−23.1%	31.7	38.1	−16.8%
5	29.9	33.5	−10.8%	28.6	41.2	−30.8%
Average	25.2	30.2	−16.6%	29.9	39.5	−24.4%

(continued)

Table 2. (continued)

Scenario 2 (Red stop)	Driver1 (ml) With	Driver1 (ml) Without	Difference	Driver2 (ml) With	Driver2 (ml) Without	Difference

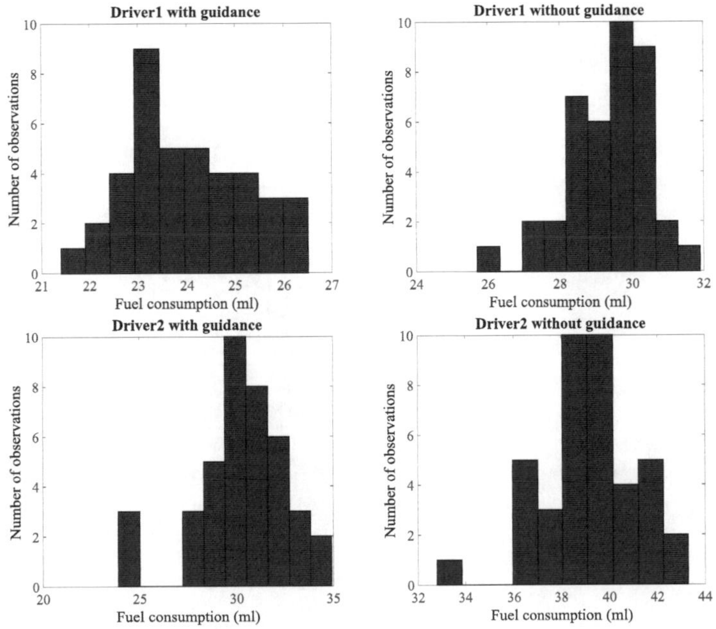

Scenario 3 (Red)	Driver1 (ml) With	Driver1 (ml) Without	Difference	Driver2 (ml) With	Driver2 (ml) Without	Difference
Field test	21.9	31.2	–29.8%	21.9	31.2	–29.8%
1	29.4	33.8	–13.2%	28.6	38.9	–26.5%
2	30.4	34.2	–11.0%	29.1	39.6	–26.6%
3	32.5	35.7	–9.1%	31.7	40.3	–21.3%
4	33.4	36.9	–9.5%	31.4	41.2	–23.9%
5	30.5	34.8	–12.3%	26.4	39.3	–32.9%
Average	31.2	35.1	–11.0%	29.4	39.9	–26.2%

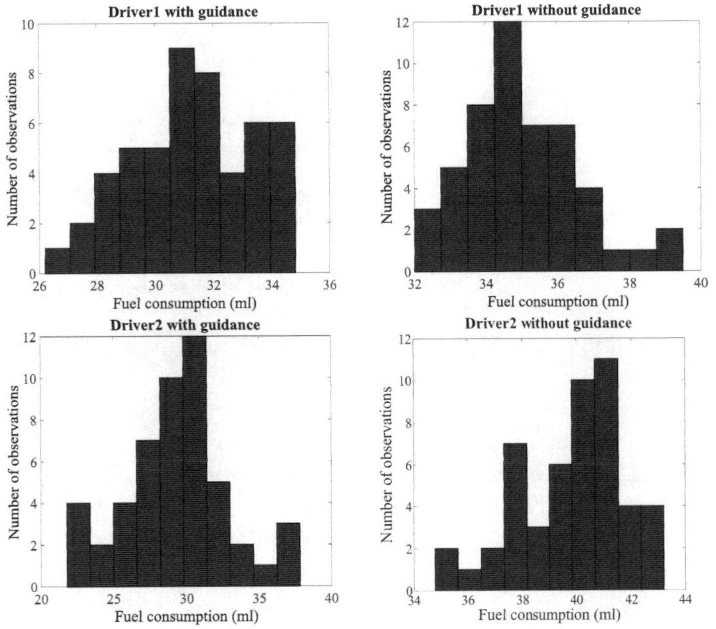

(continued)

Table 2. (continued)

Scenario 4 (EcoSignal)	Driver1 (ml) With	Driver1 (ml) Without	Difference	Driver2 (ml) With	Driver2 (ml) Without	Difference
Field test	23.9	40.7	−41.9%	23.9	40.7	−41.9%
1	24.8	36.5	−32.0%	26.0	38.4	−32.4%
2	25.3	37.0	−31.5%	25.4	37.1	−31.5%
3	25.6	36.7	−30.2%	26.1	37.6	−30.7%
4	25.6	37.1	−31.1%	26.7	36.3	−26.5%
5	26.2	37.3	−29.9%	27.3	39.0	−30.0%
Average	25.5	36.9	−30.9%	26.3	37.7	−30.2%

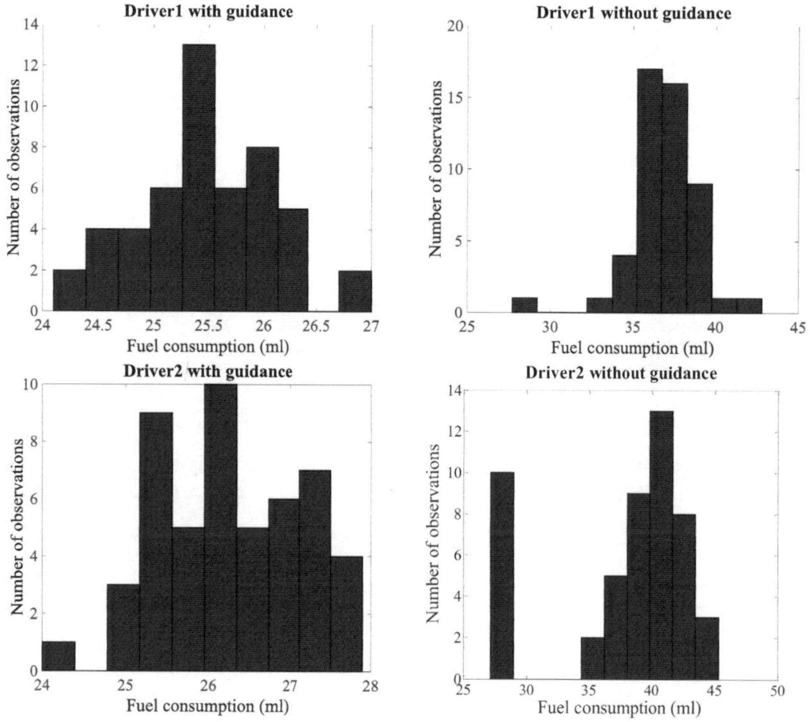

significant between with and without guidance. Scenario 4 achieved the highest fuel consumption reduction of approximately 30%, but again lower than that of the field test result.

Based on the findings of this research, the following future research directions are recommended.

- A more accurate fuel consumption model is needed as the driving simulator-based simulation results highly depend on the theoretical calculation.
- If an accurate fuel consumption model is available for all vehicle types and models, the eco-guidance algorithm can be designed by considering specific fuel consumption model for each vehicle so as to maximize the fuel savings.
- A network-wide evaluation is needed, especially for the eco-signal case, where the impact of the eco-driving system not only to the subject vehicle but also its surrounding vehicles which are affected by the subject vehicle or the eco-signal also needs to be considered.
- Scenarios that multiple vehicles follow the eco-guidance need to be considered in order to induce the vehicles to drive eco-friendly using interactions between them.
- Effect of eco-signal on the overall traffic flow needs to be further investigated because increasing green splits could affect delays of vehicles waiting for green-time signal and cause traffic congestions.

Appendix A

Fuel Consumption Results For Every Trial

Field Test Results

Scenario 1 (Green)		1	2	3	4	5	Average	Difference
Fuel consumption (ml, field)	w/	24.1	27.1	23.1	25.1	24.2	24.7	–22.2%
	w/o	34.1	32.0	33.5	28.3	30.9	31.7	
Fuel consumption (ml, VT-Micro)	w/	21.1	23.3	21.4	22.8	21.6	22.0	–10.6%
	w/o	26.5	25.2	25	23.2	23.37	24.7	
Scenario 2 (Red stop)								
Fuel consumption (field)	w/	21.8	19.8	23.7	22.6	21.9	22.0	–25.7%
	w/o	26.4	26.5	27.1	31.4	36.3	29.5	
Fuel consumption (ml, VT-Micro)	w/	21.6	20.1	22.9	22.2	22.1	21.8	–16.1%
	w/o	24.8	24.3	25.2	28.2	27.3	26.0	
Scenario 3 (Red)								
Fuel consumption (ml, field)	w/	22.3	23.9	19.0	20.3	22.1	21.5	–39.8%
	w/o	37.5	41.6	33.0	34.1	32.5	35.7	
Fuel consumption (ml, VT-Micro)	w/	22.5	22.8	20.9	20.7	22.5	21.9	–29.8%
	w/o	31.9	33.7	30.1	30.8	29.3	31.2	
Scenario 4 (Eco signal)								
Fuel consumption (ml, field)	w/	26.9	28.4	29.8	29.5	26.5	28.2	–45.0%
	w/o	50.8	49.4	52.6	49.8	54.8	51.4	
Fuel consumption (ml, VT-Micro)	w/	23.4	23.8	24	23.5	23.4	23.9	–41.9%
	w/o	41.3	39.5	41.7	39.5	41.7	40.7	

Driving Simulator Test Results

		1	2	3	4	5	6	7	8	9	10	Average	Difference
Scenario 1 (Green, ml)													
Driver1 (1-1)	w/	23.2	25.6	28.5	24.8	24.6	25.0	24.9	25.4	25.6	26.6	25.4	**–4.7%**
	w/o	28.5	26.1	27.2	27.3	26.5	26.9	26.0	26.2	25.5	26.6	26.7	
Driver2 (1-1)	w/	28.0	28.2	25.4	29.2	26.5	27.1	26.0	25.5	26.7	25.7	26.8	**–4.9%**
	w/o	30.3	28.7	26.9	27.7	28.0	27.3	26.9	27.6	29.2	29.5	28.2	
Driver1 (1-2)	w/	21.9	26.8	24.4	25.0	24.9	25.3	25.1	25.7	26.0	23.1	24.8	**–8.3%**
	w/o	29.2	26.6	27.6	26.6	26.6	26.5	27.0	27.2	26.5	26.9	27.1	
Driver2 (1-2)	w/	28.8	26.7	27.7	26.8	25.6	27.4	27.5	28.9	27.8	27.9	27.5	**–5.4%**
	w/o	29.9	29.3	27.9	28.7	30.4	27.3	28.0	30.0	29.3	30.0	29.1	
Driver1 (1-3)	w/	24.7	24.6	24.1	24.3	24.5	24.9	24.7	24.4	24.9	24.9	24.6	–1.9%
	w/o	25.4	26.1	25.4	26.5	24.2	24.1	25.2	24.9	24.5	24.5	25.1	
Driver2 (1-3)	w/	27.0	28.2	26.5	27.6	27.8	25.2	25.0	25.4	27.1	28.3	26.8	–1.3%
	w/o	25.7	26.8	27.9	27.6	27.4	27.7	26.8	26.8	27.9	26.9	27.2	
Driver1 (1-4)	w/	25.0	25.0	25.7	25.8	26.3	25.4	25.5	25.6	25.5	26.0	25.6	0.6%
	w/o	25.7	25.2	25.1	25.5	26.3	24.8	25.2	25.3	24.8	26.5	25.4	
Driver2 (1-4)	w/	27.8	25.8	27.8	26.4	26.7	26.0	25.7	25.2	25.6	25.0	26.2	**–3.7%**
	w/o	28.3	27.0	26.5	26.6	27.2	26.8	28.0	26.1	28.7	27.0	27.2	
Driver1 (1-5)	w/	25.5	25.5	25.4	25.8	25.6	26.3	25.6	26.1	25.3	26.7	25.8	0.6%
	w/o	25.7	25.7	26.3	24.8	25.1	25.2	27.0	25.3	24.9	26.2	25.6	
Driver2 (1-5)	w/	25.6	26.1	26.2	26.2	25.8	26.1	26.5	28.2	26.8	28.3	26.6	–1.6%
	w/o	28.0	27.8	27.7	26.7	26.6	26.7	25.8	27.9	27.0	26.0	27.0	
Scenario 2 (Red stop, ml)													
Driver1 (2-1)	w/	23.7	23.1	22.7	23.4	23.0	25.8	26.0	24.4	25.0	25.4	24.3	**–17.6%**
	w/o	30.0	29.5	25.7	30.5	28.7	30.1	28.2	30.2	30.2	31.1	29.4	
Driver2 (2-1)	w/	23.9	30.0	31.7	30.4	29.3	29.9	31.3	29.9	28.3	31.7	29.6	**–24.3%**
	w/o	36.8	38.5	40.3	38.7	38.0	39.4	41.1	39.1	41.3	38.2	39.1	

(continued)

(continued)

		1	2	3	4	5	6	7	8	9	10	Average	Difference
Scenario 2 (Red stop, ml)													
Driver1 (2-2)	w/	23.5	23.5	26.5	24.4	23.1	24.6	24.1	25.9	26.4	23.5	24.6	**−16.1%**
	w/o	29.9	27.5	29.1	30.2	29.1	31.0	29.6	27.6	28.7	29.8	29.3	
Driver2 (2-2)	w/	32.5	31.2	28.7	32.1	28.0	28.5	32.8	28.2	32.9	29.8	30.5	**−22.2%**
	w/o	37.8	42.1	39.3	39.8	41.3	36.3	36.5	38.4	39.6	40.4	39.2	
Driver1 (2-3)	w/	22.7	24.7	24.2	24.9	24.0	25.4	24.5	25.7	25.1	23.3	24.5	**−16.5%**
	w/o	29.4	30.0	29.6	28.6	30.4	30.2	28.6	28.2	27.5	30.4	29.3	
Driver2 (2-3)	w/	24.2	33.8	29.6	30.8	31.6	30.2	30.2	28.8	24.7	28.2	29.2	**−27.2%**
	w/o	39.0	39.9	39.2	41.5	39.7	38.1	42.3	39.4	38.9	43.3	40.1	
Driver1 (2-4)	w/	21.4	23.4	22.3	23.7	23.0	22.8	22.5	23.2	22.2	23.1	22.8	**−23.1%**
	w/o	29.8	31.9	29.1	29.8	28.7	28.1	29.0	30.0	28.9	30.6	29.6	
Driver2 (2-4)	w/	30.6	30.8	34.9	34.3	32.0	30.5	32.0	31.0	30.9	29.5	31.7	**−16.8%**
	w/o	40.5	41.4	38.1	36.0	38.4	32.8	40.1	36.5	38.7	38.0	38.1	
Driver1 (2-5)	w/	29.9	28.7	30.2	29.1	29.4	31.3	28.8	30.4	30.3	31.3	29.9	**−10.8%**
	w/o	32.1	35.6	33.2	32.3	32.1	34.4	33.7	33.7	34.0	34.2	33.5	
Driver2 (2-5)	w/	29.1	26.2	26.1	28.2	27.1	30.5	28.0	30.1	27.9	32.3	28.6	**−30.8%**
Scenario 3 (Red, ml)													
Driver1 (3-1)	w/	28.9	29.3	29.6	27.9	30.6	26.2	28.2	29.8	31.2	32.0	29.4	**−13.2%**
	w/o	34.5	34.0	34.0	33.2	33.7	32.0	34.0	38.0	32.4	32.5	33.8	
Driver2 (3-1)	w/	28.1	32.5	26.4	22.7	28.1	25.5	32.4	30.3	30.9	28.9	28.6	**−26.5%**
	w/o	35.2	37.6	40.5	39.7	37.5	40.6	40.2	41.1	38.6	38.0	38.9	
Driver1 (3-2)	w/	29.4	28.7	31.1	31.1	29.8	28.7	29.8	31.7	32.4	31.3	30.4	**−11.0%**
	w/o	33.9	34.7	34.6	34.1	33.1	35.7	32.8	35.2	33.9	33.5	34.2	
Driver2 (3-2)	w/	34.1	31.0	29.6	27.0	29.4	29.4	28.6	28.1	24.3	29.1	29.1	**−26.6%**
	w/o	39.6	41.3	41.5	40.1	37.1	37.7	39.7	40.4	39.7	38.9	39.6	
Driver1 (3-3)	w/	31.7	31.2	34.0	33.5	31.8	32.1	34.0	32.9	31.0	32.3	32.5	**−9.1%**
	w/o	35.9	34.3	33.5	36.6	35.9	35.6	36.5	36.8	35.3	36.5	35.7	
Driver2 (3-3)	w/	31.5	36.8	29.7	30.0	35.0	36.7	27.6	30.7	29.3	29.9	31.7	**−21.3%**
	w/o	41.3	42.2	34.8	38.0	41.7	40.7	42.6	40.9	40.2	40.6	40.3	
Driver1 (3-4)	w/	33.6	34.8	33.3	31.6	33.6	32.4	31.7	34.5	34.2	34.5	33.4	**−9.5%**
	w/o	38.6	34.9	37.0	37.2	34.4	38.8	36.0	36.4	39.5	36.5	36.9	
Driver2 (3-4)	w/	32.6	37.8	31.2	30.7	29.9	29.1	31.5	31.4	30.5	29.1	31.4	**−23.9%**
	w/o	41.0	41.0	42.4	41.6	40.6	36.7	42.5	41.2	41.9	43.2	41.2	
Driver1 (3-5)	w/	33.6	29.7	29.7	30.8	31.1	29.4	28.5	27.2	31.5	33.3	30.5	**−12.3%**
	w/o	34.8	34.7	34.7	34.5	34.1	34.3	35.2	34.6	35.3	35.3	34.8	
Driver2 (3-5)	w/	30.9	27.5	27.6	24.5	26.5	26.3	21.8	22.1	22.6	34.2	26.4	**−32.9%**
	w/o	40.0	36.3	38.9	39.7	39.7	38.1	38.0	41.1	40.6	40.9	39.3	
Scenario 4 (EcoSignal, ml)													
Driver1 (4-1)	w/	24.3	25.2	24.7	25.4	24.5	24.1	24.4	25.0	24.9	25.2	24.8	**−32.0%**
	w/o	38.1	36.7	39.0	35.7	39.6	33.9	36.1	34.9	35.9	34.6	36.5	
Driver2 (4-1)	w/	25.0	27.1	25.7	26.6	25.1	26.3	26.3	25.1	25.9	26.9	26.0	**−32.4%**
	w/o	39.6	42.2	40.9	41.0	38.5	40.3	36.9	36.1	40.6	28.3	38.4	
Driver1 (4-2)	w/	25.7	25.4	25.8	25.2	25.9	25.2	25.0	25.3	25.3	24.5	25.3	**−31.5%**
	w/o	37.0	35.6	36.2	37.4	37.3	38.2	35.5	37.7	35.8	39.0	37.0	
Driver2 (4-2)	w/	25.8	25.3	25.4	25.5	25.5	25.3	26.5	25.5	24.0	25.4	25.4	**−31.5%**
	w/o	27.6	42.1	38.8	39.9	27.2	38.5	37.0	40.3	41.7	37.9	37.1	
Driver1 (4-3)	w/	24.9	25.5	26.2	25.4	26.1	25.5	25.7	25.4	25.5	25.9	25.6	**−30.2%**
	w/o	38.1	39.3	38.3	36.7	37.0	33.8	36.0	36.8	33.5	37.3	36.7	
Driver2 (4-3)	w/	27.2	25.3	26.3	26.3	26.1	25.7	26.2	26.1	25.4	26.2	26.1	**−30.7%**
	w/o	41.7	27.6	28.5	42.1	45.3	40.0	39.6	41.5	41.8	28.2	37.6	
Driver1 (4-4)	w/	25.6	26.1	25.4	24.9	26.0	25.5	25.4	26.2	25.8	24.6	25.6	**−31.1%**
	w/o	39.0	38.3	35.7	38.2	38.1	37.8	35.5	35.7	36.1	36.5	37.1	
Driver2 (4-4)	w/	27.0	26.2	26.4	27.4	27.6	26.1	27.0	26.5	26.9	25.8	26.7	**−26.5%**
	w/o	43.8	37.0	41.7	27.7	39.4	27.4	41.2	38.8	27.1	39.0	36.3	
Driver1 (4-5)	w/	27.0	25.9	26.9	25.3	25.8	25.9	26.2	26.2	26.0	26.4	26.2	**−29.9%**
	w/o	42.8	40.6	36.8	27.7	39.2	36.7	37.0	37.7	39.0	35.6	37.3	
Driver2 (4-5)	w/	26.8	27.4	27.6	27.2	26.7	27.6	27.5	27.4	27.3	27.9	27.3	**−30.0%**
	w/o	27.9	36.5	39.7	42.9	40.6	41.4	41.3	39.9	43.9	36.2	39.0	

Note: The bolded cells indicate significant differences in the confidential level of 95%.

Acknowledgments

This research project was partly supported by Global Research Laboratory Program (2013K1A1A2A02078326) through NRF, DGIST Research and Development Program (CPS Global Center) funded by the Ministry of Science, ICT & Future Planning, and KAIA grant funded by the Korea Government (MOLIT). (No.15TLRP-C105654-01, Development and Verification of Signal Operation Algorithms in Local Intersection Network utilizing V2X Communication Infrastructure).

Author Contributions

The authors confirm contribution to the paper as follows: study conception and design: Park, Choi, Ko, and Cui; data collection: Ko, Cui, and Ryu; analysis and interpretation of results: all authors; draft manuscript preparation: Ko and Cui. All authors reviewed the results and approved the final version of the manuscript.

References

1. US Energy Information Administration. https://www.eia.-gov/totalenergy/data/monthly/#consumption. Accessed July 17, 2016.
2. Li, M., K. Boriboonsomsin, G. Wu, W. B. Zhang, and M. Barth. Traffic Energy and Emission Reductions at Signalized Intersections: A Study of the Benefits of Advanced Driver Information. *International Journal of ITS Research*, Vol. 7, No. 1, 2009, pp 49–58.
3. Saboohi, Y., and H. Farzaneh. Model for Developing an Eco-driving Strategy of a Passenger Vehicle Based on the Least Fuel Consumption. *Applied Energy*, Vol. 86, No. 10, 2009, pp. 1925–1932.
4. Barth, M., S. Mandava, K. Boriboonsomsin, and H. Xia Dynamic ECO-driving for arterial corridors. *IEEE Forum on Integrated and Sustainable Transportation System (FISTS)*, Vienna, Austria, 2011.
5. Rakha, H., and R. K. Kamalanathsharma. Eco-driving at Signalized Intersections Using V2I Communication. *2011 14th International IEEE Conference on Intelligent Transportation Systems*. Washington, D.C., 2011.
6. Qian, G., and E. Chung. Evaluating Effects of Eco-driving at Traffic Intersections Based on Traffic Micro-simulation. *Transport Research Forum 2011 Proceedings*, Adelaide, Australia, 2011.
7. Katwijk, R. Using I2V Communication to Reduce Traffic Emissions at Urban Signalized Intersections. *Procedia - Social and Behavioral Sciences*, Vol. 48, 2012, pp. 2385–2392.
8. Xia, H., K. Boriboonsomsin, and M. Barth. Dynamic Eco-driving for Signalized Arterial Corridors and its Indirect Network-Wide Energy/Emissions Benefits. *Journal of Intelligent Transportation Systems*, Vol. 17, No. 1, 2013, pp 31–41.
9. CIECA. International Project on 'Eco-driving' in Category B Driver Training and the Driving Test. Final Report, Brussels, 2007.
10. Stillwater, T., and K. Kurani. Field Test of Energy Information Feedback: Driver Responses and Behavioral Theory. Presented at 90th Annual Meeting of the Transportation Research Board, Washington, D.C., 2011.
11. Tulusan, J., H. Steggers, E. Fleisch, and T. Staake. Supporting Eco-driving with Eco-feedback Technologies: Recommendations Targeted at Improving Corporate Car Drivers' Intrinsic Motivation to Drive More Sustainable. *Proc., 14th ACM International Conference on Ubiquitous Computing*. Pittsburgh, PA, 2012.
12. Barth, M., and K. Boriboonsomsin. Energy and Emissions Impacts of a Freeway-based Dynamic Eco-driving System. *Transportation Research Part D: Transport and Environment*, Vol. 14, No. 6, 2009, pp. 400–410.
13. Kurani, K. S., T. Stillwater, and M. Jones. *Ecodrive I-80: A Large Sample Fuel Economy Feedback Field Test*. Final Report, ITS-RR-13-15. Oak Ridge National Laboratory, Oak Ridge, TN, 2013.
14. Kirsten, D., S. Nina, H. Richard, F. Philippus, and A. Bart. Roadside versus In-Car Speed Support for Green Wave: A Driving Simulator Study. Presented at 87th Annual Meeting of the Transportation Research Board, Washington, D.C., 2008.
15. Hiraoka, T., Y. Terakado, S. Matsumoto, and S. Yamade. Quantitative evaluation of Eco-driving on fuel consumption based on driving simulator experiments. *Proc., 16th World Congress on Intelligent Transport Systems*, Stockholm, Sweden, 2009.
16. Krause, M., and K. Bengler. Traffic Light Assistant – Driven in a Simulator. *Proc., 2012 International IEEE Intelligent Vehicle Symposium (IV-2012)*, Madrid, Spain, 2012.
17. Niu, D., and J. Sun. Eco-driving Versus Green Wave Speed Guidance for Signalized Highway Traffic: A Multi-Vehicle Driving Simulator Study. *Social and Behavioral Sciences*, Vol. 96, 2013, pp. 1079–1090.
18. Zhao, Y., A. Wagh, Y. Hou, K. Hulme, C. Qiao, and A. W. Sadek. Integrated Traffic-Driving-Networking Simulator for the Design of Connected Vehicle Applications: Eco-Signal Case Study. *Journal of Intelligent Transportation Systems*, Vol. 20, No. 1, 2014, pp. 75–87.
19. Cui, L., and B. Park. Assessing Eco-driving Behaviors Using Driving Simulator. Presented at 96th Annual Meeting of the Transportation Research Board, Washington, D.C., 2017.
20. SAE International. *J2735 Dedicated Short Range Communications (DSRC) Message Set Dictionary*. 2016.
21. Barth, M., S. Mandava, K. Boriboonsomsin, and H. Xia. Dynamic ECO-driving for Arterial Corridors. *IEEE Forum on Integrated and Sustainable Transportation System (FISTS)*, Vienna, Austria, 2011.
22. Jung, H., S. Choi, B. B. Park, and S. H. Son. Bi-Level Optimization for Eco-Traffic Signal System. *Proc., 5th*

International Conference on Connected Vehicles and Expo (ICCVE), Seattle, WA, 2016.
23. Ko, B., S. Choi, B. B. Park, and S. H. Son. Field Implementation of Eco-driving and Eco-signal System. *Proc., 3rd International Conference on Vehicle Technology and Intelligent Transport Systems*, Porto, Portugal, 2017
24. Wikipedia. https://en.wikipedia.org/wiki/Braking_distance. Accessed July 18, 2017.
25. Taoka, G. T. Brake Reaction Times of Unalerted Drivers. *ITE Journal*, Vol. 59, No. 3, 1989, pp. 19–21.
26. TASS International, TNO. *PreScan R7.2.0 Manual – A Simulation & Verification Environment for Intelligent Vehicle Systems*, September 2015.

The Standing Committee on Transportation and Air Quality (ADC20) peer-reviewed this paper (18-05021).

Article

TRR
JOURNAL OF THE TRANSPORTATION RESEARCH BOARD

Transportation Research Record
2018, Vol. 2672(25) 139–149

DOI: 10.1177/0361198118797805
journals.sagepub.com/home/trr
SAGE

Navigation Application Programming Interface Route Fuel Saving Opportunity Assessment on Large-Scale Real-World Travel Data for Conventional Vehicles and Hybrid Electric Vehicles

Lei Zhu[1], Jacob R. Holden[1], and Jeffrey D. Gonder[1]

Abstract

The green-routing strategy instructing a vehicle to select a fuel-efficient route benefits the current transportation system with fuel-saving opportunities. This paper introduces a navigation application programming interface (API), route fuel-saving evaluation framework for estimating fuel advantages of alternative API routes based on large-scale, real-world travel data for conventional vehicles (CVs) and hybrid electric vehicles (HEVs). Navigation APIs, such as Google Directions API, integrate traffic conditions and provide feasible alternative routes for origin–destination pairs. This paper develops two link-based fuel-consumption models stratified by link-level speed, road grade, and functional class (local/non-local), one for CVs and the other for HEVs. The link-based fuel-consumption models are built by assigning travel from many global positioning system driving traces to the links in TomTom MultiNet and road grade data from the U.S. Geological Survey elevation data set. Fuel consumption on a link is computed by the proposed model. This paper envisions two kinds of applications: (1) identifying alternate routes that save fuel, and (2) quantifying the potential fuel savings for large amounts of travel. An experiment based on a large-scale California Household Travel Survey global positioning system trajectory data set is conducted. The fuel consumption and savings of CVs and HEVs are investigated. At the same time, the trade-off between fuel saving and travel time due to choosing different routes is also examined for both powertrains.

Emerging intelligent transportation system technologies, such as connected and automated vehicles, could be implemented to provide positive effects on mobility, fuel consumption, and greenhouse gas emissions (*1*, *2*) in the near future. One particular interest in connected and automated vehicle technology is to provide vehicles with guidance to achieve better fuel efficiency during driving, which includes two types of tactics. One type is called "eco-driving," offering operational feedback to drivers, such as maintaining a steady speed and smoothing acceleration (*3*). The second type is guiding drivers to choose more fuel-efficient routes, referred to as "green routing." The selected ideal route considers road features and the traffic conditions, which are dominant factors in driving behavior (*4*) and fuel efficiency. A study by Nie and Li revealed that when the route is chosen, the operational tactics seem to have relatively small impacts on operating speed and fuel efficiency (*5*). Therefore, if the estimated fuel consumption of each alternative route is known before departure, the driver will potentially save fuel and reduce greenhouse gas emissions for that trip by choosing the most fuel-efficient route.

The green-routing algorithms find the most fuel-efficient route for an origin and destination (OD) pair (*6*, *7*). The fuel-saving opportunities of green-routing strategies have been extensively studied recently. A comprehensive performance study of current eco-routing methods showed that an average saving of 12.5% could be achieved under ideal assumptions (*8*). A study in Lund, Sweden, on potential reductions of fuel consumption and carbon dioxide emissions through an eco-routing navigation system (*9*) was applied to 109 trips. The results indicated an average 8.2% fuel saving by choosing the greenest route. A navigation service, "GreenGPS," uses participatory sensing data that allow

[1]National Renewable Energy Laboratory, Golden, CO

Corresponding Author:
Address correspondence to Lei Zhu: Lei.Zhu@nrel.gov

the driver to find the most fuel-efficient route (*6*), which can save about 10% of fuel. A similar tool, Eco-Routing Navigation System developed by researchers at the University of California at Riverside (*10*), requires an extensive traffic database and accurate model inputs. A green-routing environmental benefit evaluation study in the greater Buffalo–Niagara metropolitan area, New York (*11*), proposed a "green-user equilibrium" concept. The TRansportation ANalysis SIMulation System (TRANSIMS)–Motor Vehicle Emission Simulator (MOVES) framework carries out the potential environmental benefit analysis. These studies are applied either on a small travel data set (e.g., a small number of trips or mileages) or on simulation data to prove the concept. They are restricted by quality traffic and network data requirements and the accurate pre-trip fuel-consumption estimation model applicable to large-scale real-world travel data.

Therefore, assessing the route choice fuel-saving potential for a large-scale, real-world travel data set is not easy. First, hosting and maintaining a routing server for route choice is costly and requires detailed and quality traffic, network (*12–14*) and other data, such as global positioning system (GPS) trajectories (*10*, *13*). Second, a pre-trip fuel-consumption estimation model that accurately predicts the fuel consumption of a route before it is taken for different vehicle powertrains is desired.

To overcome the rigorous traffic and network data requirements, studies utilize the outputs of traffic simulation tools such as TRANSIMS (*11*) or Dynamic Urban Systems in Transportation (DynusT) (*15*). However, establishing a traffic simulation application is challenging due to the efforts of model calibration and the lack of computational resources. Also, their link cost attributes can be relatively simplistic in the area of energy estimation. Moreover, the simulation and traffic assignment results, either user equilibrium or system optimal, do not reflect traffic in the real world. Therefore, the fuel-saving analysis based on simulation solutions may not be accurate and persuasive. In that case, an efficient and effective way of applying real-world travel data to find the possible green route options is needed.

Routing application programming interfaces (APIs), such as Google Maps Directions API (*16*), provide feasible route solutions for any OD pair by considering typical traffic conditions. The API offers quality routes conveniently because of the industry-level high-quality network and real-time traffic data. Although the API routes may not be exhaustive in presenting all possible routes for an OD pair, it is reasonable that the API routes are logical alternatives when considering both ease of following the routes and travel time. They may offer a fuel-saving opportunity for an actual route by comparing its estimated fuel consumption to alternative API routes.

With actual routes and API alternative routes, pre-trip fuel-consumption estimation methods evaluate actual and API routes' fuel consumption based on the correlation of fuel consumption and influencing factors about trips, vehicles, and drivers. Macroscopic models assume experimentally fixed fuel-consumption rate values for particular powertrain models. Microscopic-level vehicle models, such as the Future Automotive Systems Technology Simulator (FASTSim) (*17*) and Autonomie (*18*), consider vehicle driving and road details to provide a fuel-consumption estimation. The mesoscopic-level model is a compromise between macroscopic and microscopic models. Mesoscopic-level models do not need complete driving cycles and consider various fuel economy impact factors, such as traffic conditions, trip road features, offering acceptable estimation results. The present mesoscopic-level studies (*19–21*) mainly rely on the average speed of a trip to estimate the trip fuel-consumption rate. Except for aggregated traffic conditions, the mesoscopic fuel-consumption estimation models also need a vehicle's powertrain information. A hybrid electric vehicle (HEV) is a type of vehicle that "combines a conventional internal combustion engine system with an electric propulsion system" (*22*). The partially electric powertrain can achieve better fuel economy than a conventional vehicle (CV).

This study applies the enhanced pre-trip fuel-consumption estimation models for CVs and HEVs. The model was trained and developed using millions of driving cycle GPS trajectory point data and considering the route traffic conditions, functional class, and road grade factors. The advantages of the models are that they do not need complete trip drive cycles and can provide accurate fuel-consumption rate estimations for both powertrains for a route that has not yet been driven.

In that case, a fuel-saving opportunity assessment framework using a routing API and the pre-trip fuel-consumption rate estimation models for large-scale, real-world travel data is proposed. The framework has the capability to quantify fuel-saving opportunities for a large-scale, real-world travel data set by comparing the API routes to the actual routes for all OD pairs, which is enabled by the automated similarity comparison between the actual route GPS trajectories and the alternative API routes. Meanwhile, the trade-off between fuel saving and travel-time saving is also investigated. The contributions of the proposed route choice fuel-saving assessment method include but are not limited to:

- Quantifying and analyzing potential fuel consumption and savings in a large-scale, real-world travel data set for both CVs and HEVs for the first time.

- The API (Google Directions) method provides alternative route options and is easy to implement and compatible with any cities that have travel data.
- The enhanced pre-trip fuel-consumption rate estimation models for CVs and HEVs evaluate the route fuel use and enable pre-trip green route selection.

Methodology

The proposed framework consists of several modules:

(1) Initialization
(2) API query
(3) Map matching
(4) Route fuel-consumption estimation
(5) Route similarity assessment
(6) Fuel saving analysis

Initialization

An initialization procedure deals with the raw GPS trajectories and abstracts the real driving routes from a large-scale, real-world travel data set (*23*, *24*) generated by GPS-instrumented vehicles within a region. A preprocessing procedure is used to cleanse raw GPS data (*25*) to remove the stationary points and outliers. The actual routes are segregated from the raw GPS trajectories by the time and distance gaps in the GPS data. Meanwhile, the actual route OD locations are extracted from the first and last GPS point coordinates of the route. The trip departure time is read directly from the timestamp of the origin GPS point. The actual route point-based speed sequence (speed profile) is used as a proxy to indicate the route's traffic condition (*26*). The actual route length is computed by summing the coordinate distances among all consecutive GPS point pairs.

API Query

The navigation API (e.g., Google Directions API) provides routes for a specific OD pair and departure time. Google Directions API responds with the route topology, length, and duration in traffic. The traffic model of Google Directions API is obtained by providing an additional input parameter, "departure_time," in the API request URL. Then, the API returns the "duration in traffic" feature in the response. The "duration in traffic" feature is the route travel time under traffic conditions at a typical future time with the same time-of-day and day-of-week information as the actual route. Although the typical future time is not the historical time when the actual route occurred, the typical traffic conditions at the time with the similar temporal feature (e.g., time of day, day of week) as the actual route time is still meaningful in general cases (except for accidents or unexpected events). The API route is composed of a sequence of shape nodes with coordinates.

However, the API cannot offer the route speed profile, which reflects detailed driving operation and traffic conditions. Nevertheless, the segmented traffic data are considered as a surrogate for the speed profile. First, the API query obtains multiple consecutive route segments (legs). Then, the second-level API query is conducted to fetch the polyline (i.e., shape node sequence), duration in traffic, and distance features for each route segment. The average speed of the segment is calculated by the distance over the duration on the segment, and then the speed is assigned to each shape node on the segment. After that, the combination of the ordered shape nodes of all segments form the route topology; the point-based speed sequences of all segments compose the route speed profile.

Map Matching

The map-matching procedure (*25*, *27*, *28*) matches all routes onto a common road network (*13*, *21*) to procure the additional road attributes for the routes according to route location. After map matching, each GPS point in the actual routes or each shape node in the API routes is associated with a link identification, which can be used to retrieve link functional class and elevation (grade). The link functional class and road grade are used for estimating route-based fuel consumption, which will be explained in the fuel estimation model.

Route Fuel Consumption Estimation

With detailed road link information, the enhanced fuel-consumption rate estimation model estimates the route fuel usage for CVs and HEVs. The model uses link-level average speed, road grade, and a local and non-local road classifier based on functional road class, rather than detailed second-by-second speed profiles, as input to provide accurate energy estimates for one route. The FASTSim model (*17*) is used in conjunction with real-world driving data, made available by the Transportation Secure Data Center at the National Renewable Energy Laboratory (*23*), to obtain FASTSim-estimated fuel economy data as the ground truth for nearly 1 million miles of driving in the United States. The driving data are used as the feedstock for developing a trip-level fuel estimation model, which does not require second-by-second drive cycles. The raw GPS points may cause drive-cycle profiles to be noisy, so the drive cycles are cleansed and filtered via the National Renewable Energy Laboratory's standard processing

(25) to make them more suitable for the powertrain simulation model. As a part of the processing, the U.S. Geological Survey Digital Elevation Model is used to append road grade to the drive-cycle data, and the GPS points are matched to a road network with the map-matching procedure.

FASTSim is run for all cleansed drive cycles to determine the second-by-second fuel consumption for the drive cycle. The FASTSim fuel-consumption results for the HEV and CV powertrain models are taken to be the ground truth for the model development. Since FASTSim returns fuel-consumption estimations for each point in the drive cycle provided as an input, the resulting resolution is too fine for the link-based energy estimation model. Therefore, the point-based FASTSim results are aggregated to the link level as the total fuel consumption on the link, defined as the sum of all point-based FASTSim fuel-consumption estimations on the link. Therefore, the link-level results include total fuel consumption, functional class (local [such as urban major/minor arterials, local streets, etc.] or non-local [such as freeways]), road grade, and average speed of travel. The functional class attribute is directly obtained from the road network attributes. The road grade is computed as a ratio of "rise" to "run," in which run is the horizontal distance and rise is the vertical distance—the result is taken as a percentage. The average speed of the link is calculated by averaging all speed points on the link.

The link-based result data are grouped into "bins" by average link speed, functional class, and road grade. Average fuel-consumption rates for each bin are calculated from the link-level FASTSim results, and this generates the estimation model. The model is a three-dimensional look-up table. Figure 1*a* shows the fuel-consumption rate (gallons per 100 miles) for each speed and functional class bin in the model. Figure 1*b* has fuel-consumption rates as a function of road grade and functional class. Each figure shows the results for the HEV and CV powertrains on local and non-local road links. Overall, the accuracy of the model is computed from the modeled to the ground-truth error in estimated fuel consumption (gallons) for a trip. The normalized averaged absolute error for all trips used to generate this model is approximately 5.7%. The fuel economy of HEVs outperforms that of CVs in all situations. In particular, Figure 1*a* indicates that the HEV has more fuel-use benefits on low-speed ranges, partially due to the HEV's fuel-efficient performance on urban streets. For example, when cruising at low speed or stopping at a traffic signal, an HEV is likely to be under full electric power and thus has a better fuel economy.

The fuel-consumption rate of a route comprised of links can be estimated by looking at the model tables through the link average speed, functional class, and road grade attributes. Link fuel consumption in gallons is computed by the link length (in miles) multiplied by the fuel consumption rate. The fuel consumption for the entire route is the sum of all links' fuel-consumption quantities on the route.

Route Similarity Assessment

Given the routes' estimated fuel consumption, if a route matches one of the API routes, the fuel consumptions of the matched API route and the actual route are assumed to be the same. Before studying the fuel-saving opportunity, the similarity relationship of the actual routes, the API routes, and the green route have to be examined. The route similarity is described by a longest common sub-sequence (LCS)-based similarity score (ranging between 0 and 1) (*25*, *29*, *30*). The LCS similarity score usually represents the overlapping level of two trajectories. The LCS model can match two sequences by allowing them to stretch without changing the order of elements in the sequences but allowing some parts to be unmatched (*29*). In that case, an API route matched to the actual route can be found when the API route similarity score is the maximum one among all alternative API routes and is greater than a predefined similarity threshold (for example, maximum score $>$ threshold of 0.7).

The predefined similarity score threshold affects the performance of route matching. For a large threshold, such as 1, the tightest constraint makes it very hard to find matching cases. A small threshold, such as 0.1, implies that almost all actual routes can find a matched API route. Neither case is favorable for the fuel-saving study in different route similarity situations. The similarity score threshold of 0.7 used in this study is an empirically derived value, obtained from a route-matching study in the literature (*25*) to define a reasonable route-matching case.

Fuel Saving Analysis

In the fuel-saving analysis module, the actual routes are first separated as "follow API" and "non-follow API," according to whether a route matches one of the API routes. For the "follow API" group, the estimated fuel consumption of the actual route is replaced by the fuel-consumption estimation of the matched API route. The reason for doing that is that the two matched routes are not fully identical due to the similarity threshold setting. A small discrepancy between two matched routes (the actual route and its corresponding API route, as "a matched route pair") leads to a fuel-consumption difference by using a common fuel estimation model.

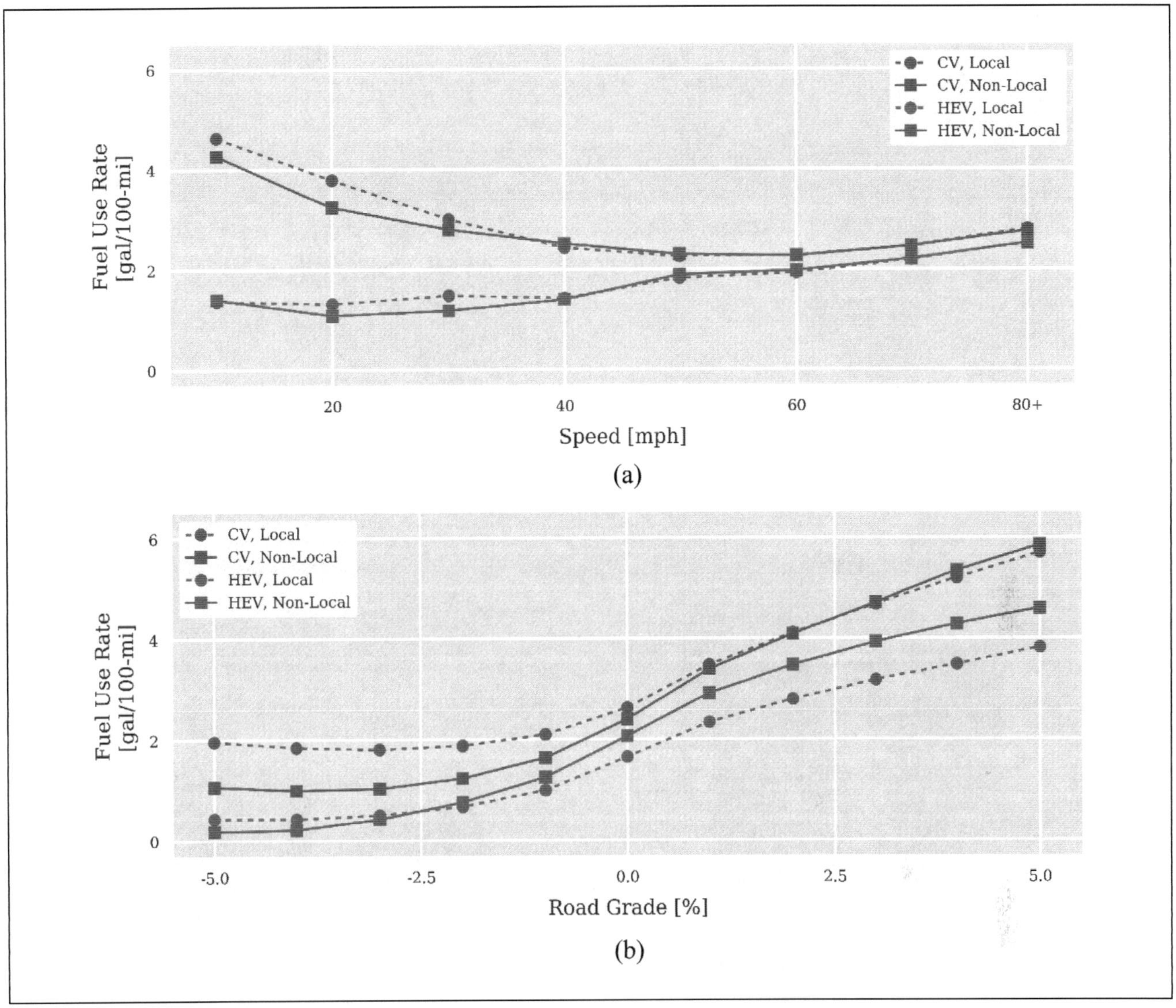

Figure 1. Fuel-consumption estimation model: (*a*) fuel-consumption rate as a function of link average speed and functional class; and (*b*) fuel-consumption rate as a function of road grade and functional class.

Therefore, it is better to choose a comparable route from the matched route pair to compare with other API routes. Since all API routes share the same Google routing engine, the matched API route is more comparable to other API routes in light of path choice logic. Using the matched API route instead of the actual route to estimate fuel saving, the adverse impact of the fuel-consumption difference between the actual route and the matched API route might be diminished in the fuel comparison procedure.

If the fuel consumption of the matched API route is higher than any other API route, the route is not the green one. Thus there is fuel-saving potential, and the actual route is categorized as having an "API saving route." Otherwise, it is denoted as "API greenest route."

For the "non-follow API" group, the estimated fuel consumption of the actual route and the API routes are compared directly. If the fuel consumption of the actual route is less than or equal to (outperforms) that of all API routes, the actual route is called "Actual outperform route." Otherwise, the actual route has fuel-saving potential, and the actual route is defined as "Actual saving route." Ultimately, the actual routes are divided into four groups: *(1) API greenest route, (2) API saving route, (3) Actual saving route*, and *(4) Actual outperform route*.

To understand the total fuel saving, the cumulative fuel consumption of those four groups is calculated. According to the definitions, the fuel-saving amount is derived from the fuel-consumption differences of the actual routes and their green API routes for the potential

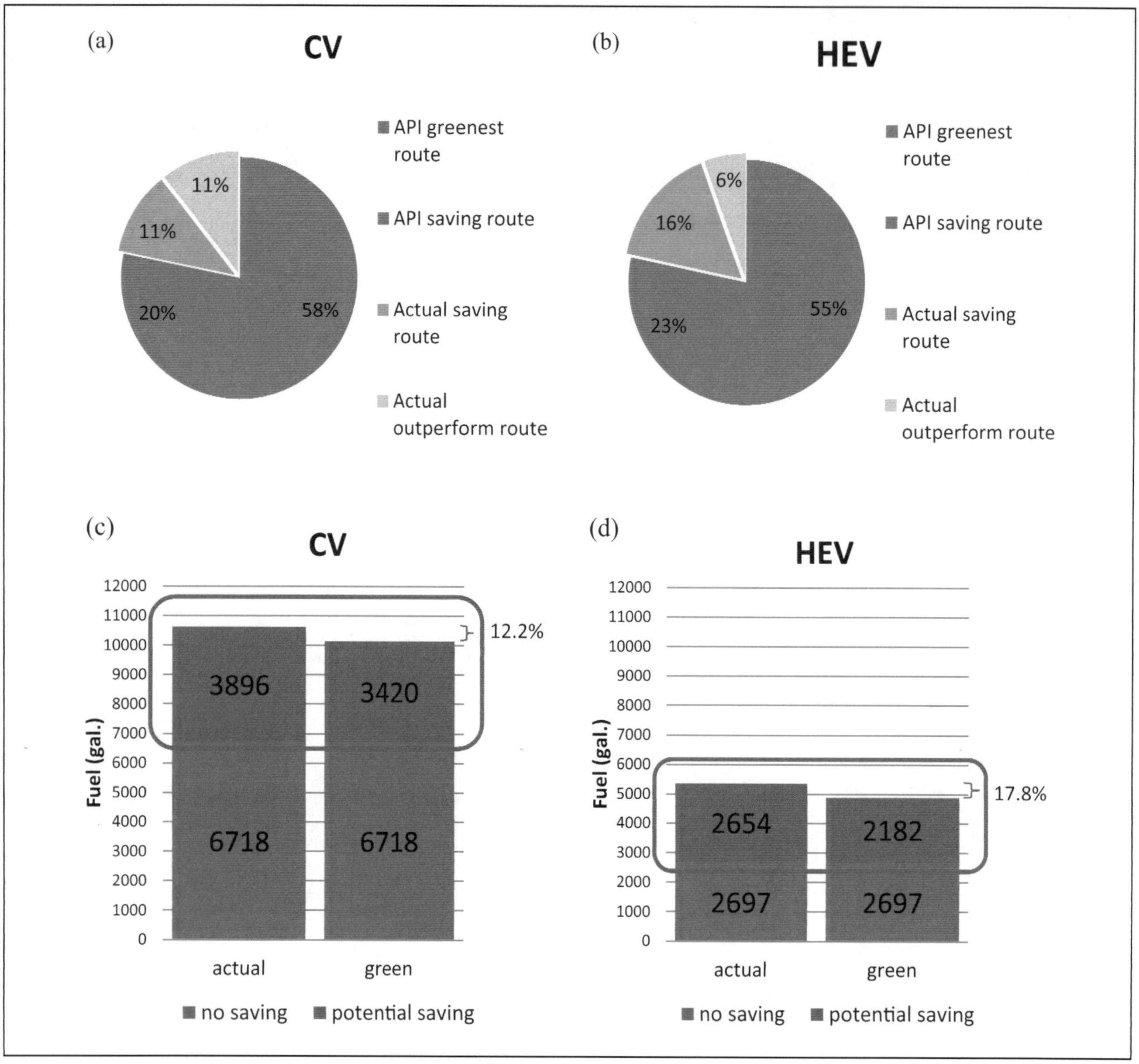

Figure 2. Ratio distribution of actual routes for: (*a*) CVs; and (*b*) HEVs; and cumulative fuel consumption for actual routes for: (*c*) CVs; and (*d*) HEVs.

fuel-saving groups, including (2) *API saving route*, and (3) *Actual saving route*.

The fuel-consumption and saving patterns of the two powertrain models—CV and HEV—are investigated respectively. In addition to the quantitative study of fuel savings, the fuel-saving and travel-time relationship study reveals the changes of fuel saving and travel time for specific potential fuel-saving routes.

Experiment and Results Discussion

The California travel data set in the Transportation Secure Data Center is used in the experiment. The data set has 44,805 OD pairs containing 4,265,064 GPS points, which are extracted from 111,096 miles of driving by Californian travelers from 2010 to 2012. There are 100,031 API routes procured by the Google Directions API queries. Using TomTom MultiNet as the underlying road network layer and U.S. Geological Survey elevation data for all routes in the map-matching procedure, the additional link attributes are appended to the routes.

The framework processing efficiency was assessed. The computational time for processing one OD pair is impacted by many factors, such as route length and internet speed. The API query takes on average about 1 second for an OD pair. The map-matching module takes about 8 seconds for an average length trip. One OD pair may have two to three API routes and one actual route. So, the map matching may take about 28 seconds. And, considering the time costs of other modules, the overall computational time for computing a green route and comparing its fuel consumption will take about 30 seconds.

Fuel Saving Quantitative Analysis

Overall Actual Route Ratio Distribution. The ratio distribution of the four actual route groups for CVs and HEVs is illustrated in Figure 2. From the figure, similar distribution patterns for CVs and HEVs are observed. In Figure 2*a*, for CVs, 31% of actual routes have fuel-saving potential (all blue slices) while 69% of actual routes (all green slices) do not. In Figure 2*b*, for HEVs, 39% of actual routes have fuel-saving potential while 61% of actual routes do not. In that case, HEVs have a larger number of potential fuel-saving routes (39%) than CVs have (31%).

For the subset of cases that follow API routes (dark blue and dark green slices, 78% of all actual routes), most of them (58% for CVs and 55% for HEVs) followed the greenest API route so do not have fuel-saving potential.

Cumulative Fuel Consumption and Saving. The cumulative fuel consumption of potential fuel-saving routes and no fuel-saving routes for CVs and HEVs are demonstrated in Figure 2*c* and *d*. The columns marked as "actual" and "green" denote the actual routes and their corresponding green routes (either actual route itself or its greener API route). The red bars in the "actual" column represent the cumulative estimated fuel consumption of potential fuel-saving actual routes. The red bars in the "green" column illustrate the cumulative fuel consumption of the green routes corresponding to the potential fuel-saving actual routes. The blue bars in the "actual" columns indicate the cumulative fuel usage of no-fuel-saving actual routes, which is identical to that of the corresponding green routes, marked with the blue bars in the "green" columns.

The cumulative fuel-consumption difference between CVs and HEVs is significant, and HEVs always outperform CVs. For the actual routes' fuel-consumption perspective, HEV cumulative fuel consumption is roughly half of that for CVs. Thus, considerable fuel-saving potential exists for HEVs compared with CVs, though this requires vehicle switching which is more costly and complicated than route switching.

For the CV case, the total potential fuel saving derived from the fuel-use difference between actual routes and their green routes is 476 gallons (3,896–3,420 gallons), which is 12.2% of the cumulative fuel consumption of the actual routes with potential fuel-saving. For the HEV case, the total fuel-saving amount is 472 gallons, which accounts for 17.8% of the potential fuel-saving actual routes' cumulative fuel consumption. Although the absolute fuel-saving amounts of CVs and HEVs are very close, the fuel-saving ratio of the HEVs is higher in this analysis.

The percentages of the local and non-local roads (functional class attribute) on an actual route may impact the estimated fuel consumption and fuel saving. For CVs, 84% of actual routes with fuel-saving potential are on local roads, and 16% of actual routes with fuel-saving potential are on non-local roads (i.e., freeway). For HEVs, 78% of the potential fuel-saving routes are on local roads, and 22% of the potential fuel-saving routes are on non-local roads. Considering the comparable ratios of potential fuel-saving actual routes for two powertrains (HEV—39% vs. CV—31%), the non-local routes are more likely to have fuel-saving potential for HEVs compared with that for CVs.

Trade-Off between Fuel Saving and Travel Time

For the actual routes with fuel-saving potential, the time penalties or savings of green alternative routes are uncertain. Scatter diagrams in Figure 3 illustrate fuel-saving and travel-time differences of potential fuel-saving actual routes: Figure 3*a* for CVs, and 3*b* for HEVs. Each actual route is represented as a colored dot, and the color of the dot describes the *actual route duration* in seconds, which ranges from 15 to 10,800 seconds (3 hours). The actual route duration minus the green route duration yields the *travel-time difference*.

The actual routes are separated into two classes according to time savings and penalties. The actual routes with both time and fuel savings are the *more desirable routes*, which are located above the zero-time difference line (dashed line). The colored dots located below the zero-time difference line with fuel saving and without travel-time saving represent *less desirable routes*. The statistics of more desirable and less desirable routes for CVs and HEVs are illustrated in Table 1. The fuel saving is described by the fuel-saving amount in gallons and as a percentage, calculated by the fuel-saving amount of the category over the total fuel saving amount. From the table, HEVs have fewer more desirable routes (as compared with CVs) that save both fuel and travel time. For the less desirable routes category, it follows that the HEV case has more actual routes than the CV case with time penalties (negative time savings) that accompany the fuel savings. These differences are also visible from the point scattering trend of Figure 3*a* and *b*. For the CV case, more dots are located above the zero-time

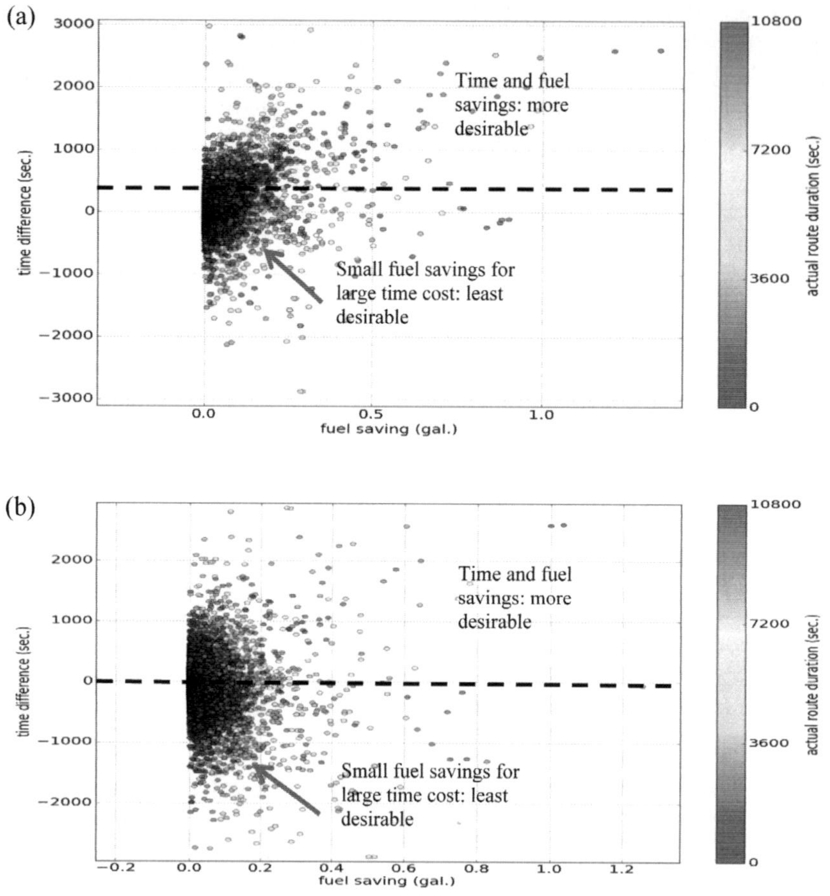

Figure 3. Fuel saving vs. time difference for potential fuel-saving actual routes for: (*a*) CVs; and (*b*) HEVs.

Table 1. Statistical Summary of Most Desirable and Less Desirable Routes

	Most desirable routes			Less desirable routes		
	# of actual routes	Fuel saving (gal./%)	Time-saving (hours)	# of actual routes	Fuel saving (gal./%)	Time-saving (hours)
CV	6,831	315/66.2%	438	7,059	161/33.8%	−263
HEV	6,005	170/36.0%	347	11,397	302/64.0%	−712

difference line, while for the HEV case more dots are below the zero line.

Table 2 summarizes statistics of actual route attributes for both the CV and HEV, including route duration (seconds), length (miles), and road classification (local/non-local), broken out between the more desirable and less desirable route designations. For the CV cases, the mean values of length and duration for the more desirable route group are statistically higher than those of the less desirable route group. On the other hand, the splits between local and non-local road classes are very similar between the more and less desirable route options for the CV—in both cases, the number of local routes is significantly larger than the number of non-local routes.

For the HEVs, the duration attribute has a similar pattern to the CV case, while the pattern for length flips (the mean length value of the most desirable route group is smaller than that of the less desirable route group). This implies that green routes that save both fuel and travel time for an HEV are more likely to be found for shorter routes with relatively longer travel times. The road classification attribute again shows more green-

Table 2. Statistical Summary of Route Attributes for CVs and HEVs

	Mean	Std.	*p*-value	*t*-value
CV				
Length				
Most desirable	8.32	14.23	.01[a]	2.62
Less desirable	7.64	16.10	–	–
Duration				
Most desirable	1,079.14	1,202.44	.0[a]	16.80
Less desirable	763.10	1,005.16	–	–
Functional class	# of local	# of non-local	Local%	Non-local%
Most desirable	5,833	1,018	85%	15%
Less desirable	5,873	1,206	83%	17%
HEV	Mean	Std.	*p*-value	*t*-value
Length				
Most desirable	7.97	13.78	.00[a]	−4.47
Less desirable	8.97	14.41	–	–
Duration				
Most desirable	1,074.55	1,171.95	.00[a]	9.74
Less desirable	903.10	962.36	–	–
Functional class	# of local	# of non-local	Local%	Non-local%
Most desirable	5,173	832	86%	14%
Less desirable	8,470	2,927	74%	26%

[a]Significance level of 0.05.

routing opportunities overall for local roads than non-local roads, though the percentage skews higher toward non-local roads for the less desirable HEV green route options.

Conclusions

The proposed navigation API route fuel-saving opportunity assessment framework provides a feasible way to assess the potential fuel saving for a large-scale, real-world travel data set for CVs and HEVs. The API methods do not need rigorous network and traffic data to provide the possible green routes for OD pairs in the target research area. The framework using large-scale, real-world travel data rather than simulation-based or small-scale travel data can better reflect the traffic patterns, and the estimated aggregate fuel-saving results of route choice are more reliable and convincing.

In the experimental data set of 44,805 OD pairs, 31% of actual routes for CVs and 39% of actual routes for HEVs show opportunities for fuel savings by choosing a different route. Of those actual routes, the total estimated fuel saving is 476 gallons for the CV case, which is 12.2% of potential fuel-saving of actual routes' fuel consumption. For the HEV case, the overall fuel-saving estimation is very similar (472 gallons), and accounts for 17.8%.

Due to the overall better fuel efficiency performance of HEVs, if HEVs drive all the routes, the total fuel consumption could be cut in half relative to the CV cases, although this would require full vehicle replacement rather than simply alternative route selection. From the perspective of fuel vs. travel-time trade-offs, the analysis indicates that a given green route alternative for a CV is more likely to save time as well as fuel, whereas a given HEV green route is more likely to trade off an increase in travel time against the decrease in fuel consumption.

The proposed framework is transferable and can be applied to any city with real-world travel data and road networks. The findings are promising and convincing, thanks to the use of the large-scale, real-world GPS trajectory data.

Acknowledgments

This report and the work described were sponsored by the U.S. Department of Energy (DOE) Vehicle Technologies Office (VTO) under the Systems and Modeling for Accelerated Research in Transportation (SMART) Mobility Laboratory Consortium, an initiative of the Energy Efficient Mobility Systems (EEMS) Program. The authors acknowledge Eric Rask of Argonne National Laboratory for leading the Connected and Automated Vehicles Pillar of the SMART Mobility Laboratory Consortium. David Anderson of the DOE Office of Energy Efficiency and Renewable Energy (EERE) provided important support in establishing the project concept, advancing implementation, and giving ongoing guidance.

Author Contributions

The authors confirm contribution to the paper as follows: study conception and design: Lei Zhu, Jeffrey D. Gonder; data collection: Lei Zhu; analysis and interpretation of results: Lei Zhu, Jacob R. Holden; draft manuscript preparation: Lei Zhu, Jacob

R. Holden, Jeffrey D. Gonder. All authors reviewed the results and approved the final version of the manuscript.

References

1. Ahn, K., and H. Rakha. The Effects of Route Choice Decisions on Vehicle Energy Consumption and Emissions. *Transportation Research Part D: Transport and Environment*, Vol. 13, No. 3, 2008, pp. 151–167. https://doi.org/10.1016/j.trd.2008.01.005
2. Manzie, C., H. Watson, and S. Halgamuge. Fuel Economy Improvements for Urban Driving: Hybrid vs. Intelligent Vehicles. *Transportation Research Part C: Emerging Technologies*, Vol. 15, No. 1, 2007, pp. 1–16. https://doi.org/10.1016/j.trc.2006.11.003
3. Boriboonsomsin, K., A. Vu, and M. Barth. *Eco-Driving: Pilot Evaluation of Driving Behavior Changes Among U.S. Drivers*. UCTC-FR-2010-20, University of California Transportation Center, Berkeley, CA, 2010.
4. Hu, X., Y.-C. Chiu, and L. Zhu. Behavior Insights for an Incentive-Based Active Demand Management Platform. *International Journal of Transportation Science and Technology*, Vol. 4, No. 2, 2015, pp. 119–133. https://doi.org/10.1260/2046-0430.4.2.119
5. Nie, Y., and Q. Li. An Eco-Routing Model Considering Microscopic Vehicle Operating Conditions. *Transportation Research Part B: Methodological*, Vol. 55, 2013, pp. 154–170. https://doi.org/10.1016/j.trb.2013.06.004
6. Ganti, R. K., N. Pham, H. Ahmadi, S. Nangia, and T. F. Abdelzaher. GreenGPS: A Participatory Sensing Fuel-Efficient Maps Application. *Proc., 8th International Conference on Mobile Systems, Applications, and Services*, Association for Computing Machinery, New York City, NY, 2010.
7. L. Zhu, J. Holden, E. Wood and J. Gender. Green Routing Fuel Saving Opportunity Assessment: A Case Study using Large-Scale Real-World Travel Data. *2017 IEEE Intelligent Vehicles Symposium (IV)*, Los Angeles, CA, 2017, pp. 1242–1248.
8. M. Kubiĉka, J. Klusacck, A. Sciaretta, A. Cela, H. Mounier, L. Thibault, and S.I. Niculeseu. Performance of Current Eco-Routing Methods. *2016 IEEE Intelligent Vehicles Symposium (IV)*, Gothenburg, Sweden, 2016, pp. 472–477.
9. Ericsson, E., H. Larsson, and K. Brundell-Freij. Optimizing Route Choice for Lowest Fuel Consumption–Potential Effects of a New Driver Support Tool. *Transportation Research Part C: Emerging Technologies*, Vol. 14, No. 6, 2006, pp. 369–383. https://doi.org/10.1016/j.trc.2006.10.001
10. Boriboonsomsin, K., M. J. Barth, W. Zhu, and A. Vu. Eco-Routing Navigation System Based on Multisource Historical and Real-Time Traffic Information. *IEEE Transactions on Intelligent Transportation Systems*, Vol. 13, No. 4, 2012, pp. 1694–1704. https://doi.org/10.1109/TITS.2012.2204051
11. Guo, L., S. Huang, and A.W. Sadek. An Evaluation of Environmental Benefits of Time-Dependent Green Routing in the Greater Buffalo–Niagara Region. *Journal of Intelligent Transportation Systems*, Vol. 17, No. 1, 2013, pp. 18–30. https://doi.org/10.1080/15472450.2012.704336
12. Zhu, L., Y.-C. Chiu, and Y. Chen. *Traffic Analysis Network Abstraction Approach: Framework and Numerical Analysis*. Transportation Research Board, Washington, D.C., 2017. https://doi.org/10.1049/iet-its.2016.0287
13. Zhu, L., and Y.-C. Chiu. Transportation Routing Map Abstraction Approach: Algorithm and Numerical Analysis. *Transportation Research Record: Journal of the Transportation Research Board*, 2015. 2528: 78–85.
14. Zhu, L. *Routing Map Topology Analysis and Application*. Dissertation. University of Arizona, Tuscon, AZ, 2014.
15. Lin, J., Y.-C. Chiu, S. Vallamsundar, and S. Bai. Integration of MOVES and Dynamic Traffic Assignment Models for Fine-Grained Transportation and Air Quality Analyses. *Proc., 2011 IEEE Forum on Integrated and Sustainable Transportation Systems (FISTS)*, IEEE, Vienna, Austria, 2011.
16. Svennerberg, G. *Beginning Google Maps API 3*. Apress, New York city, NY, 2010.
17. Brooker, A., J. Gonder, L. Wang, E. Wood, S. Lopp, and L. Ramroth. *FASTSim: A Model to Estimate Vehicle Efficiency, Cost and Performance*. SAE Technical Paper 2015-01-0973, 2015. https://doi.org/10.4271/2015-01-0973
18. Halbach, S., P. Sharer, S. Pagerit, A. P. Rousseau, and C. Folkerts. *Model Architecture, Methods, and Interfaces for Efficient Math-Based Design and Simulation of Automotive Control Systems*. SAE Technical Paper 2010-01-0241, 2010. https://doi.org/10.4271/2010-01-0241
19. Barth, M., F. An, J. Norbeck, and M. Ross. Modal Emissions Modeling: A Physical Approach. *Transportation Research Record: Journal of the Transportation Research Board*, 1996. 1520: 81–88.
20. Chen, Y., L. Zhu, J. Gonder, S. Young, and K. Walkowicz. Data-Driven Fuel Consumption Estimation: A Multivariate Adaptive Regression Spline Approach. *Transportation Research Part C: Emerging Technologies*, Vol. 83, 2017, pp. 134–145. https://doi.org/10.1016/j.trc.2017.08.003
21. Zhu, L., Y.-C. Chiu, and Y. Chen. Road Network Abstraction Approach for Traffic Analysis: Framework and Numerical Analysis. *IET Intelligent Transport Systems*, Vol. 11, No. 7, 2017, pp. 424–430. https://doi.org/10.1049/iet-its.2016.0287
22. Hybrid Electric Vehicle. *Wikipedia*, 2017. https://en.wikipedia.org/w/index.php?title=Hybrid_electric_vehicle&oldid=789980728.
23. Zhu, L., J. Gonder, and L. Lin. Prediction of Individual Social-Demographic Role Based on Travel Behavior Variability Using Long-Term GPS Data. *Journal of Advanced Transportation*, Vol. 11, No. 7, 2017, pp. 424–430. https://doi.org/10.1155/2017/7290248
24. Holden, J., E. W. Wood, L. Zhu, and J. D. Gonder. *Development of a Trip Energy Estimation Model Using Real-World Global Positioning System Driving Data: Preprint*. National Renewable Energy Laboratory (NREL), Golden, CO, 2017.
25. Zhu, L., J. R. Holden, and J. D. Gonder. Trajectory Segmentation Map-Matching Approach for Large-Scale, High-Resolution GPS Data. *Transportation Research*

Record: Journal of the Transportation Research Board, 2017. 2645: 67–75.

26. Hu, X., Y.-C. Chiu, Y.-L. Ma, and L. Zhu. Studying Driving Risk Factors Using Multi-Source Mobile Computing Data. *International Journal of Transportation Science and Technology*, Vol. 4, No. 3, 2015, pp. 295–312. https://doi.org/10.1260/2046-0430.4.3.295
27. Wang, M., X. Bao, L. Zhu, and Y. L. Bao. A Map-Matching Method Using Intersection-Based Parallelogram Criterion. *Advanced Materials Research*, Vol. 403, 2012, pp. 2746–2750. https://doi.org/10.4028/www.scientific.net/AMR.403-408.2746
28. Zhu, L., Y. Bao, S.-G. Wang, Q. Zhou, and X. Bao. Map-Matching Compatible with Junction Adjusting in Vehicle Navigation System. In *Recent Advances in Computer Science and Information Engineering*, Springer, New York City, NY, 2012, pp. 451–457.
29. Vlachos, M., G. Kollios, and D. Gunopulos. Discovering Similar Multidimensional Trajectories. *Proc., 18th International Conference on Data Engineering, 2002*, IEEE, San Jose, CA, 2002. https://doi.org/10.1109/ICDE.2002.994784
30. Kim, J., and H. S. Mahmassani. Spatial and Temporal Characterization of Travel Patterns in a Traffic Network Using Vehicle Trajectories. *Transportation Research Part C: Emerging Technologies*, Vol. 59, 2015, pp. 375–390. https://doi.org/10.1016/j.trc.2015.07.010

The Standing Committee on Transportation and Air Quality (ADC20) peer-reviewed this paper (18-05302).

Article

TRR
JOURNAL OF THE TRANSPORTATION RESEARCH BOARD

Transportation Research Record
2018, Vol. 2672(25) 150–162

DOI: 10.1177/0361198118793244
journals.sagepub.com/home/trr
SAGE

Modeling Urban Bus Fuel Consumption in Shanghai, China, Based on Localized MOVES

Wenjian Jia[1], Xiaohong Chen[1], and Xiaonian Shan[1]

Abstract

In China, urban bus energy consumption is an increasing concern due to system expansion and poor energy efficiency due to frequent stopping and starting by buses. This study develops a mesoscopic bus energy consumption model based on the U.S. Environment Protection Agency's Motor Vehicle Emission Simulator (MOVES). To localize MOVES, link operating mode distribution is calculated by bus GPS data collected from nine routes in Shanghai, China. A comparison of bus fuel economy between the U.S.A. and China is conducted to determine the model years in U.S.A. and China which have similar fuel consumption performance for buses with a certain weight. After MOVES localization, link energy consumption factors are estimated, and then the impacts of average speed, vehicle stops, acceleration, and road facility on link energy consumption factors are explored. Based on this exploration of influential variables, this study develops link-level bus energy consumption factor look-up tables for a variety of bus types. Model validation indicates that using link-level indicators to estimate bus energy consumption can achieve acceptable accuracy, and that the link type classification method can influence the accuracy of the mesoscopic bus energy consumption model. This study is useful to estimate bus energy consumption when instantaneous speed data is unavailable. This study also explores the extended application of MOVES by offering a procedure for applying MOVES to develop a bus energy consumption model in regions beyond the U.S.A.

Urban bus systems have been developing rapidly in China in recent years thanks to the priority given to public transport policy. Taking Shanghai for example, the number of bus routes, total length of routes, and length of reserved bus lanes increased by 48%, 10%, and 94%, respectively, from 2006 to 2015 (*1*). With the expansion of urban bus systems, bus energy consumption has been an increasing concern. In Shanghai, the energy consumption, PM emissions, and NOx emissions from buses account for 6.4%, 18%, and 14.5% of those from all motor vehicles respectively, though the ratio of number of buses to number of all motor vehicles is only 0.53% (*2*). Thus, reducing bus energy consumption is an important means to achieve the goal of energy saving and emissions reduction in the transport sector.

Some bus improvement measures have the potential to reduce energy consumption, such as Intelligent Transportation System (ITS) strategies (*3*, *4*) and reserved bus lanes (*5*). To evaluate the energy saving effectiveness of such improvement measures, a bus energy consumption model is needed. Over the years, numerous energy consumption models have been developed. Macroscopic models, like EMFAC (*6*), can evaluate network-wide energy consumption using aggregated network parameters, but they are not well suited for assessing the energy impacts of bus improvement measures (e.g., bus signal priority) at vehicle or trip level. In contrast, microscopic models, such as VT-Micro (*7*) and CMEM (*8*), can estimate energy consumption much more accurately, but they require instantaneous vehicle trajectory data which can be difficult to obtain.

Considering the limitations of macroscopic and microscopic models, mesoscopic models based on link-level parameters represent an acceptable alternative for estimation of bus energy consumption. To develop a mesoscopic model, researchers mainly use two kinds of energy consumption data, as shown in Table 1. The first category (*9–11*) uses real-world energy consumption data collected by Portable Emission Measurement System (PEMS) to explore the relationship between energy

[1]Key Laboratory of Road Traffic Engineering of the Ministry of Education, Tongji University, Shanghai, P. R. China

Corresponding Author:
Address correspondence to Xiaohong Chen: chenxh@tongji.edu.cn

Table 1. Studies on the Development of Mesoscopic Emissions/Energy Consumption Models

Energy data source	Researchers	Research object	Link definition	Method	Major conclusions	Limitations
Portable Emission Measurement System (PEMS)	Sonntag and Gao, 2009 (9)	Link particle number emission rates of diesel buses	The segment between intersections for the urban Farmington route	Using linear mixed models to study the relationship between operating parameters and diesel transit buses' particle emissions at the link-level	The model, which includes road type, vehicle speed, and average grade, can explain 92.9% of the variability of link-level particle number emissions.	Data collection is costly and often interrupts daily bus operations during the installation of on-board equipment
	Zhai et al., 2008 (*10*)	Link average emission rates of diesel transit buses	The segment of road between two consecutive bus stops	Categorizing road links by facility type and average link speed, then calculating the average emission rates of each road link category	Link average emission rates are sensitive to link mean speed, but not to facility type. To individual vehicle trip emissions estimates, average errors are $\pm$ 12% for CO_2 emissions.	
	Song, et al. 2013 (*11*)	Link emission and fuel consumption factor	Traveling fragment by different time interval	Using statistical techniques to fit the relationship between emission (fuel consumption) factor and average speed	The mesoscopic model can be used to evaluate the environmental effectiveness of ITS strategies with acceptable estimation accuracy.	
Output from MOVES	Alam, et al., 2016 (*12*)	Link greenhouse gas (GHG) emission factor	The road segment between two successive bus stops	Grouping road links into 25 speed categories; For each speed category, a representative driving cycle was constructed and then imported into MOVES database	Bus emissions could be largely under/overestimated when the default distributions of MOVES are used. The developed local opMode distributions embedded within MOVES can be used to estimate bus emissions when second-by-second speed data is unavailable.	Without considering the localization of bus base emission rates stored in MOVES database
	Alam et al., 2015 (*13*)	Link GHG and $PM_{2.5}$ emission factor	The road segment between two successive bus stops	Multiple linear regression model	Link average speed, grade, and acceleration affect link emissions significantly. The regression model can be used to estimate emissions when second-by-second speed data is unavailable.	

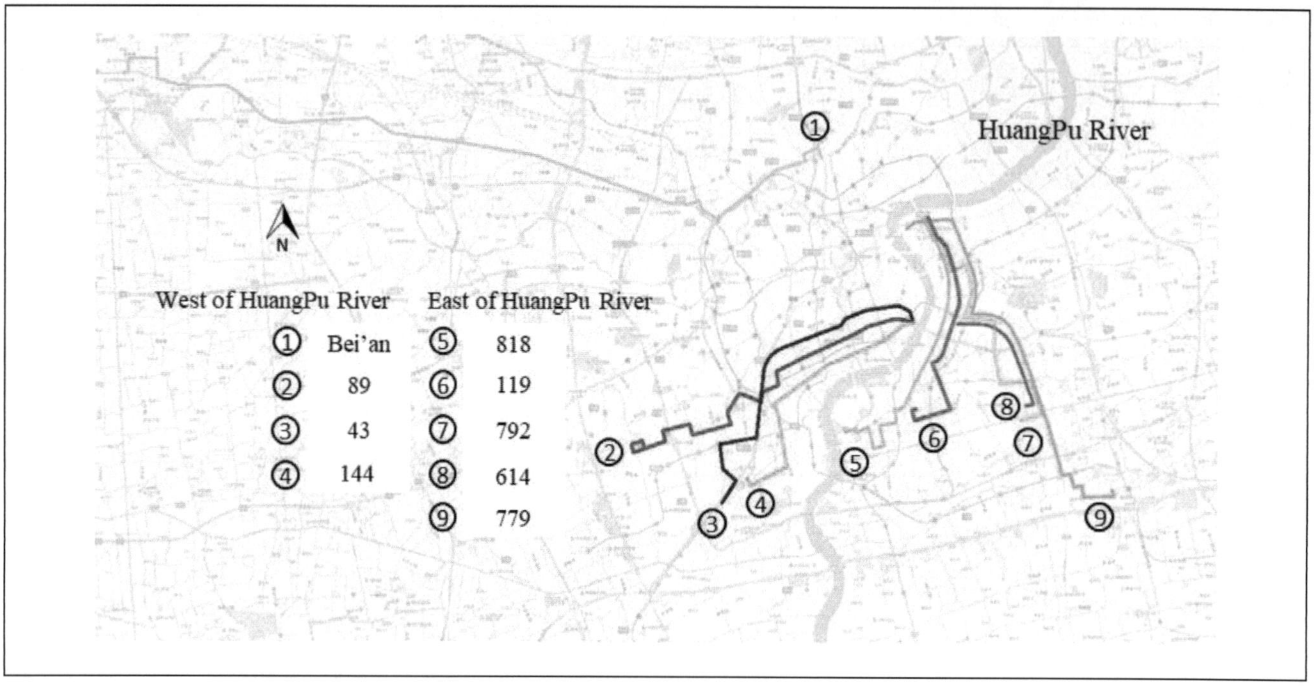

Figure 1. Selected bus routes for data collection.

consumption factor and vehicle activities on a link basis. Such data collection, however, is costly and often interrupts daily bus operation due to the installation of on-board equipment.

Instead of using PEMS, the second category (*12*, *13*) uses an established microscopic model platform (e.g. MOVES project level) to output energy consumption data. The localization of the microscopic model is vital for studies from the second category. Though some have considered the differences in bus driving characteristics in different regions, the difference in bus fuel economy is often ignored. Thus, it is necessary to propose a procedure for microscopic model localization to develop a mesoscopic energy consumption model for urban buses.

In addition to energy consumption data sources, another important issue in development of a mesoscopic model is classification of link type. In the studies listed in Table 1, a link is often categorized by average speed or facility type of the link. Other indicators, like the number of vehicle stops, are seldom discussed. Performance of these mesoscopic models varies, however, according to how the links are classified. To the authors' knowledge, few studies have systematically explored the impacts of link type classification on the accuracy of mesoscopic bus energy consumption models.

MOVES, developed by the U.S. Environment Protection Agency, is widely used for emission and fuel consumption model development application thanks to its open database structure (*14*). This study aims to: (a) estimate link energy consumption factors after MOVES localization; (b) explore the impacts of vehicle activity and road facility on link energy consumption factors; (c) develop look-up tables for link energy consumption factors for different bus types; (d) evaluate the impacts of classification of link type on the accuracy of the mesoscopic bus energy consumption model. The rest of the paper is organized as follows: the next section presents data collection, followed by a section on the model development framework. The fourth section presents MOVES localization. The development of link energy consumption factor look-up tables is presented in the fifth section, followed by model validation in the sixth section. The final part concludes this paper with recommendations.

Data Collection and Preparation

From the major bus corridors in Shanghai, nine bus routes were selected for data collection covering varied road facilities, directions, and regions. As shown in Figure 1, four routes (bei'an, 89, 43, and 144) are operated west of the HuangPu River and in an east–west direction. The other five routes (818, 119, 792, 614, and 779) are operated east of the HuangPu River and in a north–south direction. Except for routes bei'an, 89, and 119, the routes have reserved bus lanes on parts of their route segments.

After route selection, data was collected on-board in June 2016 for two parameters: (a) bus speed, instantaneously recorded by GPS software on smartphones over the entire route, and (b) the timing of bus arrival and departure from each stop, recorded by survey assistants. Parameter (b) was used to clean data through comparing time records collected by survey assistants and by GPS software. For each route, data were collected for morning peak, off-peak, and evening peak periods in both directions. Thus, a total of six trips for each route and 54 trips for all routes were collected, whereby a trip was defined from the start to the end of one route in a single direction.

Data cleaning was then conducted. Note that a link is defined as the road segment between two successive bus stops. Trip-level instantaneous speed profiles were broken down into link-level speed profiles, and each link-level speed profile consisted of the initial dwell time at the upstream bus stop and the running time before stopping at the downstream stop. Thus, link average speed was calculated based on the instantaneous speed between successive stops without the idling portion at the downstream stop. Samples were removed if: (a) GPS records were missing on that particular link, (b) the dwell time at a bus stop recorded by survey assistants did not match the record by GPS data, or (c) link average speed was lower than 1 km/h, due to a road accident or other reasons. Finally, 904 links were kept for subsequent analysis. The average length of a link was 0.65 km and the average speed was 16.73 km/h.

Method

This paper develops a link-level bus energy consumption model based on MOVES. First, localization of MOVES is conducted by: (a) calculating local link-level bus operating mode distributions; (b) comparing bus fuel economy between the U.S.A. and China to determine the corresponding base energy rates stored in the MOVES database for different bus types in China. Bus types in this paper are classified by gross vehicle weight (GVW) and model year; (c) localizing other parameters, such as fuel supply, meteorology, and road types, etc.

After localization of MOVES, link energy consumption factors are estimated. The impacts of average speed, vehicle stops, acceleration, and road facility on link energy consumption factor are then explored. Based on the exploration of such influential variables, link energy consumption factor look-up tables are developed for different bus types.

The last step is model validation. Through comparing link energy consumption factors estimated by instantaneous speed and by link-level indicators, the accuracy of using link-level variables to estimate bus energy consumption was discussed. Further, the mesoscopic bus energy consumption model developed in this paper is also validated using real-world bus fuel consumption data.

MOVES Localization

Local Link opMode Distributions

For heavy-duty vehicles, MOVES classifies vehicle operating modes into 23 opMode categories based on scaled tractive power (STP) and speed. STP is calculated by Equation 1 (*15*). After determining the opMode for each second, we calculate link opMode distribution, which describes the fractions of time that a bus has spent under each opMode category on that link. Setting bus total mass to be 16 ton in Equation 1, Figure 2 shows an example for link opMode distribution for a link with an average speed of 13.43 km/h and a length of 973 m:

$$STP = \frac{A \times v + B \times v^2 + C \times v^3 + m \times v \times (\alpha + g \sin \theta)}{f_{scale}}$$
$$A = m \times 0.0643$$
$$B = 0$$
$$C = 0.0032 + 5.06 \times 10^{-5} \times m \tag{1}$$

where

A = the rolling resistance coefficient (kW·sec/m)
B = the rotational resistance coefficient (kW·sec^2/m^2)
C = the aerodynamic drag coefficient (kW·sec^3/m^3)
m = mass of bus (metric ton)
f_{scale} = fixed mass factor (17.1)
v = instantaneous vehicle velocity (m/s)
α = instantaneous vehicle acceleration (m/s^2)
g = the acceleration due to gravity (9.8 m/s^2)
$\sin\theta$ = the (fractional) road grade

Bus Fuel Economy Comparison between U.S.A. and China

MOVES stores an inventory of base energy rates of U.S. urban buses for model years from 1960 to 2050. For model years 1960 to 2013, MOVES treats model year as having little impact on energy rates. In contrast, the base energy rates are revised for model year 2014 and later to reflect the impact of the 2014 Medium and Heavy-Duty Greenhouse Gas Rule (*16*). Besides, MOVES does not classify the energy rates by vehicle age.

Considering the difference in bus fuel economy between the U.S.A. and China, for buses with a certain weight, it is necessary to determine the corresponding model years in both countires which have similar fuel consumption performance. To achieve this purpose, adapted world transient vehicle cycle (C-WTVC) was simulated in MOVES for U.S. buses with varied mass and model year. Requirements of C-WTVC simulation environments include: temperature should be in the range of 5° C to 35° C, relative humidity should be lower than 95%, and atmospheric pressure should be in the range from 91 kPa to 104 kPa (*17*).

Simulated energy consumption values of each bus type in MOVES are then compared with the limit value from China's bus fuel consumption standards. Note that the

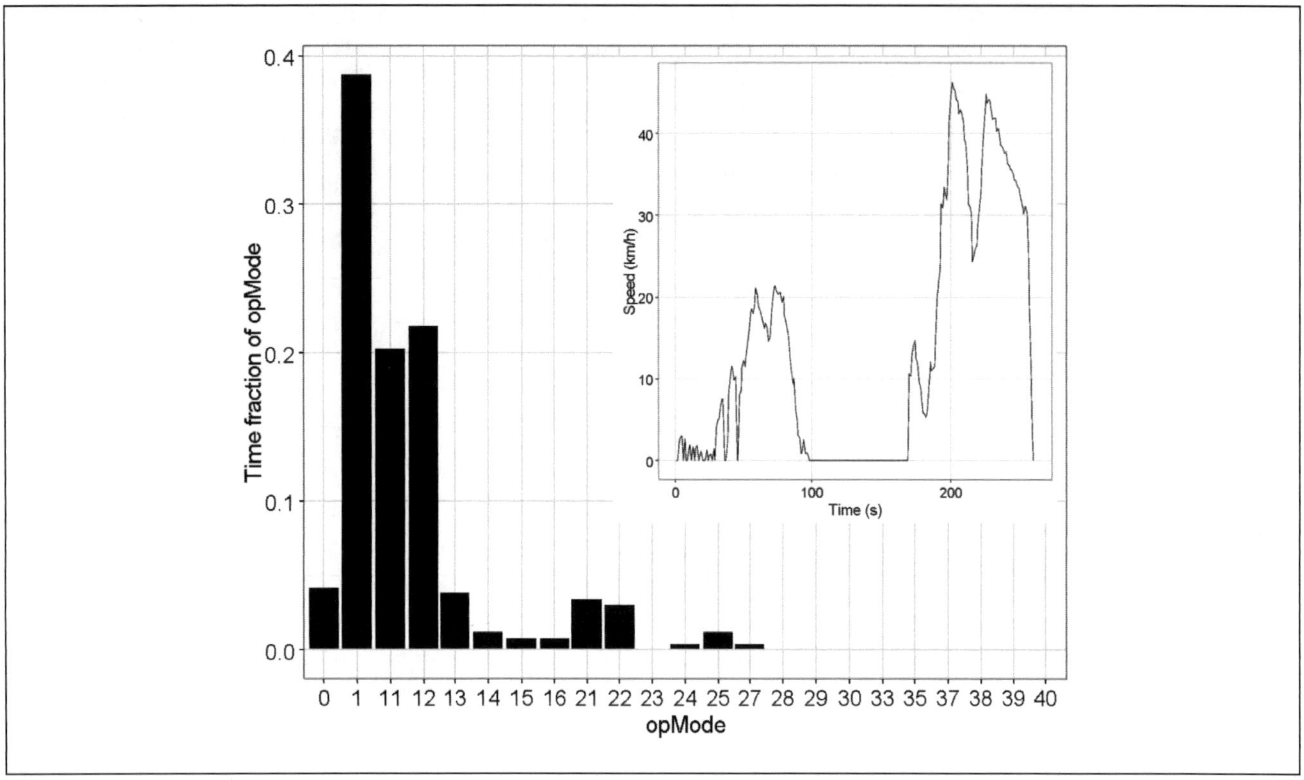

Figure 2. An example for link operating mode distribution.

limit value from China's bus fuel consumption standard is also developed based on C-WTVC (*17*). Figure 3 shows the corresponding model years in the U.S.A. and China which have the most similar fuel economy for buses with a certain mass. For example, for buses with a total mass of 17 ton, the base energy rates of a 2014 model-year bus in China are equivalent to a model-year 2017 bus in the U.S.A. Besides, since China has no bus fuel consumption limit standard for buses with model years prior to 2014, we assume their base energy rates are similar to model-year 1960 to 2013 buses in the U.S.A.

Other Parameters

In addition to bus operating mode distribution and vehicle information (bus total mass and model year), MOVES also requires: (a) fuel supply including its type and formulation; (b) meteorology including temperature and relative humidity; (c) link length. Based on the sulfur content regulated by local standard on diesel fuel in Shanghai (*18*), we set FuelTypeID = 2, FuelSubtypeID = 20, and fuelFormulationID = 20 (the sulfur content for this fuelFormulationID in MOVES is closest to the sulfur content for diesel in Shanghai). Since MOVES does not consider the impact of ambient temperature on energy consumption, we set a normal temperature of 24° C. Furthermore, link length was calculated by GPS data. Finally, road type was set to be "urban unrestricted road" (*19*).

Link Energy Consumption Factor Modeling

After MOVES localization, link energy consumption of different bus types was estimated. Link energy consumption factor was then calculated by Equation 2:

$$LinkFactor_{ij} = \frac{LinkEnergy_{ij}}{LinkLen_j} \quad (2)$$

where

$LinkFactor_{ij}$ = link energy consumption factor for bus type i on link j (10^4 kJ/km)

$LinkEnergy_{ij}$ = link energy consumption for bus type i on link j (10^4 Kj)

$LinkLen_j$ = length for link j (km)

Taking a 16-ton, 2012 model-year bus as an example, the Influential Variables Analysis section below presents the relationship between link energy consumption factor and link average speed, number of stops per kilometer, average positive acceleration, and road facility type.

Influential Variables Analysis

Link Average Speed. Figure 4 shows that link average speed has a non-linear relationship to link energy consumption

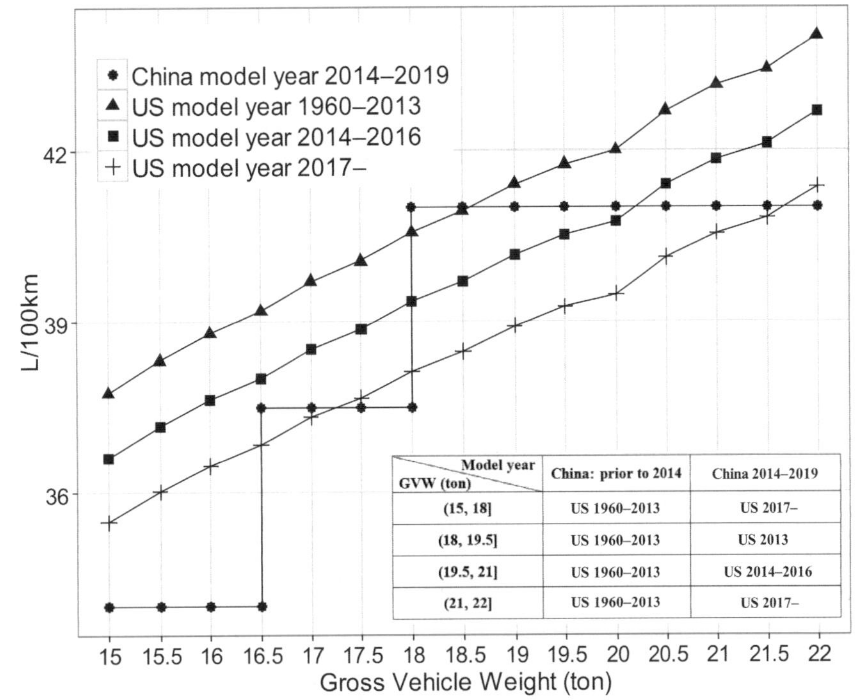

GVW (ton) \ Model year	China: prior to 2014	China 2014–2019
(15, 18]	US 1960–2013	US 2017–
(18, 19.5]	US 1960–2013	US 2013
(19.5, 21]	US 1960–2013	US 2014–2016
(21, 22]	US 1960–2013	US 2017–

Figure 3. Comparison of bus fuel consumption performance between U.S.A. and China based on adapted world transient vehicle cycle.

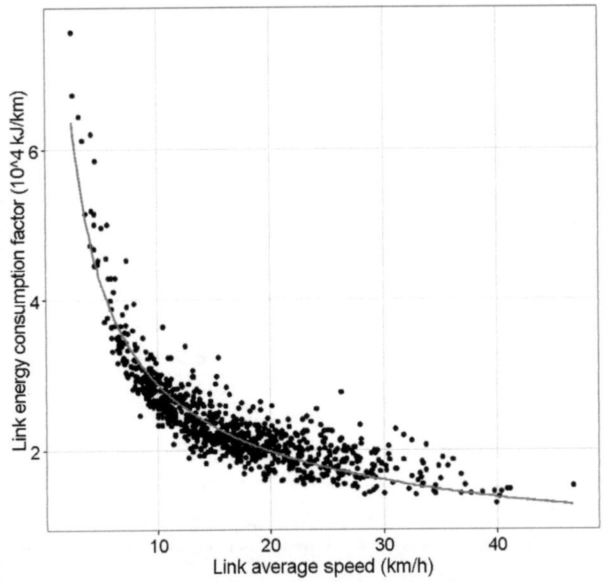

Figure 4. Relationship between link energy consumption factor and average speed.

factor. When average speed increased from 10 km/h to 20 km/h, link energy consumption factor decreased by 31%. When average speed increased from 20 km/h to 30 km/h, link energy consumption factor decreased by 19%. Therefore, giving priority to improving average speed on highly congested routes can gain more significant energy saving benefits.

Number of Stops per Kilometer. Figure 5 indicates that, in addition to link average speed, number of stops per kilometer also has a an important effect on the link energy consumption factor. Within the same speed interval, link energy consumption factor increases with the increasing number of stops per kilometer. For example, in the 10–15 km/h speed category, when number of stops per kilometer were <3, 3–6, 6–9, 9–12, and >12, average link energy consumption factors were 2.19, 2.39, 2.54, 2.68, 2.95 (10^4 kJ/km) respectively.

Link Average Positive Acceleration. Figure 6 indicates that link energy consumption factor increases with the increase in link average positive acceleration. When link average positive acceleration increases from 0.5 m/s^2 to 0.9 m/s^2, average link energy consumption factor increases by 28.95%. However, note that the distribution of link energy consumption factors is scattered. Compared with link average speed and number of stops per kilometer, link average positive acceleration has less impact on link fuel consumption factor.

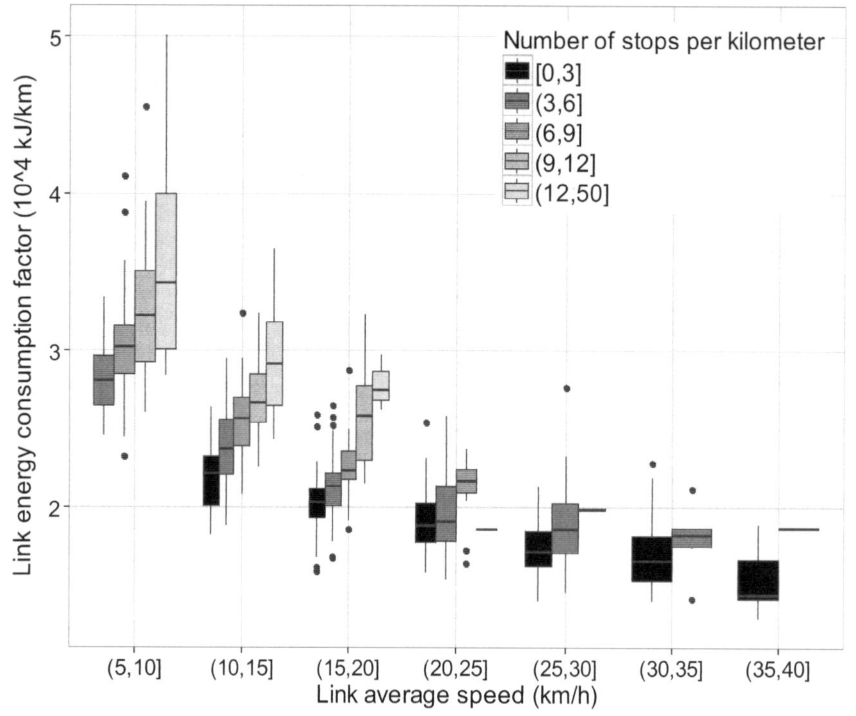

Figure 5. Relationship between link energy consumption factor and number of stops.

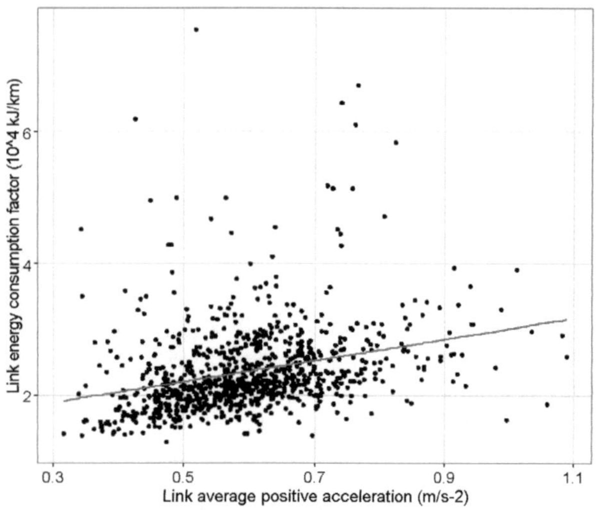

Figure 6. Relationship between link energy consumption factor and link average positive acceleration.

Road Facility Type. Figure 7*a* shows the difference in link energy consumption factors between links with and without reserved bus lane. When average speed is lower than 10 km/h, average energy consumption factor of links with bus lane is lower than that of links without bus lane. However, when average speed is higher than 10 km/h, for each speed category, the presence of a bus lane has little impact on link energy consumption factors. Similarly, Figure 7*b* presents the variation of energy consumption factors among principal arterial road, minor arterial road, and local road. When average speed is lower than 10 km/h, link average energy consumption factors of principal arterial road, minor arterial road, and local road increase successively. However, when average speed is higher than 10 km/h, to each speed category, no statistically significant difference was observed among the link energy factors of principal arterial road, minor arterial road, and local road.

Therefore, road facility matters only when link average speed is relatively low (specifically, lower than 10 km/h). Taking the 5–10 km/h speed category for example, Figure 8*a* shows that number of stops per kilometer of links with bus lane is significantly lower than that of links without bus lane. Figure 8*b* shows that number of stops per kilometer of principal arterial road, minor arterial road, and local road increase successively. Thus, when average speed is relatively low, in the same speed category, fewer stops associated with links with reserved bus lane or arterial road can account for their relatively low fuel consumption factors.

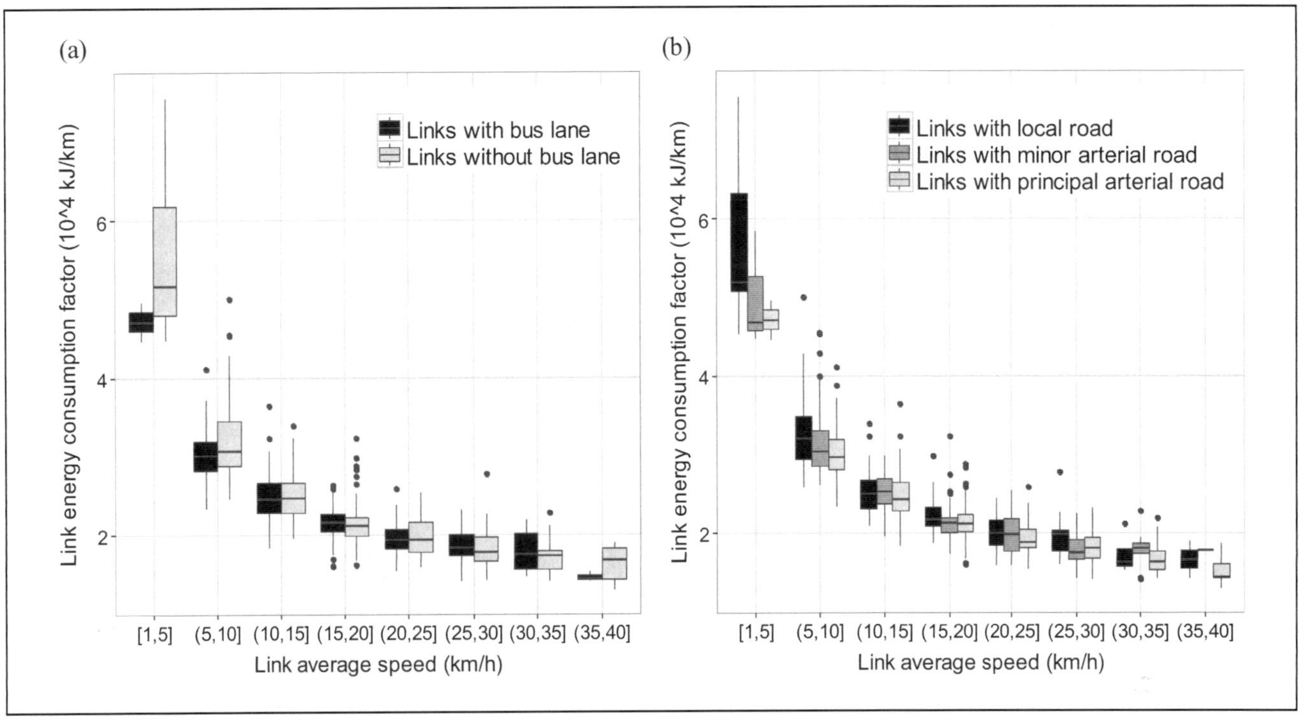

Figure 7. Relationship between link energy consumption factor and road facility type: (*a*) links with or without bus lanes; (*b*) links with local, minor arterial, or principal road.

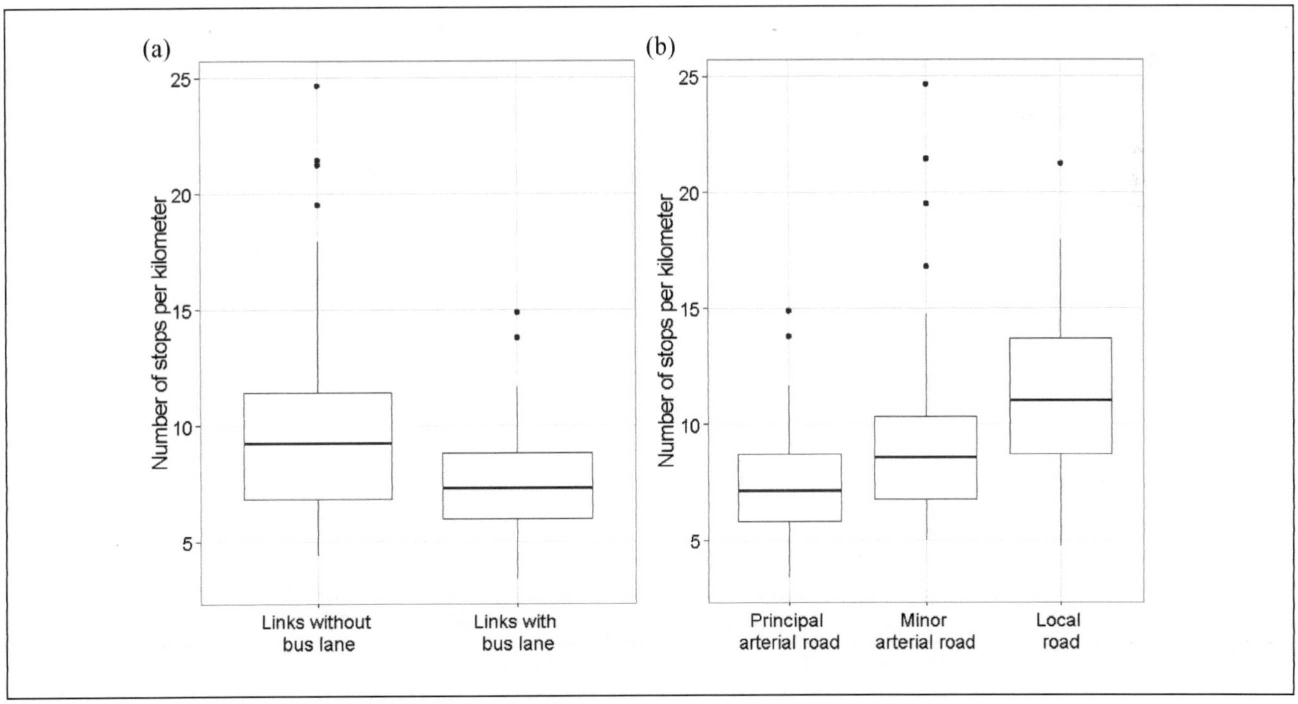

Figure 8. Relationship between road facility type and number of stops per kilometer (link average speed category = 5–10 km/h): (*a*) links with or without bus lanes; (*b*) links with local, minor arterial, or principal road.

Link Energy Consumption Factor Look-up Tables

Based on results presented in the previous section, links are categorized according to average speed, number of stops per kilometer, and road facility type. The combination of these three indicators formulates different link classification methods, as shown in Table 2. The average

Table 2. Link Type Classification Methods

Link type classification method	Indicators of link type classification: Link average speed (5 km/h interval)	Link average speed (2 km/h interval)	Links in principal arterial / minor arterial/local road	Links with bus lane / without bus lane	Number of stops per kilometer
Method 1	√				
Method 2	√		√		
Method 3	√			√	
Method 4	√				√
Method 5		√			
Method 6		√	√		
Method 7		√		√	
Method 8		√			√

Table 3. Average Link Energy Consumption Factor of a Bus with 16-ton Total Mass (10^4 kJ/km)

	Model year									
	Prior to 2014					2014–2019				
	Number of stops per kilometer									
Average speed (km/h)	[0,3]	(3,6]	(6,9]	(9,12]	>12	[0,3]	(3,6]	(6,9]	(9,12]	>12
[1,4]					6.4544					6.2608
(4,6]			3.7907	3.9200	4.7496			3.6769	3.8024	4.6071
(6,8]		3.0030	3.2524	3.4652	3.6704		2.9129	3.1548	3.3613	3.5603
(8,10]		2.8061	2.8975	2.9651	2.9611		2.7219	2.8105	2.8761	2.8723
(10,12]		2.5273	2.6843	2.7839	2.8605		2.4515	2.6038	2.7003	2.7746
(12,14]	2.3213	2.3265	2.5267	2.6304	3.1845	2.2517	2.2567	2.4509	2.5515	3.0889
(14,16]	2.1237	2.2107	2.3368	2.6563	2.9790	2.0600	2.1444	2.2667	2.5766	2.8896
(16,18]	2.0606	2.1283	2.2451	2.5154	2.6880	1.9988	2.0645	2.1777	2.4399	2.6074
(18,20]	1.9550	2.1069	2.1644	2.2231		1.8963	2.0436	2.0995	2.1564	
(20,22]	1.9312	2.0100	2.1794	1.8644		1.8733	1.9497	2.1140	1.8084	
(22,24]	1.8868	1.8849	2.0713			1.8302	1.8284	2.0092		
(24,26]	1.8526	1.8992	2.1581			1.7970	1.8423	2.0934		
(26,28]	1.7899	1.9186	1.9898			1.7362	1.8611	1.9301		
(28,30]	1.6863	1.7498	1.9977			1.6357	1.6973	1.9378		
(30,32]	1.9087	1.7253				1.8515	1.6736			
(32,34]	1.6620	2.1183				1.6122	2.0547			
(34,36]	1.6163	1.7914				1.5678	1.7377			
(36,38]	1.5108	1.8707				1.4654	1.8146			
(38,40]	1.3894					1.3477				

link energy consumption factors of different bus types were calculated based on each link classification method. Finally, eight look-up tables of energy consumption factors were developed. For example, as part of the look-up table related to method 8 (using average speed and number of stops per kilometer to categorize links), Table 3 shows the link energy consumption factors of a 16-ton bus. Due to space limitation, link energy consumption factors of other bus types are not exhibited. When second-by-second speed data is unavailable, bus energy consumption can be estimated by thumbing link energy consumption factors in the corresponding look-up table.

Model Validation

Estimates by Link-Level Indicators versus Estimates by Instantaneous Speed

Route 56 was chosen randomly to test the validity of using link-level indicators to estimate bus energy consumption. Instantaneously recorded data on bus speed on two trips from route 56 were collected, taking a 16-ton, 2012 model-year bus as validation object. After data cleaning, as described above, 29 links were kept. First, the operating mode distribution of 29 links were calculated based on their instantaneous speed profile. By

Table 4. Estimation Performance of Link-Level Approach for Eight Link Type Classification Methods

Link	Instantaneous speed approach (10^4 kJ/km)	Link-level indicator approach: estimation performance for eight link type classification methods							
		Method 1 (%)	Method 2(%)	Method 3 (%)	Method 4 (%)	Method 5 (%)	Method 6 (%)	Method 7 (%)	Method 8 (%)
1	2.1588	14.82	15.49	14.85	10.82	13.12	13.42	14.66	7.77
2	2.6815	−7.56*	−7.02	−7.54	−10.78	−8.93	−8.69	−7.69	-13.24
3	2.4353	1.78	2.38	1.81	−1.76	7.89	6.64	6.25	3.78
4	2.7454	−9.71	−9.18	−9.69	−12.86	−4.30	−5.40	−5.75	-7.94
5	2.9793	5.26	7.14	7.63	2.16	−2.93	−1.27	−3.05	-2.75
6	2.8272	10.92	12.90	13.43	7.66	2.29	4.04	2.17	2.49
7	2.2766	8.88	9.51	8.91	5.08	7.26	7.55	8.72	2.19
8	2.6124	20.04	22.19	22.75	24.70	10.70	12.59	10.57	13.50
9	1.7756	20.81	20.74	20.70	13.59	16.39	15.63	14.52	10.10
10	2.7822	−10.90	−10.39	−10.88	−8.47	−5.56	−6.66	−7.00	-3.52
11	2.1356	16.07	16.74	16.09	12.02	5.80	4.66	6.60	3.51
12	3.7770	−16.97	−13.06	−15.10	−13.75	15.78	19.05	17.04	3.78
13	2.3545	−8.89	−6.53	−8.97	−9.53	−8.60	−6.42	−8.23	-9.61
14	3.8776	−19.13	−21.90	−17.30	−21.51	12.78	3.07	14.00	-2.24
15	1.7878	19.99	18.79	19.88	19.14	20.37	19.32	20.86	19.05
16	3.0367	3.27	5.11	5.60	0.23	−4.76	−3.14	−4.88	-4.58
17	3.5053	−10.54	−8.94	−8.52	−7.07	−3.55	−2.02	−1.18	-1.14
18	2.2712	9.14	11.32	9.17	5.33	7.52	13.55	8.98	2.44
19	2.4567	−12.68	−10.42	−12.76	−7.93	−8.03	−6.38	−7.33	-4.88
20	2.5871	21.21	26.93	23.95	9.78	11.79	12.69	11.65	8.46
21	1.7080	14.78	16.58	15.06	11.80	10.66	16.00	10.39	8.46
22	2.4620	0.68	2.70	0.71	−2.83	6.72	7.77	5.10	2.65
23	2.1007	−6.67	−5.36	−6.45	−9.10	−10.03	−11.07	−10.25	-11.81
24	1.4049	23.48	27.42	21.52	22.28	20.80	20.04	20.62	18.30
25	2.1318	0.63	3.23	0.54	−0.08	0.95	3.36	1.36	-0.16
26	1.8065	8.52	10.22	8.79	5.71	6.04	10.14	7.33	4.44
27	1.6082	13.74	21.81	13.29	17.26	17.52	23.19	17.24	18.10
28	1.7297	13.34	15.12	13.62	12.77	16.25	13.98	15.90	16.20
29	1.7711	10.69	12.25	10.96	7.82	6.71	5.47	6.46	4.60
On average (%)		−11.45~11.90	−10.31~13.93	−10.80~12.46	−8.81~11.07	−6.30~10.87	−5.67–11.61	−6.15~11.02	−5.63~8.32

Note: − = estimate of link energy consumption factor by link-level indicator approach is 7.56% lower than estimate by instantaneous speed approach.

inputting opMode distribution of each link into MOVES, the link energy consumption factor was estimated. This approach is called the instantaneous speed approach. Second, based on each link classification method, as shown in Table 2, the link type of each of the 29 links was determined. For each link type, the energy consumption factor was extracted from the look-up tables developed in this paper. This approach is called link-level indicator approach. Finally, the estimation results between instantaneous speed approach and link-level indicator approach is compared.

As shown in Table 4, the link-level indicator approach produced acceptable estimates for link energy consumption factors, with average estimation error ranging from −5.6% to ~13.9% depending on link classification method: (a) when average speed was the only indicator to classify links, the average difference between instantaneous speed approach and link-level indicator approach is −6.3%~10.87% for method 5 (2 km/h speed interval), while that difference is −11.45%~11.9% for method 1 (5 km/h speed interval); (b) using link average speed (2 km/h speed interval) and number of stops to categorize links, method 8 shows the best estimation performance among the eight link classification methods; (c) compared with using only link average speed to categorize links, adding link facility indicator to categorize links has little impact on the estimation performance of the link-level indicator approach.

Therefore, when link average speed and number of stops per kilometer can be obtained, using these two indicators to categorize links and then to estimate link energy consumption factor will achieve the best estimation performance for link-level models. When link average speed can only be obtained, using a smaller speed interval to categorize links will also produce higher estimation accuracy. However, adding the road facility indicator to

Table 5. Bus Fuel Consumption Estimation Error of Different Link Classification Methods

Observations			Daily fuel consumption estimation error for varied link classification methods					
Date	Vehicle code	Real-world daily fuel consumption (L)	Method 1(%)	Method 2(%)	Method 3(%)	Method 5(%)	Method 6(%)	Method 7(%)
2015/3/7	S2F-001	80.02	−2.07	−2.28	−1.91	−1.46	−5.02	−1.29
2015/3/10	S2F-001	75	22.25	21.70	22.39	22.65	21.08	22.80
2015/4/4	S2F-001	79.98	13.19	13.12	13.15	12.62	11.23	13.15
2015/4/7	S2F-001	68	16.55	16.28	17.10	16.47	13.96	17.10
2015/4/20	S2F-001	60.03	27.01	25.56	26.74	26.03	22.96	26.74
2015/4/21	S2F-001	80	−4.85	−4.98	−3.92	−4.46	−5.62	−3.92
2015/4/22	S2F-001	80	16.30	15.54	16.36	15.66	14.05	16.36
2015/4/25	S2F-025	80	−7.03	−7.73	−6.30	−6.47	−7.26	−6.30
2015/4/25	S2F-029	80.01	−5.97	−6.80	−5.58	−5.93	−6.61	−5.58
2015/4/25	S2F-030	70.04	10.00	9.62	9.72	9.57	8.80	9.72
2015/4/26	S2F-026	50	−15.19	−16.82	−15.57	−15.38	−17.00	−15.57
2015/4/26	S2F-027	70.01	1.07	0.45	−0.99	1.23	−0.90	−0.99
2015/4/26	S2F-030	70	11.98	12.01	11.62	11.35	10.15	11.62
2015/4/26	S2F-033	90	−16.96	−16.93	−16.90	−17.09	−17.76	−16.90
2015/4/27	S2F-026	89.2	−17.64	−19.02	−17.92	−17.91	−18.63	−17.92
2015/4/27	S2F-027	57.34	5.92	4.75	6.12	5.91	4.47	6.12
2015/4/27	S2F-028	90	3.03	2.43	2.70	2.54	1.38	2.70
2015/4/29	S2F-003	75	3.81	3.56	4.44	3.98	2.27	4.44
2015/4/29	S2F-025	90	2.82	2.04	3.86	3.40	2.00	3.86
2015/4/30	S2F-033	90	−12.77	−12.80	−12.38	−12.99	−14.55	−12.38
Average estimation error (%)	-	–	−10.30~11.16	−10.92~10.59	−9.05~12.20	−10.21~10.95	−10.37–10.21	−8.98~12.24

categorize links scarcely improves the performance of link-level energy consumption models.

Estimates by Mesoscopic Model versus Real-World Fuel Consumption

Route 9 was used to validate the mesoscopic energy consumption model based on real-world bus fuel consumption data. The 2011 model-year buses operating on route 9 weigh 15 ton (curb weight plus passenger weight). For each bus, four kinds of data on March and April, 2015 were collected from Pudong Public Transport Corporation: (a) the timing of bus arrival at each stop and departure from each stop recorded by vehicle monitoring system, which can be used to calculate link average speed; (b) link facility type obtained by field observation; (c) daily operating mileage; (d) daily fuel consumption measured by the amount of fuel required to completely fill the tank when it accomplishes its daily operating schedule.

Because the real-world bus fuel consumption data is collected on a daily basis, all the links which a bus has traveled all day were aggregated. First, for each link which a bus has passed on a certain day, the road facility type and average speed were determined. When calculating link average speed, some links may not be recorded by vehicle monitoring systems due to missing signal. Considering this situation, bus daily operating mileage estimation based on records of vehicle monitoring system was compared with the real-world operating mileage. If the difference is less than 5%, this observation can be used for model validation. After this selection, only 20 daily observations were kept. Second, link fuel consumption is estimated by thumbing the link energy consumption factor for a 15-ton, 2011 model-year bus from look-up tables. Daily fuel consumption is estimated by summing the fuel consumption of all links which a bus has passed on that day. Finally, the estimated daily fuel consumption was validated against real-world daily fuel consumption.

Additionally, the impact of link classification methods listed in Table 2 on model accuracy was discussed. Because the number of stops cannot be obtained by vehicle monitoring system records, estimation performance of method 4 and method 8 cannot be evaluated. Table 5 presents estimation errors for different link classification methods. When link average speed was the only indicator to classify links, the average daily fuel consumption estimation error of method 5 (2 km/h speed interval) is −10.21%~10.95%. This estimation error is close to the error of method 1 (5 km/h speed interval). This is probably because the aggregation of links on a daily basis offsets the error variation of link fuel consumption estimation. Thus, real-world fuel consumption data on a link basis collected by PEMS are needed for future validation.

Conclusion and Recommendations

This study develops a mesoscopic bus energy consumption model based on MOVES. First, based on bus GPS data, local bus operating mode distribution is calculated as the input for MOVES. Then, fuel economy of buses in the U.S.A. and China was compared. Because the test cycle and performance measure are different between the U.S.A. and China's bus fuel consumption standards, it is difficult to compare bus fuel economy directly according to fuel consumption limit values released by the two countries. So, this study uses buses with different mass and model years in MOVES to simulate C-WTVC from China's bus fuel consumption standard. The simulated fuel consumption values of each bus type in MOVES are compared with the limit values from China's bus fuel consumption standards to determine the corresponding model years in the U.S.A. and China which have similar energy consumption performance for buses with a certain weight.

After MOVES localization, link energy consumption factors are estimated, and the influential variables of link energy factor are explored. With the increase in link average speed, link energy consumption factor decreases exponentially. Thus, giving priority to improving the speed of buses on highly congested routes would gain significant energy saving benefits. In the same average speed range, link energy consumption factor increases with the increasing number of stops per kilometer. Thus, bus signal priority measures, which have the potential to reduce vehicle stops, can achieve significant bus energy saving benefits.

Based on the various characteristics of the link energy consumption factor, link energy consumption factor look-up tables are developed for different bus types. Therefore, when instantaneous speed data is unavailable, bus energy consumption can be estimated using this link-level model. Validation tests indicate that using easily-accessible link-level indicators to estimate bus energy consumption can achieve acceptable accuracy. Furthermore, link type classification method affects the accuracy of the mesoscopic model. When link average speed and number of vehicle stops can be obtained, using these two indicators to categorize links will achieve better model estimation performance. When only link average speed is obtained, using smaller speed intervals to categorize links will also result in higher model estimation accuracy. However, adding road facility type to categorize links scarcely improves the estimation accuracy of the mesoscopic energy consumption model.

The model developed in this paper could be used to evaluate energy impacts of measures to improve bus services when second-by-second speed data is unavailable, and this research can be used as a reference for MOVES application outside the U.S.A., especially in China. However, there are some shortcomings: (a) the effect of road grade is ignored. The authors note that when applying the methodology to regions with significant variations in road terrain, road grade should be considered as an influential factor; (b) because we did not obtain link-level energy consumption data, daily bus fuel consumption is used to validate model. However, aggregation of links on a daily basis can offset the variation of links estimation error. Future research should collect link-level energy consumption data using PEMS to validate the mesoscopic model.

Acknowledgments

The authors thank Shanghai Pudong Public Transport Corporation for providing related data used in this paper; and students from School of Transportation Engineering, Tongji University for their assistance in field data collection.

Author Contributions

The authors confirm contribution to the paper as follows: study conception and design: Wenjian Jia, Xiaohong Chen; data collection: Wenjian Jia; analysis and interpretation of results: Wenjian Jia, Xiaohong Chen, Xiaonian Shan; draft manuscript preparation: Wenjian Jia. All authors reviewed the results and approved the final version of the manuscript.

References

1. Shanghai Municipal Transportation Commission. *Investigation Report on Bus Passenger Flow in Shanghai*. 2015.
2. Tongji University. *Study on the Effect of Shanghai Automobile Energy Structure Optimization on Traffic Energy Saving and Emission Reduction*. 2013.
3. Li, J. Q., S. D. Gupta, L. Zhang, K. Zhou, and W.-B. Zhang. Evaluate Bus Emissions Generated Near Far-Side and Near-Side Stops and Potential Reductions by ITS: An Empirical Study. *Transportation Research Part D: Transport and Environment*, Vol. 17, No. 1, 2012, pp. 73–77.
4. Qian, Y. *Research on Urban Diesel Bus Emissions Modelling and Application*. Doctoral dissertation. Southeast University, 2016.
5. Alam, A., E. Diab, A. M. El-Geneidy, and M. Hatzopoulou. A Simulation of Transit Bus Emissions along an Urban Corridor: Evaluating Changes under Various Service Improvement Strategies. *Transportation Research Part D: Transport and Environment*, Vol. 31, 2014, pp.189–198.
6. CARB. *EMFAC2014 Web Database*, 2016.

7. Ahn, K., H. Rakha, A. Trani, and M. Van Aerde. Estimating Vehicle Fuel Consumption and Emissions based an Instantaneous Speed and Acceleration Levels. *Journal of Transportation Engineering*, Vol. 128, No. 2, 2002, pp.182–190.
8. Barth, M., F. An, T. Younglove, G. Scora, C. Levine, M. Ross, and T. Wenzel. *Development of a Comprehensive Modeal Emissions Model.* NCHRP Project 25-11 Final Report, National Academy Press, Washington, D.C., 2000.
9. Sonntag, D. B., and H. O. Gao. Developing Link-Based Particle Number Emission Models for Diesel Transit Buses Using Engine and Vehicle Parameters. *Transportation Research Part D: Transport and Environment*, Vol. 14, No.4, 2009, pp.240–248.
10. Zhai, H., H. C. Frey, and N. M. Rouphail. A Vehicle-Specific Power Approach to Speed-and Facility-Specific Emissions Estimates for Diesel Transit Buses. *Environmental Science & Technology*, Vol. 42, No. 21, 2008, pp. 7985–7991.
11. Song, Y., E. Yao, T. Zuo, et al. Emissions and Fuel Consumption Modeling for Evaluating Environmental Effectiveness of its Strategies. *Discrete Dynamics in Nature and Society*, Vol. 2013, 2013, 9p.
12. Ahsan, A., and H. Marianne. Embedding Locally Derived Operating Mode Distributions into the MOVES Database to Estimate Transit Bus Emissions across an Urban Network. Presented at 95th Annual Meeting of the Transportation Research Board, Washington, D.C., 2016.
13. Alam, A., J. Xu, and M. Hatzopoulou. An Analysis of Instantaneous Speed Distributions and Emissions for Transit Buses across an Urban Network. Presented at 94th Annual Meeting of the Transportation Research Board, Washington, D.C., 2015.
14. *Motor Vehicle Emission Simulator (MOVES) User Guide for MOVES2014.* U.S. Environmental Protection Agency, Assessment and Standards Division Office of Transportation and Air Quality, Washington, D.C., 2014.
15. *Exhaust Emission Rates for Heavy-Duty On-road Vehicles in MOVES2014.* US Environment Protection Agency, Assessment and Standards Division Office of Transportation and Air Quality, Washington, D.C., 2015.
16. *Greenhouse Gas Emissions Standards and Fuel Efficiency Standards for Medium- and Heavy-Duty Engines and Vehicles.* US Environmental Protection Agency, National Highway Traffic Safety Administration (NHTSA), and Department of Transportation (DOT), Washington, D.C., 2011.
17. *Fuel Consumption Test Methods for Heavy-Duty Commercial Vehicles.* GB/T 27840-2011. National Standards of People's Republic of China, 2011.
18. *Diesel Fuel for Motor Vehicle.* DB31/428-2009. Shanghai Local Standard, 2009.
19. *Population and Activity of On-road Vehicles in MOVES2014.* US Environmental Protection Agency, Assessment and Standards Division Office of Transportation and Air Quality, Washington, D.C., 2016.

The Standing Committee on Transportation and Air Quality (ADC20) peer-reviewed this paper (18-05666).

TRR
JOURNAL OF THE TRANSPORTATION RESEARCH BOARD

Article

Transportation Research Record
2018, Vol. 2672(25) 163–173

DOI: 10.1177/0361198118798244
journals.sagepub.com/home/trr
SAGE

Efficiently Simulating Personal Vehicle Energy Consumption in Mesoscopic Transport Models

Zachary A. Needell[1,2] and Jessika E. Trancik[1,3]

Abstract

Mesoscopic transport models can efficiently simulate complex travel behavior and traffic patterns over large networks, but simulating energy consumption in these models is difficult with traditional methods. As mesoscopic transport models rely on a simplified handling of traffic flow, they cannot provide the second-by-second measurement of vehicle speeds and accelerations that are required for accurately estimating energy consumption. Here we present extensions to the TripEnergy model that fill in the gaps of low-resolution trajectories with realistic, contextual driving behavior. TripEnergy also includes a vehicle energy model capable of simulating the impact of traffic conditions on energy consumption and CO_2 emissions, with inputs in the form of widely available calibration data, allowing it to simulate thousands of different real-world vehicle makes and models. This design allows TripEnergy to integrate with mesoscopic transport models and to be fast enough to run on a large network with minimal additional computation time. We expect it to benefit from and enable advances in transport simulation, including optimizing traffic network controls to minimize energy, evaluating the performance of different vehicle technologies under wide-scale adoption, and better understanding the energy and climate impacts of new infrastructure and policies.

A variety of transport simulation models allow researchers and policy makers to predict the behavior of travelers in great detail under different policy, infrastructure, and travel demand scenarios. Producing energy-use and emissions estimates with these models opens up an even larger set of possible research and policy-making applications—from estimating the emissions implications of different policies or traffic control strategies to evaluating the expected usage patterns of emerging vehicle technologies. However, barriers still remain to widespread application of linked transport and energy/emissions models. In this paper, we present a new, flexible energy model calibrated on a large sample of GPS driving behavior data and capable of linking with a mesoscopic transport model to efficiently produce energy estimates for personal vehicles on large transport networks.

To be widely applicable to transport simulation, an energy model must balance efficiency and specificity (*1*). Increases in computational power allow city-sized transport networks to be simulated at the agent level—capturing the travel-related decisions and movements of millions of travelers at a second's time-scale. These models can simulate congestion, trip-making decisions, and traveler responses to information and incentives to a degree that is fundamentally impossible with simpler models. The massive scale of these simulations requires that any energy model at the vehicle scale operate very quickly and with minimal memory overhead. However, the large number of different vehicle types on the road and the variety of operating conditions they face during real-world use also require that any energy model either depends on vast amounts of calibration data or incorporates simplifications. While striking this balance, a useful model must capture the relationship of vehicle energy use to the congestion, driver behavior, and external conditions that motivate the use of detailed transport models in the first place.

Here we present an extension of the existing TripEnergy model (*2*), allowing for the efficient linking of a mesoscopic transport simulation model with an energy model calibrated to reproduce real-world energy use for a

[1]Institute for Data, Systems, and Society, Massachusetts Institute of Technology, Cambridge, MA
[2]Department of Civil and Environmental Engineering, Massachusetts Institute of Technology, Cambridge, MA
[3]Santa Fe Institute, Santa Fe, NM

Corresponding Author:
Address correspondence to Jessika E. Trancik: trancik@mit.edu

wide range of personal vehicles. By relying primarily on empirical observations and simple physical approximations, this energy model uses fewer tunable parameters than other methods. It neither requires high-resolution simulation of individual vehicle motion, nor does it use traffic flow theory to fill in the gaps of the moderate-resolution vehicle trajectories produced by more computationally efficient mesoscopic models. Both of these alternative approaches require calibration against disaggregate data that, though available for traffic counts and speeds, are far more difficult to incorporate when the error to be minimized is against energy consumption (*3*). Instead, the model presented here draws from a database of over 100,000 GPS speed histories taken from various statewide travel surveys of real-world drivers (*2, 4–6*) and fills in the gaps of moderate-resolution simulated trajectories with representative examples of realistic high-resolution driving behavior. This energy model also does not require detailed emission rates as inputs, instead using a simple vehicle energy model that can be calibrated on widely available regulatory data but that still reproduces energy consumption accurately enough for most purposes, allowing thousands of individual vehicle types to be simulated with minimal additional calibration. It achieves sufficient computational efficiency to run on large-scale networks through a novel separation of the driving behavior and vehicle performance steps of the energy estimation procedure.

This paper outlines the extensions to TripEnergy that allow it to efficiently process mesoscopic transportation model outputs, describes a validation procedure for the matching method and energy estimates, and presents results of simulation on a test network. We describe the energy estimation procedure in two steps—a vehicle model that translates characteristics of a high-resolution trajectory into an energy estimate, and a driving behavior model that links moderate-resolution simulated vehicle trajectories to a database of high-resolution trajectories. Model validation is performed by downsampling drive cycles to moderate resolution and then application of the driving behavior and vehicle models, comparing results to energy estimates given full knowledge of the trajectory and a much more detailed vehicle simulation (*7*) and dynamometer results (*8*). The model is then linked to the SimMobility mesoscopic transport simulation model (*9*) and results are examined for a toy road network, reproducing expected energy performance for a range of different real-world vehicle models.

Background

Traffic Network Modeling

Transport models work by considering the relationship between transportation supply and demand. The characteristics of the road and transit network such as travel times and costs are considered in the supply model, whereas the demand model considers the destinations, departure times, modes, and route choices for individuals' trips. For supply, dynamic traffic assignment (DTA) moves vehicles through the traffic network based on real-time traffic conditions (*10*). Implementations of this method are able to capture certain effects, such as gridlock, queuing, and rerouting, which cannot be easily modeled using static methods. Some vehicle-following models directly simulate acceleration, lane changing, and gap-acceptance, and, given extensive calibration, can reproduce the realistic high-resolution vehicle trajectories necessary to estimate energy use. However, even if such calibration data exist, which is often not the case, it remains infeasible to run such detailed models on large networks. As an alternative, "mesoscopic models" are able to capture many of the complex aspects of transport supply while still achieving sufficient computational efficiency to run on large networks. Mesoscopic models also follow individual agents across the network, but they contain a simplified treatment of vehicle movement and often operate at a lower temporal resolution, typically on the order of 5 to 15 s. Though sacrificing some details such as variability in driving style and complex interactions between vehicles, such mesoscopic methods have been successfully applied to large networks (*11*) and can run quickly enough to optimize controls (*12*). Mesoscopic traffic supply is included in many transport simulation models, such as SimMobility Midterm (*9*), DTAlite (*13*), and MATSIM (*14*).

For demand, most state-of-the-art models are "activity based." Unlike more traditional methods, activity-based models simulate individual travelers' trip-making decisions in terms of how they allow travelers to complete a daily sequence of activities. Because they track traveler location and choices throughout the day, activity-based models can capture more realistic feedback between transportation network performance and travel demand (*15*). Energy modeling is more closely related to the supply side of the modeling framework, but certain questions can be fully answered only by integrating energy calculations directly into the supply/demand interaction. For example, the limited range of existing battery-electric vehicles means that charging requirements will likely become a greater component of travel decision-making, motivating development of energy models that can run in concert with transport supply and demand calculations.

Vehicle Modeling

Transport models implicitly or explicitly simulate vehicle movement, and there exist various software tools to turn simulated trajectories into energy estimates given a

Table 1. Summary of Selected Methods to Calculate Energy Based on Transport Network Simulations

Model	Resolution needed	Energy method
COPERT-micro (Samaras et al. [*16*])	Link	Speed
EMFAC (California Air Resource Board [*17*])	Trip	Speed
ADVISOR (Markel et al. [*7*])	High-resolution	Microsimulation
VT-Micro (Ahn and Rakha [*18*])	High-resolution	Speed/Acceleration
MOVES County (Assessment and Standards Division, Office of Transportation and Air Quality [*19*])	Link	Vehicle-specific power
MOVES Project (Assessment and Standards Division, Office of Transportation and Air Quality [*19*])	High-resolution	Vehicle-specific power
MOVESlite (Zhou et al. [*20*])	High-resolution	Vehicle-specific power
VT-Macro (Zegeye et al. [*21*])	Link	Speed/Acceleration
Zhou et al. (*20*)	Mesoscopic	Vehicle-specific power

vehicle type. These characterizations of vehicle movement can vary in terms of their temporal resolution and the fidelity with which they can be expected to represent real driving behavior, leading to different scales at which energy models must operate. The characteristics of selected models are summarized in Table 1.

Most traditional transport simulations model supply based on average speeds and densities of road network links. These link-level speeds can be turned into energy estimates by models such as COPERT (*16*) that rely on empirical measurements of typical energy and emissions rates but do not directly capture the local effects of accelerations on energy use. Link speeds can also be aggregated into beginning-to-end trip average speeds, which can be used to estimate energy consumption by models such as EMFAC (*17*).

If a network simulation produces high-resolution trajectories (typically 1 Hz or greater), those trajectories can be fed into detailed vehicle simulations such as ADVISOR (*7*). These models simulate the mechanics of vehicles in great detail and are best equipped to simulate complicated processes such as criterion pollutant emissions or engine performance under extreme operating conditions. However, these detailed models are slow to run, making them impractical to run on the output of large network simulations.

To achieve greater efficiency while still capturing the detail provided by high-resolution trajectories, researchers have developed simpler models to approximate vehicle performance. Some estimate energy consumption rates based on polynomial fits to instantaneous speed (*16*) and others use fits based on speed and acceleration, including the widely-used VT-Micro model (*22*). To better incorporate factors such as road grade, a number of models estimate energy and emissions rates using vehicle-specific power (VSP), a measure of the relative energy intensity of the vehicle motion. One commonly used VSP-based model is CMEM (*23*), which relies on measurements from a variety of different vehicle types that cover the breadth of models in the on-road vehicle fleet. Other vehicle models, some requiring less calibration data, have been proposed and evaluated as well (*24*).

One particularly comprehensive and widely used energy and emissions model is MOVES (*19*), which uses a database of emissions rates for a large number of different vehicle types over a variety of speed and VSP bins known as operating modes. The distribution of time spent in each of these operating mode bins is estimated based on a combination of disaggregate speed, temperature, travel behavior, and vehicle fleet data. MOVES can be cumbersome to run at its most detailed settings, so for added efficiency and flexibility others have developed a simplified version, MOVESLite (*25*), with a more limited but still comprehensive set of vehicle types and operating modes.

In all of these cases, there is an underlying difficulty: efficient simulation of large networks requires some sort of simplification in the treatment of traffic flow, but energy-consumption modeling requires a high-resolution representation of vehicle movement. This tradeoff is apparent in the various available levels of detail in MOVES. At the "County" level, MOVES populates its operating mode distribution table with results from a small set of high-resolution drive cycles. It then calculates emissions by weighting and aggregating the emissions rates of these operating modes based on traffic patterns. This method, however, is not capable of directly simulating individual vehicle consumption or taking advantage of the full level of detail present in simulated mesoscopic trajectories. At the higher-detail "Project" level, however, MOVES requires drive schedules or operating mode distributions for each road link. To provide these inputs, it is necessary either to run a full high-resolution simulation of the network or to rely on default values that do not reflect local traffic conditions. This means that there is no widely used energy estimation methodology that is capable of taking advantage of both the computational efficiency of mesoscopic models as well as their ability to capture individual vehicle trajectories at a higher resolution than link average speeds. More recently, some researchers have proposed inferring detailed trajectory information from the moderate-resolution trajectories produced by mesoscopic network simulations. This approach has been implemented (*20, 21*) using traffic flow theory to model the acceleration behavior of drivers and then inferring VSP from these interpolated trajectories. The method we present has three main differences. First, TripEnergy fills in the details of mesoscopic trajectories with empirical, real-world behavior, requiring fewer calibration parameters. Second, it is designed to simulate the energy consumption of a large number of real-world vehicle types without needing large amounts of calibration data for each one. And, third, it uses a simpler model of instantaneous vehicle energy use, trading the ability to simulate emission of non-CO_2 pollutants and engine performance under extreme operating conditions to allow for more flexible and efficient implementation using a linear model.

Methods

Vehicle Model

Drive Energy. This method of calculating energy consumption from a high-resolution trajectory relies on a simplification that allows the important aspects of the trajectory to be stored in a vehicle-independent manner but allows energy consumption for any vehicle type to be efficiently calculated from those stored values. The basis for the energy model is the version presented in a previous paper (*2*) and applied to questions of range constraints in battery-electric vehicles (*26*), but reformulated here to depend only on instantaneous quantities. Unlike other pollutants such as CO and particulate matter, CO_2 emissions from a gasoline-powered vehicle are almost precisely linearly related to the energy content of the fuel, allowing this method to also estimate CO_2 emissions from gasoline-powered vehicles.

The basis for the energy calculations is the tractive power, a function of the vehicle's instantaneous speed v, acceleration a, and road grade θ, as well as its mass m, a set of polynomial coastdown coefficients A, B, and C, and a rotational inertia fraction ϵ, set to be 0.05 here (*27*):

$$\begin{aligned}\mathcal{P}_{tr}(t) &= Av(t) + Bv(t)^2 + Cv(t)^3 \\ &\quad + (1+\epsilon)mv(t)a(t) + mg\sin(\theta(t))v(t).\end{aligned} \quad (1)$$

In this paper, final energy quantities are written in script, whereas pre-conversion energy values (measured at the battery or gas tank) are written in normal font.

The rate of change of stored energy in the gas tank or in the battery consists of power going to the wheels, returning via the brakes (if regenerative braking exists), and going toward auxiliary electronics such as climate control. All of these processes involve conversion losses, taken into account with loss functions $P = L(\mathcal{P})$. The rate of change of stored energy is the power going toward motion $P_{accel} = L_{accel}(\mathcal{P}_{tr})$ minus the power sourced from regenerative brakes $P_{brake} = L_{brake}(\mathcal{P}_{tr})$, plus power going to the auxiliaries $P_{aux} = L_{aux}(\mathcal{P}_{aux})$, which depends on climate control and dashboard settings.

For clarity, we can define an indicator function for whether tractive power is positive (and the engine is active) or negative (and the brakes are active):

$$PTF(t) = \begin{cases} 1 & \mathcal{P}_{tr}(t) > 0 \\ 0 & \mathcal{P}_{tr}(t) \leq 0 \end{cases}. \quad (2)$$

Finally, we make the further simplifying assumption that all loss functions are linear and that auxiliary power is constant over the course of a trip. The linear approximation has been shown to be reasonably accurate for internal combustion engine vehicles (*28*), and the framework we present here can easily be adjusted to allow for higher-order polynomial terms, such as those tested by Saerens et al. (*24*). Using the linear approximation we have $L_{accel}(\mathcal{P}_{tr}(t)) = P_{idle} + \frac{\mathcal{P}_{tr}(t)PTF(t)}{\eta_{max}}$, $L_{brake}(\mathcal{P}_{tr}(t)) = \eta_{brake}\mathcal{P}_{tr}(t)(1 - PTF(t))$, and $L_{aux} = \frac{\mathcal{P}_{aux}}{\eta_{aux}}$. P_{idle} is the rate of energy use at no tractive power (including rest and coasting), effectively accounting for lower engine efficiency at low power. η_{max} is the slope of the $\mathcal{P}_{tr}$-P_{accel} line, effectively giving the engine efficiency approached at higher power. η_{brake} is the portion of energy going through the brakes re-cycled into the vehicle's battery and then back to the wheels. Defining the acceleration energy and braking energy as:

$$\begin{aligned}\mathcal{E}_{accel} &= \int_{t=0}^{T} \mathcal{P}_{tr}(t)PTF(t)dt; \\ \mathcal{E}_{brake} &= \int_{t=0}^{T} -\mathcal{P}_{tr}(t)(1 - PTF(t))dt,\end{aligned} \quad (3)$$

we can simplify the total energy equation:

$$E_{use} = \frac{\mathcal{E}_{accel}}{\eta_{max}} - \eta_{brake}\mathcal{E}_{brake} + \left(P_{idle} + \frac{\mathcal{P}_{aux}}{\eta_{aux}}\right)T. \quad (4)$$

The function for $\mathcal{E}_{accel}$ is dependent on the distance traveled and the average speed of the trip. Here we separate out the effects of these quantities by defining unitless variables based on t, $v(t)$ and $a(t)$:

$$\tau = (t - t_0)/T,\ u(\tau) = v(\tau)/\bar{v},\ \text{and}\ \alpha(\tau) = a(\tau)/a_0$$

where T is the total duration of the drive cycle; $\bar{v}$ is its average speed; and a_0 is a characteristic acceleration, defined as $0.15\ m/s^2$.

We can express the equation for $\mathcal{E}_{accel}$ by defining a set of "moments" consisting of the time-integrals of these unitless quantities:

$$\mu_1 = \int_{\tau=0}^{1} u(\tau)PTF(\tau)d\tau \quad \mu_2 = \int_{\tau=0}^{1} u(\tau)^2 PTF(\tau)d\tau$$
$$\mu_3 = \int_{\tau=0}^{1} u(\tau)^3 PTF(\tau)d\tau \quad \mu_a = \int_{\tau=0}^{1} \alpha(\tau)u(\tau)PTF(\tau)d\tau. \qquad (5)$$

These definitions allow us to express the total acceleration energy without any integrals:

$$\mathcal{E}_{accel} = D(A\mu_1 + B\bar{v}\mu_2 + C\bar{v}^2\mu_3 + (1+\epsilon)m\ \alpha_0\mu_a + mg\Delta Z\mu_1) \qquad (6)$$

where D is total distance. For the road grade component, and we assume that road grade is relatively constant over each time interval, leaving the energy impact dependent on the total elevation change ΔZ. Braking energy is calculated similarly for $PTF(t)>0$, for which we define the comparable moments as ν_i:

$$\mathcal{E}_{brake} = -D\big(A\nu_1 + B\bar{v}\nu_2 + C\bar{v}^2\nu_3 + (1+\epsilon)m\ \alpha_0\nu_a + mg\Delta Z\nu_1\big). \qquad (7)$$

Equations 7 and 8 and an estimate of auxiliary power yield total energy consumption via Equation 4. Pulling these average quantities outside of the integral effectively treats total energy consumption as a linear combination of these average quantities multiplied by correction factors that depend on the specific details of the true speed history. The final expression for energy consumption is a linear combination of 10 terms relating to the vehicle and its use:

$$\mathbb{V} = \begin{bmatrix} A/\eta_{max} \\ B/\eta_{max} \\ C/\eta_{max} \\ (1+\epsilon)m/\eta_{max} \\ A\eta_{brake} \\ B\eta_{brake} \\ C\eta_{brake} \\ (1+\epsilon)m\eta_{brake} \\ P_{idle} \\ 1/\eta_{max} \end{bmatrix}; \ \mathbb{U} = D \begin{bmatrix} 1 \\ \bar{v} \\ \bar{v}^2 \\ 1 \\ -1 \\ -\bar{v} \\ -\bar{v}^2 \\ -1 \\ 1/\bar{v} \\ \mathcal{P}_{aux}/\bar{v} \end{bmatrix} \circ \begin{bmatrix} \mu_1 \\ \mu_2 \\ \mu_3 \\ \alpha_0\mu_a \\ \nu_1 \\ \nu_2 \\ \nu_3 \\ \alpha_0\nu_a \\ 1 \\ 1 \end{bmatrix};$$
$$E_{drive} = \sum_{k=1}^{10} V_k U_k. \qquad (8)$$

These moments are a helpful way of expressing the total energy because they are almost independent of vehicle type. This is a simplification only because the function $PTF(t)$ varies slightly from vehicle to vehicle, but these differences end up having little effect on the total energy and are ignored here. The moments, calculated for a partial or full vehicle trajectory, can serve a similar purpose to the link-specific driving schedules or operating mode distributions required as an input to MOVES, defining the factors of a vehicle's use that determine energy requirements. Rather than querying a detailed set of vehicle-specific emissions rates or evaluating integrals, all that is needed to evaluate this energy function is to access these stored parameters that can be calculated in advance, allowing for faster computation.

Drive Efficiency Calibration. The vehicle-specific component of the total energy in Equation 8, $\mathbb{V}$, depends in part on physical vehicle road load characteristics A, B, C, m that are publically available (*29*). The remaining three parameters, P_{idle}, η_{max}, and η_{brake} are estimated by fitting to energy consumption over a set of U.S. Environmental Protection Agency (EPA) test cycle results, similar to methods used elsewhere (*2*, *24*, *26*). We minimize mean squared error between the predicted ($MPGest$) and measured (MPG) fuel economy values for a set of test cycles DC with known fuel economy:

$$\underset{P_{idle},\eta_{max},\eta_{brake}}{\operatorname{argmin}} \sum_{i \in DC} (MPG_i - MPGest_i(P_{idle}, \eta_{max}, \eta_{brake}))^2. \qquad (9)$$

For light-duty vehicles, these drive cycles can consist of the U.S. Environmental Protection Agency (EPA) city (federal test procedure) and highway (Highway Fuel Economy Test, HWFET) drive cycles, as well as for the high speed (US06) cycle for the vehicles with it available. For this paper, we calibrate these parameters for 3,784 vehicle types with model years from 2010 to 2016. During calibration, P_{brake} is fixed to 0 for vehicles without regenerative braking. For vehicles with regenerative braking but only two test cycles available, P_{idle} is fixed—to 1409 W for hybrid-electric vehicles and 741 W for fully electric vehicles—to ensure a unique solution. The value for hybrid-electric vehicles is chosen by taking the median fit value for P_{idle} of the hybrid vehicles for which three unique fuel economy estimates are available. The value for fully electric vehicles is fixed to the value for P_{idle} found by calibrating the 2013 Nissan Leaf on additional drive cycles, whose energy consumption is provided in the Downloadable Dynamometer Database (*8*).

Each vehicle being simulated in the transport model linked to TripEnergy must be assigned a specific vehicle type and associated energy parameters. Accurately assigning each simulated agent with a correct vehicle type is likely too burdensome to be worthwhile for some applications, and in such cases a representative vehicle can be chosen for different vehicle categories. For cases in which more detail in the vehicle fleet is required, vehicle types can be assigned to households by a discrete choice model as part of population synthesis. This functionality is especially useful for emerging integrated models where household location and vehicle ownership decisions are tied to the utilities of different transportation alternatives (*9*).

Drive-Cycle Matching

We frame the process of estimating energy consumption from simulated trajectories as a prediction problem. Given limited information about a vehicle *j*'s characteristics: $\mathcal{V}_j$, and about its trajectory: $\mathcal{U}_j$, we estimate the expected value of a vehicle's energy consumption over timestep *i* under those conditions: $E[E_{ij}|\mathcal{U}_j, \mathcal{V}_j]$. In this case, the information known is the vehicle's cumulative distance traveled every advance interval, and vehicle characteristics. We assume that the vehicle could have followed any speed profile over each timestep, but that the measurements at advance intervals are accurate. Given Equation 8, the goal becomes to predict the components of the moment vector $\mathbb{U}$ for each timestep given the vehicle's trajectory $\mathcal{U}$:

$$E\left[E_{ij}|\mathcal{U}_j, \mathcal{V}_j\right] = E\left[\left(\sum_{k=1}^{10} U_{ijk} V_{jk}\right)|\mathcal{U}_j, \mathcal{V}_j\right] = \sum_{k=1}^{10} E\left[U_{ijk}|\mathcal{U}_j\right] V_{jk}. \quad (10)$$

We estimate $E[\mathbb{U}_{ij}|\mathcal{U}_j]$ by searching a database of real-world drive cycles for intervals that are similar to the portion in advance interval *i* being estimated. The vector of moments can be calculated for each matched trajectory segment and then averaged, producing an estimate $\bar{\mathbb{U}}_{ij}$ of those moments for the unknown partial trajectory, which can then be combined with the vehicle parameters $\mathbb{V}_j$ to estimate E_{ij}. This matching and averaging procedure has several advantages. It can be implemented efficiently, it does not require that detailed energy-use measurements be stored for multiple vehicle types, and it is robust to modeling decisions about driving style that must be made for short-term simulations—a useful property because trajectories simulated by short-term models might not be entirely accurate for energy-consumption purposes (*3*). This solution does not directly incorporate elevation change. For networks in which elevation change is expected to greatly effect energy consumption, separate moment databases can be created for different grades and accessed based on the instantaneous grade of the simulated vehicle at each timestep.

To efficiently implement this matching procedure, we create a lookup table of averaged moments that can be accessed based on binned properties of the medium-resolution trajectory. For instance, the lookup table for a simple speed-based matching scheme can be implemented by breaking each GPS drive-cycle into segments of the same duration as the advance interval, assigning each segment to a bin based on its average speed, calculating the mean moments for each bin, and then using those mean moments to calculate $E[\mathbb{U}_{ij}|\mathcal{U}_j]$. The drive-cycle database used for matching consists of over 100,000 trips with 1-Hz speed measurements and has been described in detail (*2, 4–6*).

Various matching methods are possible. A one-dimensional matching scheme ($\bar{v}_t$) was tested, using only the average speed over the current advance interval. Because training data are ample, this will approximate the expected value for a vehicle's energy consumption over a timestep given only current average speed, assuming the driving behavior in the area being modeled is relatively similar to that captured in the driving behavior database and that bins are reasonably small. Two two-dimensional schemes were tested: one based on the current and previous average speeds ($\bar{v}_{t-1}, \bar{v}_t$) and one based on the average speed and average acceleration of the current advance interval ($\bar{v}_t, \bar{a}_t$), with average acceleration determined based on the instantaneous speeds at the beginning and end of the advance interval (accurate instantaneous speeds is a possible output of some but not necessarily all mesoscopic traffic simulations). Two three-dimensional matching schemes were tested: one based on the average speed over the current advance interval as well as the instantaneous speeds its the beginning and end ($\bar{v}_t, v_i, v_f$), and one based on the average speed over the preceding, current, and following advance intervals ($\bar{v}_{t-1}, \bar{v}_t, \bar{v}_{t+1}$).

As explained below, the most accurate was found to be matching based on a trajectory's average speed over a given advance interval, the previous interval, and following interval. Here, we divide speeds into 38 bins in each of the three dimensions, leading to 54,872 individual bins, each of which needs to store eight different moments—a data structure that can be accessed quickly and uses a non-substantial amount of memory. This database is sufficiently populated if it provides a reasonably accurate expected value for the moments in each bin, a condition that requires a large set of drive cycles. Indeed, 84.5% of the bins for which there are at least 10 observations in the GPS dataset are not observed at all in the 49 drive schedules provided in the MOVES database. These behaviors not observed in the MOVES database account for 29% of the total driving time in the GPS dataset, showing the value of incorporating this additional driving behavior.

A visualization of this lookup table is shown in Figure 1. The average speed over the advance interval

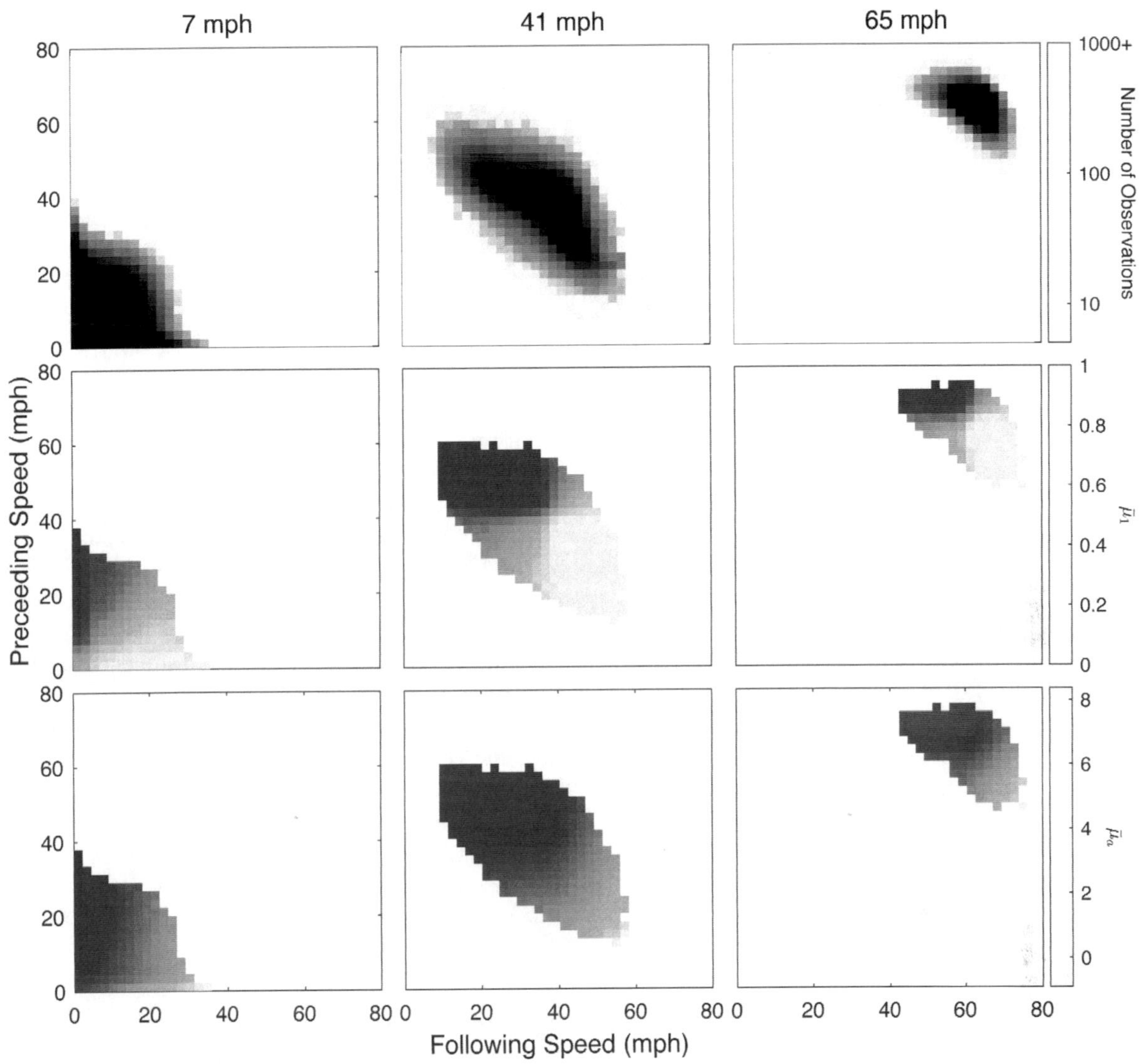

Figure 1. Visualization of the lookup table of drive-cycle moments produced by matching on three consecutive average speeds, for a 5-s advance interval.

being estimated is fixed for each of the three columns, the average speed for the previous time interval determines the y-position and the average speed for the next time interval determines the x-position within each subfigure. For example, the pixel at $Y = 35$, $X = 45$ in the middle (41 mph) column of Figure 1 represents drive-cycle segments averaging 41 mph and accelerating—in which the previous interval averaged 35 mph and the following interval will average 45 mph. The first row of subfigures shows the number of observations in each bin, with much of the range of plausible vehicle movements having over 1,000 distinct 5-s observations. The second row shows the average value of μ_1, a moment relating to the portion of time during which the engine was active. The third row shows the average value of μ_a, a moment related to the amount of work the engine does toward acceleration.

Model Validation

Microsimulation

The preferred matching method was chosen based on the root mean square (RMS) error of fuel economy predictions, ease of applicability to mesoscopic simulations, and robustness to uncertainty in mesoscopic trajectories. Estimated energy values were compared with "ground truth" values given by microsimulation, when the entire high-resolution trajectory is known. Twenty percent of approximately 100,000 available 1-Hz GPS trajectories

Table 2. Performance of Different Matching Methods, Evaluated Against ADVISOR

	No noise		± 1 mph noise		± 2 mph noise	
Estimation method	RMS (mpg)	Total energy	RMS (mpg)	Total energy	RMS (mpg)	Total energy
Spline	1.08	−2.20%	0.96	−1.20%	1.13	1.98%
$(\bar{v}_t)$	2.47	−0.78%	2.47	−0.73%	2.46	−0.55%
$(\bar{v}_{t-1}, \bar{v}_t)$	1.33	−0.65%	1.44	−0.19%	1.92	2.19%
$(\bar{v}_t, \bar{a}_t)$	0.93	−0.36%	1.08	0.50%	1.64	3.40%
$(\bar{v}_t, v_i, v_f)$	0.88	−0.40%	1.33	1.01%	2.39	4.12%
$(\bar{v}_{t-1}, \bar{v}_t, \bar{v}_{t+1})$	0.9	−0.45%	0.99	−0.06%	1.47	2.24%
Vehicle model	0.71	−0.28%	na	na	na	na

Note: na = not applicable, as "Vehicle model" assumes perfect knowledge of the vehicle trajectory. Vehicle average fuel economy is 37.0 miles per gallon.

were chosen as a test set. These were converted into simulated mesoscopic trajectories by downsampling them into 5-s resolution average speeds, consistent with a mesoscopic simulation with a 5-s advance interval. An estimate for the true energy consumption was produced by simulating vehicle performance with the ADVISOR software (*7*). The vehicle simulated was a compact internal combustion engine vehicle, with physical characteristics modeled after a 2014 Ford Focus. The vehicle model described above was then calibrated based on ADVISOR outputs—coastdown coefficients *A*, *B*, and *C* were estimated based on a polynomial fit to instantaneous road load, and drive-cycle fuel economy measurements used for model calibration were produced by running the simulated vehicle through the EPA Highway Fuel Economy Test (HWFET, highway) and Urban Dynamometer Driving Schedule (UDDS, city) drive cycles. The final vehicle model parameters used were A : 172.8 N, B : -4.676 Ns/m, C : 0.5516 Ns^2/m^2, m : 1406 kg, P_{idle} : 11.50 kW, η_{max} : 0.4081. No additional auxiliary energy use is assumed.

The remaining 80% of trips were used to generate a database of binned driving behavior moments as described above, using different binning methods. The robustness of the matching methods to errors in mesoscopic trajectories is tested by adding different degrees of uncorrelated random noise to individual timesteps in the medium-resolution trajectories.

Models were compared based on RMS estimation error for trip average fuel economy and bias in estimating the total energy consumption of the entire test set of trips, compared with ADVISOR simulations of the full high-resolution trajectory. Two other models were simulated as a comparison, neither of which is expected to be as scalable to large-scale simulations. The "Spline" model interpolated the mesoscopic trajectory to a high-resolution one using cubic splines. "Vehicle Model" assumes access to the true high-resolution trajectory but uses the simplified vehicle model described above, useful as a bound on the error introduced by the matching algorithm. Given accurate mesoscopic trajectories, the methods $(\bar{v}_t, \bar{a}_t)$, $(\bar{v}_t, v_i, v_f)$, and $(\bar{v}_{t-1}, \bar{v}_t, \bar{v}_{t+1})$ all perform roughly equivalently, with RMS errors of trip-averaged fuel economy of approximately 0.9 mpg, compared with an RMS error of 0.71 mpg given high-resolution knowledge of the trajectory and 1.08 for the spline method, and an error for the estimated total energy of all trips within 0.5% of the true value. Under increased uncertainty, the method based on three consecutive average speeds outperforms those incorporating instantaneous measurements. At high uncertainty, a spline-based method outperforms the methods based on binned moments, but all other binned methods outperform the method $(\bar{v}_t)$ based on only average speeds. A summary of these results is shown in Table 2. Given these results, the three-dimensional $(\bar{v}_{t-1}, \bar{v}_t, \bar{v}_{t+1})$ method is chosen.

Aggregate Performance Measures

Additional evaluation was performed by running a simulation of a traffic network containing a variety of vehicle types, comparing fuel economy measurements for the different vehicles with EPA estimates. The transport network simulation was performed using SimMobility Midterm (*9*), a simulator linking an activity-based demand simulator with a mesoscopic, DTA supply model. TripEnergy was implemented in C++ and run concurrently with SimMobility, leading to an increase in simulation time and memory use of less than 5%. The tests were performed on the "Virtual City" network, containing 94 nodes and 254 links, with constant elevation. The simulation was run for a full day, during which vehicles made a total of 174,594 individual trips. Four vehicle types were simulated—a 2013 Nissan Leaf (a battery-electric sedan), a 2016 Toyota Prius (a hybrid-electric sedan), a 2014 Honda Accord plug-in hybrid operating in charge-depleting mode (a sedan), and a 2010 Ford F150 (a pickup truck). Additional auxiliary energy consumption of 1000 W was assumed to account for typical

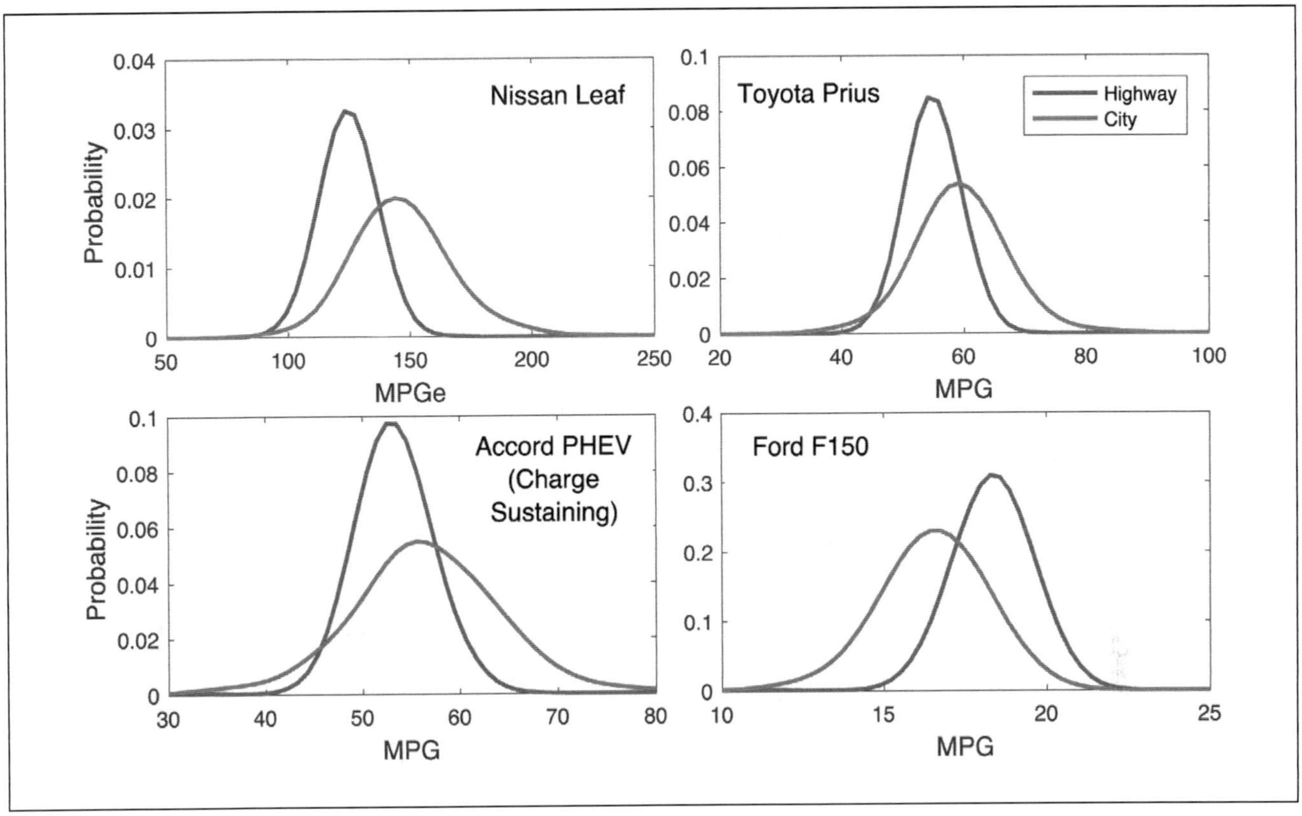

Figure 2. Distribution of trip-averaged fuel economies for different vehicle types, simulated with SimMobility over the "Virtual City" road network, measured in miles per gallon (MPG) for gasoline-powered vehicles and miles per gallon equivalent (MPGe) for electric vehicles.

auxiliary use. A charging efficiency of 83.7% was assumed for the Leaf (*30*).

Vehicle fuel economies were then compared with EPA-adjusted fuel economy estimates. Each trip was classified as "Highway" (more than 50% of the time spent at greater than 55 mph), or "City" for the remaining trips. Expected results were observed, with the gasoline internal combustion engine-powered F150 achieving lower fuel economy for city driving, and the hybrid-electric Prius and fully electric Leaf typically achieving lower fuel economy for highway trips. Probability distributions for tripwise fuel economy over different trip types are shown in Figure 2.

Average fuel consumption values for each vehicle over each class of trip were compared with EPA-published "City" and "Highway" ratings, generally showing close agreement. EPA-adjusted city/highway fuel economies for the Leaf, Prius, Accord PHEV in charge-sustaining mode, and the F150 are 126/101, 58/53, 49/45, and 15/21, respectively. In this simulation, average stop-and-go/highway fuel economies for the different vehicle types are 143/125, 59/55, 56/53, and 17/18, respectively. EPA unadjusted fuel economy ratings are used in the calibration of the efficiency parameters in the TripEnergy vehicle model, but EPA-adjusted values (typically lower) are produced through a complicated process not factored into the TripEnergy model (*31*). Any agreement between TripEnergy results and EPA-adjusted values are the result of the less-energy-efficient driving and additional auxiliary use assumed in TripEnergy and present in the real world but not the unadjusted EPA tests.

Discussion

This paper links an existing mesoscopic transport simulation with an energy model that can estimate the energy consumption and CO_2 emissions from personal vehicles. Given moderate-resolution vehicle trajectories from a mesoscopic traffic model, TripEnergy can efficiently estimate energy consumption more accurately than simple speed and acceleration-based methods. TripEnergy offers various improvements over more widely used methods such as MOVES: it utilizes the full detail on vehicle movements provided by mesoscopic simulations; it is computationally efficient enough to be run concurrently with a transport model; and it produces energy estimates that can be traced back to individual vehicles. In addition, because it relies primarily on independent measurements of high-resolution driving behavior, this method is less reliant than alternative models on tunable parameters governing vehicle movement, which are not typically calibrated to reproduce trajectories that are accurate from an energy perspective.

Because this model was developed to take advantage of the strengths of state-of-the-art transport simulations, we expect it to prove increasingly useful as simulations continue to evolve and be applied to new questions and research directions. These include problems in which the energy consumption of vehicles needs to be integrated into agents' travel choice framework, and in which energy consumption and CO_2 emissions are primary targets of study. In many of these cases, the vehicles being studied will not have detailed energy-use measurements beyond EPA-mandated tests. Specific use-cases include the development and evaluation of tools targeted at reducing individual travelers' energy use, understanding the effects of fuel economy on long-term vehicle purchasing decisions, and studying electric vehicle range and charging behavior. Additionally, the ability to calculate energy consumption from limited-resolution trajectories will be useful for contexts outside of traditional transport modeling, including suggesting energy-efficient routes given estimates of the congestion state of a traffic network, or estimating energy consumption of a trip given trajectory information captured by a smartphone.

In future work, several extensions to this analysis will be pursued. Further work is needed to quantify the degree to which mesoscopic models capture realistic traffic movement. The impacts of user-specific driving style on energy use and the degree to which the database of driving behavior data captures this variability will also be investigated further. Finally, the vehicle model could be augmented through the inclusion of additional, higher-order moments. The additional energy estimation accuracy allowed by this modification will be assessed using onboard diagnostic recorders in on-road vehicles, and the gains in accuracy will be weighed against the complexity added to the vehicle model.

Acknowledgments

The research presented here was funded by the U.S. DoE's Advanced Research Projects Agency-Energy (ARPA-E) TRANSNET Program, Award Number DE-AR0000611, and by the New England University Transportation Center at MIT under DOT grant No. DTRT13-G-UTC31.

Author Contributions

The authors confirm contribution to the paper as follows: study conception and design: ZAN, JET; model implementation: ZAN; analysis: ZAN, JET; manuscript preparation: ZAN, JET. All authors reviewed the results and approved the final version of the manuscript.

References

1. Casas, J., J. Perarnau, and A. Torday. The Need to Combine Different Traffic Modelling Levels for Effectively Tackling Large-Scale Projects Adding a Hybrid Meso/Micro Approach. *Procedia-Social and Behavioral Sciences*, Vol. 20, 2011, pp. 251–262.
2. McNerney, J., Z. A. Needell, M. T. Chang, M. Miotti, and J. E. Trancik. TripEnergy: Estimating POV Energy Consumption Given Limited Travel Survey Data. *Transportation Research Record: Journal of the Transportation Research Board*, 2017. 2628: 58–66.
3. da Rocha, T. V., L. Leclercq, M. Montanino, C. Parzani, V. Punzo, B. Ciuffo, and D. Villegas. Does Traffic-Related Calibration of Car-Following Models Provide Accurate Estimations of Vehicle Emissions? *Transportation Research Part D: Transport and Environment*, Vol. 34, 2015, pp. 267–280.
4. Department of Energy's National Renewable Energy Laboratory. *California Household Transportation Survey*. DOE-019-9889603959. 2012.
5. Atlanta Regional Commission. *Regional Travel Survey*. Final Report. 2011.
6. Texas Department of Transportation. *2002-2011 Regional Travel Surveys with GPS Data for Abilene, Austin, El Paso, Houston Galveston, Laredo, Rio Grande Valley, San Antonio, Tyler Longview, and Wichita Falls*. Texas Department of Transportation, Austin, TX, 2011.
7. Markel, T., A. Brooker, T. Hendricks, V. Johnson, K. Kelly, B. Kramer, M. O'Keefe, S. Sprik, and K. Wipke, ADVISOR: A Systems Analysis Tool for Advanced Vehicle Modeling. *Journal of Power Sources*, Vol. 110, No. 2, 2002, pp. 255–266.
8. Advanced Powertrain Research Facility. *Downloadable Dynamometer Database*. Argonne National Laboratory, Lemont, IL, 2011.
9. Adnan, M., F. C. Pereira, C. M. L. Azevedo, K. Basak, M. Lovric, S. Raveau, Y. Zhu, J. Ferreira, C. Zegras, and M. Ben-Akiva. SimMobility: A Multi-Scale Integrated Agent-Based Simulation Platform. Presented at 95th Annual Meeting of the Transportation Research Board, Washington, D.C., 2016.
10. *Transportation Research Circular E-C153: Dynamic Traffic Assignment: A Primer*. Transportation Research Board of the National Academies, Washington, D.C., 2011.
11. De Palma, A., and F. Marchal. Real Cases Applications of the Fully Dynamic METROPOLIS Tool-Box: An Advocacy for Large-Scale Mesoscopic Transportation Systems. *Networks and Spatial Economics*, Vol. 2, No. 4, 2002, pp. 347–369.
12. Ben-Akiva, M. E., S. Gao, Z. Wei, and Y. Wen. A Dynamic Traffic Assignment Model for Highly Congested Urban Networks. *Transportation Research Part C: Emerging Technologies*, Vol. 24, 2012, pp. 62–82.
13. Zhou, X., and J. Taylor. DTALite: A Queue-Based Mesoscopic Traffic Simulator for Fast Model Evaluation and Calibration. *Cogent Engineering*, Vol. 1, No. 1, 2014, p. 961345.

14. Charypar, D., K. W. Axhausen, and K. Nagel. *An Event-Driven Queue-Based Microsimulation of Traffic Flow*. Eidgenossische Technische Hochschule Zurich, Institut fur Verkehrsplanung und Transportsysteme, Zurich, Switzerland, 2006.
15. Bowman, J. L., and M. E. Ben-Akiva. Activity-Based Disaggregate Travel Demand Model System with Activity Schedules. *Transportation Research Part A: Policy and Practice*, Vol. 35, No. 1, 2001, pp. 1–28.
16. Samaras, C., L. Ntziachristos, and Z. Samaras. COPERT Micro: A Tool to Calculate the Vehicle Emissions in Urban Areas. *Proc., Transport Research Arena (TRA) 5th Conference: Transport Solutions from Research to Deployment*, Paris, France, 2014.
17. California Air Resource Board. *2.30 User Guide: Calculating Emission Inventories for Vehicles in California*. 2006. https://www.arb.ca.gov/msei/onroad/downloads/docs/user_guide_emfac2007.pdf.
18. Ahn, K., and H. Rakha. The Effects of Route Choice Decisions on Vehicle Energy Consumption and Emissions. *Transportation Research Part D: Transport and Environment*, Vol. 13, No. 3, 2008, pp. 151–167.
19. Assessment and Standards Division, Office of Transportation and Air Quality. *Population and Activity of On-Road Vehicles in MOVES2014*. Draft Report EPA-420-D-15-001. U.S. Environmental Protection Agency, Washington, D.C., 2015.
20. Zhou, X., S. Tanvir, H. Lei, J. Taylor, B. Liu, N. M. Rouphail, and H. C. Frey. Integrating a Simplified Emission Estimation Model and Mesoscopic Dynamic Traffic Simulator to Efficiently Evaluate Emission Impacts of Traffic Management Strategies. *Transportation Research Part D: Transport and Environment*, Vol. 37, 2015, pp. 123–136.
21. Zegeye, S., B. De Schutter, J. Hellendoorn, E. Breunesse, and A. Hegyi. Integrated Macroscopic Traffic Flow, Emission, and Fuel Consumption Model for Control Purposes. *Transportation Research Part C: Emerging Technologies*, Vol. 31, 2013, pp. 158–171.
22. Rakha, H., K. Ahn, and A. Trani. Development of VT-Micro Model for Estimating Hot Stabilized Light Duty Vehicle and Truck Emissions. *Transportation Research Part D: Transport and Environment*, Vol. 9, No. 1, 2004, pp. 49–74.
23. Scora, G., and M. Barth. *Comprehensive Modal Emissions Model (CMEM), Version 3.01*. User Guide. Centre for Environmental Research and Technology, University of California, Riverside, CA, 2006.
24. Saerens, B., H. Rakha, K. Ahn, and E. Van Den Bulck. Assessment of Alternative Polynomial Fuel Consumption Models for Use in Intelligent Transportation Systems Applications. *Journal of Intelligent Transportation Systems*, Vol. 17, No. 4, 2013, pp. 294–303.
25. Frey, H. C., and B. Liu. Development and Evaluation of Simplified Version of MOVES for Coupling with Traffic Simulation Model. Presented at 92nd Annual Meeting of the Transportation Research Board, Washington, D.C., 2013.
26. Needell, Z. A., J. McNerney, M. T. Chang, and J. E. Trancik. Potential for Widespread Electrification of Personal Vehicle Travel in the United States. *Nature Energy*, Vol. 1, No. 9, 2016, p. 16112.
27. Larminie, J., and J. Lowry. *Electric Vehicle Technology Explained*. West Sussex, UK: John Wiley & Sons, 2004.
28. Post, K., J. Kent, J. Tomlin, and N. Carruthers. Fuel Consumption and Emission Modelling by Power Demand and a Comparison with Other Models. *Transportation Research Part A: General*, Vol. 18, No. 3, 1984, pp. 191–213.
29. U.S. Environmental Protection Agency. *Certified Vehicle Test Results Report Data*. 2015 (April 1, 2016). http://www.epa.gov/otaq/crttst.htm. Accessed April 1, 2016.
30. Forward, E., K. Glitman, and D. Roberts. *An Assessment of Level 1 and Level 2 Electric Vehicle Charging Efficiency*. Vermont Energy Investment Corporation Transportation Efficiency Group, Burlington, VT, 2013.
31. *Fuel Economy Labeling of Motor Vehicle Revisions to Improve Calculation of Fuel Economy Estimates*. Office of Transportation and Air Quality, U.S. Environmental Protection Agency, Washington, D.C., 2006.

The Standing Committee on Transportation and Air Quality (ADC20) peer-reviewed this paper (18-05678).

Article

TRR
JOURNAL OF THE TRANSPORTATION RESEARCH BOARD

Transportation Research Record
2018, Vol. 2672(25) 174–186

DOI: 10.1177/0361198118798988
journals.sagepub.com/home/trr
SAGE

Estimating Vehicular Emission Impact of Nighttime Construction with VISSIM and Different MOVES Emission Estimation Approaches

Chaoyi Gu[1], Reza Farzaneh[2], Geza Pesti[1], Gabriel Valdez[3], and Andrew Birt[1]

Abstract

Shifting work zones from daytime to nighttime is a potential solution to air quality issues on roadway with high traffic volume and where it is undesirable to close lanes during peak hours. The expected benefit of such shifting is to reduce total fuel consumption and on-road vehicle emissions. However, the magnitude of emission reductions and air quality impacts has not been examined comprehensively at work zones. The study presented in this paper investigated the traffic-related emission impacts of work zones using an urban freeway case study. A VISSIM test bed combined with the Environmental Protection Agency's MOVES emission model was used to estimate total emissions assuming daytime and nighttime lane-closure scenarios. Vehicle emissions were estimated using a link-based method and operating mode-based method. The results from both methods demonstrated that nighttime construction has a significant impact on both traffic speeds and vehicle emissions, primarily as a result of reductions in vehicle miles traveled. In addition, a horizontal comparison between the results from the two methods was made to assess the impact of different emission estimation approaches. The outcomes from the comparison highlight the potential importance of the operating mode-based approach for accurately estimate total traffic emission quantities when data or simulations are available.

Roadway construction projects under regional transportation improvement programs reflect the awareness of transportation planners in adapting to expected fast-growing traffic demand and travel needs. Different roadway closure strategies, such as full road closure, partial road closure, and lane closure, serve as effective ways to accelerate the completion of these roadway construction projects. Among all these strategies, the lane-closure strategy is designed to keep traffic moving at the pattern closest to the no-build situation, rather than being detoured or being moved to the other side of the roadway. However, even with a lane-closure strategy, traffic is interrupted during the construction period. The effects, such as reduced speed, extended travel time, reduced safety, and increased fuel consumption and engine combustion emissions, have drawn public concern, especially when construction takes place during peak hours (*1*). Despite potential savings in vehicle emissions, construction machinery may also generate emissions in the form of fugitive dust that affects air quality. Other factors such as noise, safety, and light pollution from construction equipment may also affect the decision when selecting the time of construction activities.

In many urban areas in Texas, especially on high-volume roadways, work zone and construction activities are commonly conducted at night to reduce the impacts on congestion and mobility. Shifting roadway construction from daytime to nighttime reduces potential traffic delays caused by construction in daytime while potentially increasing traffic delays in nighttime. Considering that the traffic volume during the daytime is significantly higher than it is during the nighttime, the trade-off will benefit vehicle average speeds and total emissions in a daily cycle. From both public health and regulatory perspectives, assessing emissions from vehicle traffic in a

[1]Texas A&M Transportation Institute, College Station, TX
[2]Texas A&M Transportation Institute, Austin, TX
[3]Texas A&M Transportation Institute, El Paso, TX

Corresponding Author:
Address correspondence to Chaoyi Gu: c-gu@tti.tamu.edu

work zone is important. The public health impacts of concentrated traffic pollutants are well established, and many urban areas must comply with federal air quality regulations designed to protect air quality and public health. This suggests that in many cases, undertaking roadway work zone and construction activities during the nighttime rather than the daytime could have considerable health benefits for the public and transportation agencies.

Although there are intuitive air quality benefits of undertaking construction at night, few studies have explicitly examined the benefits of switching construction from day to night. To make effective decisions about whether to undertake nighttime construction or not, public agencies require a series of systematic approaches including traffic impact study, emission estimation, dispersion modeling, and even an extensive human health study to account for potential air quality and health impacts comprehensively. Among all the steps of assessment, the results from emission estimation provide the most effective understanding about the air quality impact of on-road traffic, by showing the amount of emissions. However, the emission impacts associated with moving construction activities from the daytime to the nighttime have not been studied in detail.

The goal of this study is to apply emission models to quantify the emission impacts of construction projects using traffic simulation from a state-funded nighttime construction study with two different emission estimation approaches—a link-based method, and a second-by-second operating mode-based method. An additional goal of the study is to evaluate these two different emission estimation processes by a horizontal comparison. The results are discussed in the context of:

- the air quality impact of daytime construction versus nighttime construction; and
- the accuracy of different approaches on estimation of total emission quantities.

Literature Review

Emissions generated within a roadway construction project can be broken down into emissions from changes in traffic activities and additional emissions from equipment used in construction and maintenance activities. Several previous studies have highlighted the emissions from roadway construction with life-cycle assessment or simulation-based approaches, or benefit-cost analysis.

Wang et al. conducted a study to characterize traffic conditions in highway work zones to model the environmental impact of work zone traffic delays (*2*). In this research, a demand-capacity model was used to calculate traffic delay. Inputs to the model were daily traffic volume, hourly traffic distribution, highway configuration, and different lane-closure strategies. They found that energy consumption and greenhouse gas (GHG) emissions impacts were relatively small in rural areas. The impact on energy savings and GHG emission reductions were negligible when no queue develops. Hanson and Noland conducted an analysis of construction project-related GHG emission from road construction (*3*). In the analysis, project-related emissions included the direct and upstream emissions associated with materials, construction equipment, mobilization of resources, traffic disruption, and maintenance over lifetime. The results from the study showed that with full roadway closure, the emissions by traffic disruption were about 26% of total life-cycle emissions, whereas with intermittent roadway closure, the emissions by traffic disruption only accounted for 2%. The emissions by traffic disruption in the intermittent roadway closure scenario were reduced by 94% compared with the full roadway closure scenario. Huang et al. used a life-cycle assessment model to investigate the additional fuel consumption and emissions caused by traffic delayed by road maintenance activities (*4*). Their results showed that reducing construction time from eight days to five days resulted in substantially lower traffic-related emissions.

González and Echaveguren developed a discrete-event simulation to describe the sustainability in roadway constructions (*5*). They simulated a hypothetical project and estimated fugitive emissions (volatile organic compounds [VOCs], $PM_{2.5}$, PM_{10}) and exhaust emissions (CO, NO_x, SO_2, VOC) generated by construction operation and traffic conditions. They used a traffic microsimulation model to evaluate the traffic delays outside of the construction site caused by these truck activities. Their results showed that the total fugitive and exhaust emissions from a site could be significantly affected by changes by constructions in truck schedules.

Avetisyan et al. conducted a study to quantify the impact of vehicle technology, traffic volume, and work zones on on-road traffic emissions (*6*). They used microscopic traffic simulations to investigate the traffic-related emissions impact of single-lane closure work zones. They found that lane closures resulted in decreases in the speed of vehicles and an increase in per-mile emissions rates. However, they observed that because the work zone caused a reduction in overall traffic through the simulated corridor, the total emissions from traffic were also reduced. Zhang et al. used a microscopic approach to compare vehicle emissions in work zone under rush hour congestions to emissions under free-flow traffic conditions (*7*). They used vehicle speed and acceleration outputs as inputs to the Comprehensive Modal Emissions Model to calculate emissions. They found that CO_2, CO, hydrocarbon (HC), and NO_x distance-based emissions rates reach their highest level during a transitional period when traffic changed from free-flow to congested or congested to free-flow conditions. They attributed these findings to increased acceleration or deceleration activities during the transition period.

A research study by Texas A&M Transportation Institute calculated traffic delays and used road-user costs to quantify characteristics of traffic flow in work zones with case study sites in Dallas, Texas and Houston, Texas (*8*). The impact of schedule and number of lanes closed were presented in the quantity of NO_x, CO, and VOC graphs for different facility demand, annual average daily traffic and work zone schedule. The results indicated that emissions and road-user costs of all scenarios are lowest with overnight schedule compared with either peak-hour schedule or off-peak-hour schedule.

In addition to studies that investigate the emission impacts of road construction activities, several continuous studies have examined the impact of work zone emissions from a human health perspective, as understanding the quantity of emissions alone is not enough to review health effects and further studies such as dispersion modeling or emission concentration study provide another level of insight. Epidemiological studies have indicated a link between vehicular emissions and adverse health impacts such as premature deaths and increased hospital admissions, as well as lost productivity through missed work or school days (*9*). Emissions concentrations from on-road and non-road sources peak near the emission source (i.e., roadways and work zones) and decay quickly within a few hundred meters to the background concentration levels (*10*, *11*).

Methodology

The paper focuses on assessing traffic-related emission impacts of shifting construction activities from daytime to nighttime. The research team used a simulation-based methodology to estimate the changes in traffic operations and the simulation results were applied to emission estimation of vehicles at both daytime and nighttime work zones through the following steps:

1. Simulate traffic by VISSIM with selection of case study sites and different lane-closure scenarios between daytime and nighttime.
2. Estimate on-road vehicle emissions by the MOtor Vehicle Emission Simulator (MOVES) with simulated vehicle activities aggregated by two approaches: a link-based approach and an operating mode-based approach.
3. Analyze the outputs from traffic simulation to estimate traffic impacts and the outputs from emission estimation to assess emission impacts.
4. Compare the emission estimation between two different estimation processes.

VISSIM Modeling

The research team developed a simulation test bed using the traffic microsimulation software VISSIM for a construction project recently undertaken on I-10 corridor in El Paso, Texas. I-10 is the southernmost cross-country transportation corridor in the Interstate Highway System from east coast to west coast. The case study work zone is located on the central-west side portion of the border region where the roadway network experiences heavy traffic congestion during peak hours caused by the proximity of a college campus, the central business district, and shopping centers. The segment of I-10 considered for this study has three lanes in each direction and no frontage road or direct connectivity to other alternative routes such as N. Mesa Drive or W. Paisano Drive. The site is currently undergoing construction to build collector-distributor lanes and frontage roads (5.94 mile total) between N. Mesa Drive and Executive Center Boulevard to improve traffic flow on I-10. The expected completion year for the project is 2019.

Figure 1 shows the case study road network. The main corridor simulated in the study is an approximately 8-mile section of the freeway. Within this corridor, a 0.9-mile work zone was simulated which involved the closure of a single lane. In the VISSIM model, 23 links were created to represent the simulated road system.

Two vehicle types—passenger cars, representing all light-duty vehicles, and heavy-duty trucks, representing all heavy-duty vehicles—were defined in the simulation with attributes that are most typical to Texas vehicles and drivers. Figure 2 shows an example of model parameters that were adjusted to represent realistic driving behaviors; in this case, acceleration functions were derived from data collected in a recent TxDOT study (*12*). These fine-scale model behaviors are important because they can affect both traffic behavior and emissions.

To run the simulations with realistic traffic conditions, the research team used a regional dynamic traffic assignment (DTA) to determine the travel demand matrix and to extract the network layout and characteristics of the study area from the broader El Paso network to capture all re-routing activities. To achieve this, a mesoscopic-to-microscopic tool was utilized to transfer link geometry, paths, speeds, and traffic volumes to VISSIM. All paths captured from the DTA model were transformed to time-dependent static routes and vehicle flows to represent current traffic conditions in the study area.

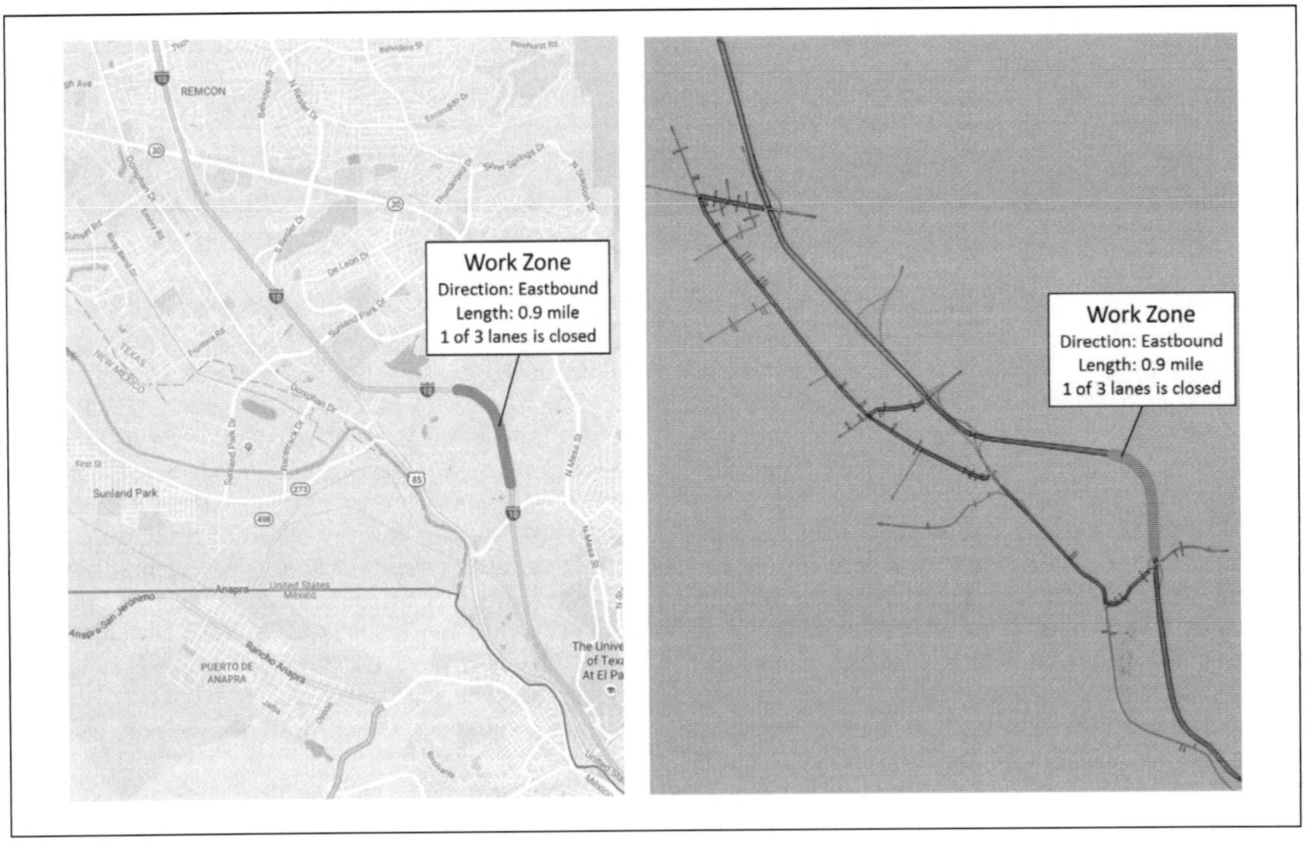

Figure 1. Horizontal layout of simulation network.

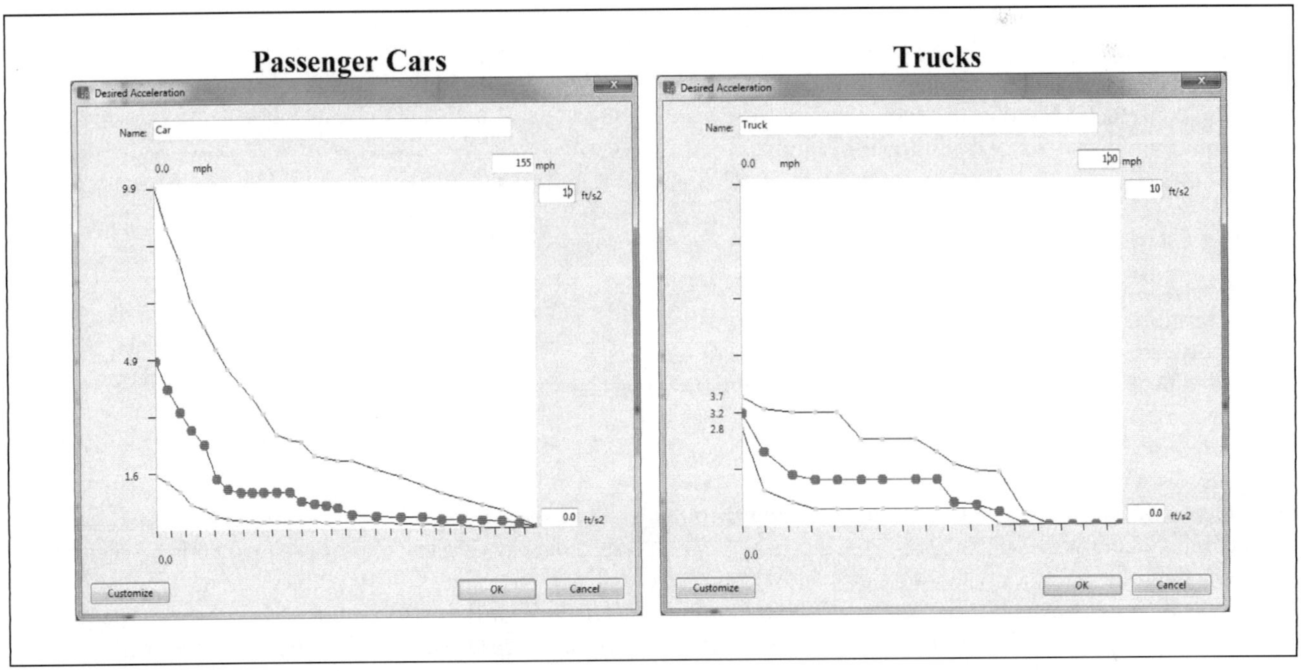

Figure 2. Texas-specific desired acceleration functions.

Table 1. Simulation Scenario Matrix

Case study sites	Daytime work zone scenarios	Daytime base scenarios	Nighttime work zone scenarios	Nighttime base scenarios
I-10 EB, El Paso	1 of 3 lanes closed 06:00–19:00	Non-lane closed 06:00–19:00	1 of 3 lanes closed 19:00–06:00	Non-lane closed 19:00–06:00

Selection of lane-closure scenarios at construction zone were based on common practices in Texas. In Texas, nighttime construction and maintenance activities on freeways typically start between 7 p.m. and 9 p.m. and finish by 6 a.m. or 7 a.m. Daytime construction could be performed at any time between 6 a.m. and 7 p.m. Four simulation scenarios for the case study site were determined (Table 1). Table 1 defines the combinations of lane-closure durations and work zone lane-closure configurations (no lane closure [base scenario], one of three lanes closed [work zone scenario]). The scenario matrix guided all simulation runs for both daytime and nighttime constructions. For each scenario, 10 replicates were simulated using different random number seeds to account for the stochastic nature of the model. An additional 30 minutes' warm-up period was added to each replicate to make sure the traffic on simulated network reaches normal level.

Second-by-second vehicle trajectory data (i.e., location, speed, and acceleration of each simulated vehicle at each time step) of each link segment were outputs from the microsimulation model. These output data were summarized to evaluate the traffic impact of daytime and nighttime construction and were used as inputs for emission estimation, discussed in the following section.

Emission Estimation

MOVES is the most recent on-road emission factor model developed by the U.S. EPA. It is a database-driven software with a disaggregate emission algorithm that provides flexibility for both inputs and outputs in the emission estimation process (*13*). It gives the options to use local meteorology data, fuel, or even local operating mode distributions as inputs to replace default data when local data are available to measure local emission impacts more accurately.

In MOVES, two distinct types of vehicle activity summaries lead to two different emission estimation approaches: a link-based method and an operating mode-based method. When only link average speeds are provided to MOVES project-level data manager, the MOVES model will use default drive cycles and its interpolation algorithm to generate default operating mode distributions. When a specific operating mode distribution is provided with link average speed at the same time, the MOVES model will use the user-input to overwrite the default operating mode distribution, and the user-specific operating mode distribution and link average speed are directly imported to the MOVES project-level panel. The operating mode-based emission estimation approach requires the real-world operating mode distribution directly converted from second-by-second vehicle trajectory data. It is considered the most accurate method, as all the vehicle activities are captured without any compromising to MOVES default drive cycles. The link average speed-based method is a mesoscopic approach that is commonly and easily used in emission estimations as only link average speed data are required.

The VISSIM model provides detailed vehicle activities in the form of second-by-second vehicle trajectory data. The second-by-second data provide the possibility to be converted to either operating mode distribution or link-level vehicle activity summaries, such as link average speed and vehicle miles traveled (VMT) on each link before emission modeling with MOVES. Therefore, both methods were applied in this research to estimate the emissions of nighttime construction versus daytime construction and the results were compared. Vehicle activities were calculated from VISSIM output accordingly, and the following two sets of vehicle activity inputs were provided to MOVES model:

- Vehicle activity inputs for the link-based method: link average speeds and VMT of all links.
- Vehicle activity inputs for the operating mode-based method: operating mode distributions, average speeds, and number of vehicle records of all links.

Vehicle Activity Calculation. For each case study scenario, to calculate average hourly vehicle speeds and VMT on each link within the simulated network, second-by-second vehicle speed data need to be summarized. Instead of using link length, the calculation focused on second-by-second vehicle distances traveled and aggregated them to total VMT on each link. For each scenario, the simulation generated millions of rows of second-by-second data; custom code written in SAS/

STAT® software was used to calculate the raw data into link average speeds and VMT for each hour of scenario with the following equations:

$$VMT_{\text{link, hour}} = \sum_{i} speed_i(mph) \times \frac{1sec}{3600(\frac{sec}{h})} \quad (1)$$

$$Average\ Speed_{\text{link, hour}} = \frac{\sum_{i=0}^{n} speed_i(mph)}{n} \quad (2)$$

Where,

$VMT_{\text{link, hour}}$ is average hourly VMT for each link of the modeled case study scenario (either composite heavy-duty or composite light-duty vehicle types);

$Average\ Speed_{\text{link, hour}}$ is hourly average speeds for each link of the modeled case study scenario (either heavy-duty or light-duty vehicle types);

$speed_i$ is the instantaneous speed of observation i for each link of the modeled case study scenario during each hour (either heavy-duty or light-duty vehicle types); and

n is the total number of observations for each link of the modeled case study scenario during each hour (either heavy-duty or light-duty vehicle types).

Meanwhile, second-by-second speed and acceleration data were calculated to vehicle specific power (VSP) using Equations 3 and 4. In the MOVES model, scaled tractive power (STP) is used as a combination measurement of instantaneous speed, acceleration, road grade, and road load for medium-duty and heavy-duty vehicles. STP is calculated and derived from VSP by Equation 4 (*13*). Each data record was subdivided into operating modes that produce unique energy consumption and emission rates. The definition of each operating mode is based on a range, which is a bin in MOVES, of VSP values and instantaneous speed values. Table 2 lists the operating mode bin definitions for most running emissions and fuel consumptions in MOVES. Similar to the VMT calculation, custom codes were written to calculate the operating modes of millions of simulation data with the following equations:

$$VSP = \frac{P}{m} = \frac{Av + Bv^2 + Cv^3 + mva}{m}. \quad (3)$$

$$STP = \frac{VSP \times m}{f_{\text{scale}}}. \quad (4)$$

Where,
P = power of the vehicle;
v = instantaneous speed;
a = instantaneous acceleration;
A = rolling resistance term;
B = rotating resistance term;
C = aerodynamic drag term;
m = mass of the vehicle;
VSP = vehicle specific power;
STP = scale tractive power; and
f_{scale} = scale factor.

Table 2. Operating Mode Bin Definitions for Most Running Emissions in MOVES

Braking (Bin 0)			
Idle (Bin 1)			
VSP (kW/tonne) \instantaneous speed (mph)	0–25	25–50	>50
<0	Bin 11	Bin 21	-
0 to 3	Bin 12	Bin 22	-
3 to 6	Bin 13	Bin 23	-
6 to 9	Bin 14	Bin 24	-
9 to 12	Bin 15	Bin 25	-
12 and greater	Bin 16	-	-
<6			Bin 33
6 to 12	-	-	Bin 35
12 to 18	-	Bin 27	Bin 37
18 to 24	-	Bin 28	Bin 38
24 to 30	-	Bin 24	Bin 39
30 and greater	-	Bin 25	Bin 40

MOVES Inputs. Besides vehicle activities, MOVES requires other input data such as source type, model year, calendar year, temperature, and humidity in project-level analysis. The case study uses the following project-level parameters for a typical winter day analysis:

- Calendar year: 2016
- Region: El Paso County, Texas
- Month: January
- Date and Time: Weekday, depending on scenario hours
- Temperature: El Paso local average temperature
- Humidity: El Paso local average humidity
- Fuel: El Paso specific fuel and fuel formulation
- Inspection and maintenance (I/M) strategy: Default 2016 MOVES I/M data
- Sourcetype: passenger cars (MOVES sourcetype 21) representing passenger cars in simulation and combination short-haul trucks (MOVES sourcetype 61) representing trucks in simulation
- Roadway type: Urban freeway

Custom batch mode MOVES run spec codes were written to set up and to operate the groups of MOVES runs. Two sets of typical emission outputs were generated from different emission estimation approaches. The typical emissions considered in this research in construction sites include total hydrocarbons (THC), carbon monoxide (CO), carbon dioxide (CO_2), oxides of nitrogen (NO_X), particulate matters under 10 um (PM_{10}), and particulate matters under 2.5 um ($PM_{2.5}$).

Results and Analysis

This section summarizes the results of all the case scenarios with different emission estimation approaches. The results include the traffic impact analysis and emission estimations of the following scenarios:

- Daytime base scenario with link average speed only,
- Daytime base scenario with link average speed and operating mode distribution,
- Daytime work zone scenario with link average speed only,
- Daytime work zone scenario with link average speed and operating mode distribution,
- Nighttime base scenario with link average speed only,
- Nighttime base scenario with link average speed and operating mode distribution,
- Nighttime work zone scenario with link average speed only, and
- Nighttime work zone scenario with link average speed and operating mode distribution.

Traffic Impact

The speed summary figure indicates that hourly vehicle average speeds during daytime scenarios are significantly different to the speeds during nighttime (Figure 3). The hourly average speeds of daytime work zone scenario are within the range of 10 mph and 15 mph, which is about 50% lower than the congested speeds during peak hours in daytime base scenario. The differences in hourly average speeds between nighttime work zone scenario and nighttime base scenario are minimal. The comparison indicates that vehicles moving through work zone during the nighttime experienced much lower interruption than when they are moving during daytime, from an average speed perspective. The drops in speed at the beginning and near the end of nighttime construction happen when the high traffic volume appears in the morning peak hours on the roadway network.

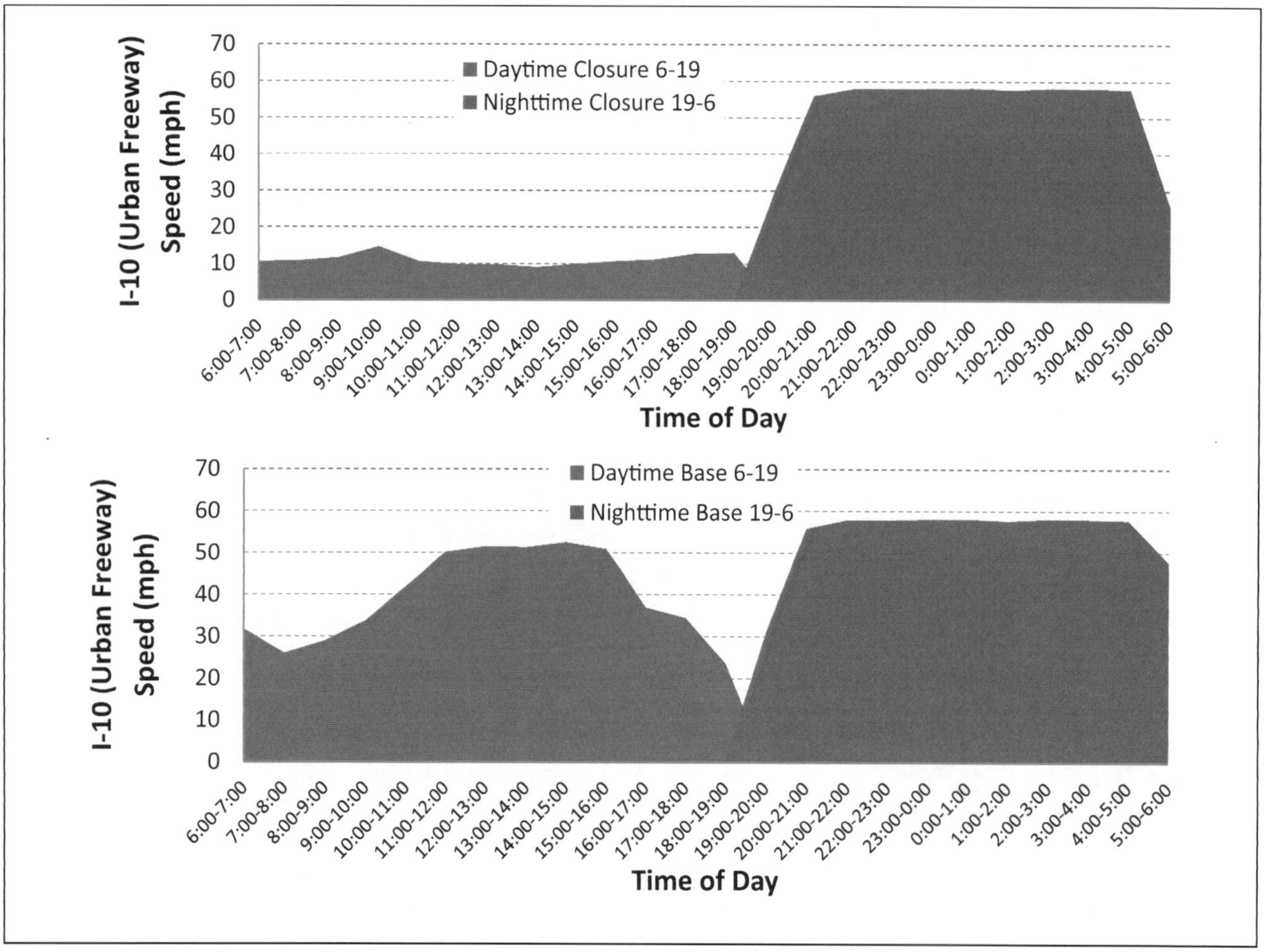

Figure 3. Vehicle speeds—nighttime vs. daytime work zones.

Emission Impact

The emission summary table illustrates the total emissions calculated from link-based estimation and operating mode-based estimation (Table 3). It also demonstrates the different emission impacts of different types of vehicles. The percentage reduction table shows a direct impact between nighttime and daytime with a percentage comparison by emission estimations from the two different estimation methodologies (Table 4).

The results from Tables 3 and 4 indicate that:

- Nighttime construction resulted in significantly lower emissions when directly compared with daytime construction in both estimations.
- The magnitude of the emission impacts was primarily determined by the significant reductions in VMT during nighttime compared with daytime.
- The magnitude of the emission reduction was more than the magnitude of the reduction in VMT. This means that the nighttime construction resulted in benefits for individual vehicles from the perspective of emission rates (gram/VMT). It indicates the individual vehicle emission rates were less during nighttime than daytime at construction sites.
- In terms of calculating percentage reduction of total emissions from daytime construction to nighttime construction, link-level emission estimations are about 6%, 6%, 4%, 1%, 1% lower than operating mode emission estimations in CO, NO_X, CO_2, $PM_{2.5}$, and PM_{10}, respectively.
- The percentage reduction in THC calculated from link-level emission estimation is similar to the reduction from operating mode emission estimation.

Emission Estimation Comparison

The comparison between link-based emission estimation and operating mode-based emission estimation of all scenarios was conducted with the following equation:

$$Percentage = \frac{Emission_{\mathrm{OPMODE}}}{Emission_{\mathrm{Link}}} \times 100\% \qquad (5)$$

It is clear in Table 5 that the emission estimations of the typical pollutants in link-based methods are greater than the estimations by the operating mode-based method, except for CO_2 (Table 5). For each pollutant, the difference in emission estimations comes from the difference in operating mode distributions. The operating mode distributions of all the case scenarios (Figure 4) are consistent with the implication of average speeds in Figure 3. In addition, the operating mode distributions in Figure 4 indicate that there are significant differences between simulated operating mode distributions and MOVES default operating mode distributions at construction sites for both work zone scenarios and base scenarios. The most significant distributional difference in daytime work zone scenario is in operating mode 1 (idling). The difference implies that the MOVES default operating mode distribution suggests more vehicles are idling when the link average speed is low (between 10 mph and 15 mph), whereas the simulation indicates fewer vehicles are idling and more vehicles are low-speed cruising (in operating mode 12, 13, and 14). For the other three control groups, the most significant difference is in operating mode 33. The difference shows that the MOVES default operating mode distribution reveals a more aggressive driving pattern with more vehicles high-speed cruising with higher accelerations (in operating mode 35 and 37), whereas the simulation shows a more conservative driving pattern with more vehicles high-speed cruising with lower accelerations (in operating mode 33).

The emission rates of each operating mode could be extracted from the MOVES database (Table 6). In this table, it is clear that the emission rates of operating mode 1 (idling) are close to emission rates of low-speed cruising operating mode 12, 13, and 14. This gives the explanation that although there is a significant difference in operating mode distributions between daytime scenarios, the difference between link-based emission estimation and operating mode-based estimation is very limited. Meanwhile, the emission rates of high-speed cruising operating mode 33 are significantly different from emission rates of operating mode 35 and 37. This supports the finding in this research that a greater difference in operating mode 33 would produce a greater discrepancy in emission estimation. The findings from operating mode emission rates are consistent with the findings of emission estimation of nighttime scenarios, shown in Table 5.

Conclusion

This paper presented a traffic simulation-based quantitative methodology to estimate the traffic emission impacts of nighttime construction versus daytime construction. The emission estimation was performed using two methods: a link average speed-based method and operating mode-based method. The results of the analysis demonstrated that nighttime construction results in improved traffic operations and decreased emissions compared with daytime construction. A preliminary analysis showed that most of these impacts are directly attributable to the considerably lower traffic volumes during nighttime compared with daytime. However, important effects also occur because of different average speed and different driving patterns (operating mode distributions)

Table 3. Emission Summary of All Case Scenarios

Total emissions estimated by link average speed method			VMT	Emissions (kg)					
				THC	CO	NO_X	CO_2	PM_{10}	$PM_{2.5}$
Daytime	Work zone	Total	296098	105.7	1386.0	1184.1	239652.5	48.0	44.0
		Ave. Rate (g/VMT)		0.36	4.68	4.00	809.37	0.16	0.15
	Base	Total	296098	44.4	806.7	757.2	145014.5	27.8	25.5
		Ave. Rate (g/VMT)		0.15	2.72	2.56	489.75	0.09	0.09
Nighttime	Work zone	Total	101854	14.1	271.5	236.9	47379.3	7.8	7.2
		Ave. Rate (g/VMT)		0.14	2.67	2.33	465.17	0.08	0.07
	Base	Total	101854	12.6	219.9	210.9	41645.4	7.3	6.7
		Ave. Rate (g/VMT)		0.12	2.16	2.07	408.87	0.07	0.07
Total light-duty emissions estimated by link average speed method			VMT	Emissions (kg)					
				THC	CO	NO_X	CO_2	PM_{10}	$PM_{2.5}$
Daytime	Work zone	Total	258906	35.6	1060.1	124.1	139808.6	3.6	3.2
		Ave. Rate (g/VMT)		0.14	4.09	0.48	540.00	0.01	0.01
	Base	Total	258906	17.3	654.1	102.3	82358.2	2.5	2.2
		Ave. Rate (g/VMT)		0.07	2.53	0.40	318.10	0.01	0.01
Nighttime	Work zone	Total	90182	6.0	226.5	36.1	28476.3	0.9	0.8
		Ave. Rate (g/VMT)		0.07	2.51	0.40	315.77	0.01	0.01
	Base	Total	90182	4.9	177.6	29.4	24364.8	0.8	0.7
		Ave. Rate (g/VMT)		0.05	1.97	0.33	270.17	0.01	0.01
Total heavy-duty emissions estimated by link average speed method			VMT	Emissions (kg)					
				THC	CO	NO_X	CO_2	PM_{10}	$PM_{2.5}$
Daytime	Work zone	Total	37192	70.1	325.8	1060.0	99843.9	44.3	40.8
		Ave. Rate (g/VMT)		1.89	8.76	28.50	2684.53	1.19	1.10
	Base	Total	37192	27.1	152.6	654.9	62656.3	25.3	23.3
		Ave. Rate (g/VMT)		0.73	4.10	17.61	1684.66	0.68	0.63
Nighttime	Work zone	Total	11672	8.1	45.1	200.8	18902.9	7.0	6.4
		Ave. Rate (g/VMT)		0.69	3.86	17.21	1619.51	0.60	0.55
	Base	Total	11672	7.7	42.3	181.5	17280.5	6.5	6.0
		Ave. Rate (g/VMT)		0.66	3.62	15.55	1480.51	0.56	0.51
Total emissions estimated by operating mode-based method			VMT	Emissions (kg)					
				THC	CO	NO_X	CO_2	PM_{10}	$PM_{2.5}$
Daytime	Work zone	Total	296098	102.9	1328.7	1173.3	242857.4	47.3	43.4
		Ave. Rate (g/VMT)		0.35	4.49	3.96	820.19	0.16	0.15
	Base	Total	296098	41.0	629.8	615.7	124029.8	25.5	23.3
		Ave. Rate (g/VMT)		0.14	2.13	2.08	418.88	0.09	0.08
Nighttime	Work zone	Total	101854	13.2	191.7	160.3	38349.8	7.0	6.4
		Ave. Rate (g/VMT)		0.13	1.88	1.57	376.52	0.07	0.06
	Base	Total	101854	11.0	143.0	125.4	30056.1	5.6	5.2
		Ave. Rate (g/VMT)		0.11	1.40	1.23	295.09	0.06	0.05
Total light-duty emissions estimated by operating mode-based method			VMT	Emissions (kg)					
				THC	CO	NO_X	CO_2	PM_{10}	$PM_{2.5}$
Daytime	Work zone	Total	258906	34.7	1022.5	110.8	143293.2	3.1	2.8
		Ave. Rate (g/VMT)		0.13	3.95	0.43	553.46	0.01	0.01
	Base	Total	258906	14.6	487.4	73.4	70472.8	2.1	1.9
		Ave. Rate (g/VMT)		0.06	1.88	0.28	272.19	0.01	0.01
Nighttime	Work zone	Total	90182	4.8	149.3	23.9	24308.0	0.8	0.7
		Ave. Rate (g/VMT)		0.05	1.66	0.26	269.54	0.01	0.01
	Base	Total	90182	3.7	107.1	17.4	18787.7	0.7	0.6
		Ave. Rate (g/VMT)		0.04	1.19	0.19	208.33	0.01	0.01
Total heavy-duty emissions estimated by operating mode-based method			VMT	Emissions (kg)					
				THC	CO	NO_X	CO_2	PM_{10}	$PM_{2.5}$
Daytime	Work zone	Total	37192	68.1	306.2	1062.5	99564.2	44.2	40.7
		Ave. Rate (g/VMT)		1.83	8.23	28.57	2677.01	1.19	1.09
	Base	Total	37192	26.4	142.4	542.3	53557.0	23.4	21.5
		Ave. Rate (g/VMT)		0.71	3.83	14.58	1440.00	0.63	0.58
Nighttime	Work zone	Total	11672	8.4	42.5	136.4	14041.8	6.2	5.7
		Ave. Rate (g/VMT)		0.72	3.64	11.69	1203.03	0.53	0.49
	Base	Total	11672	7.3	35.9	108.0	11268.4	5.0	4.6
		Ave. Rate (g/VMT)		0.63	3.08	9.25	965.42	0.43	0.39

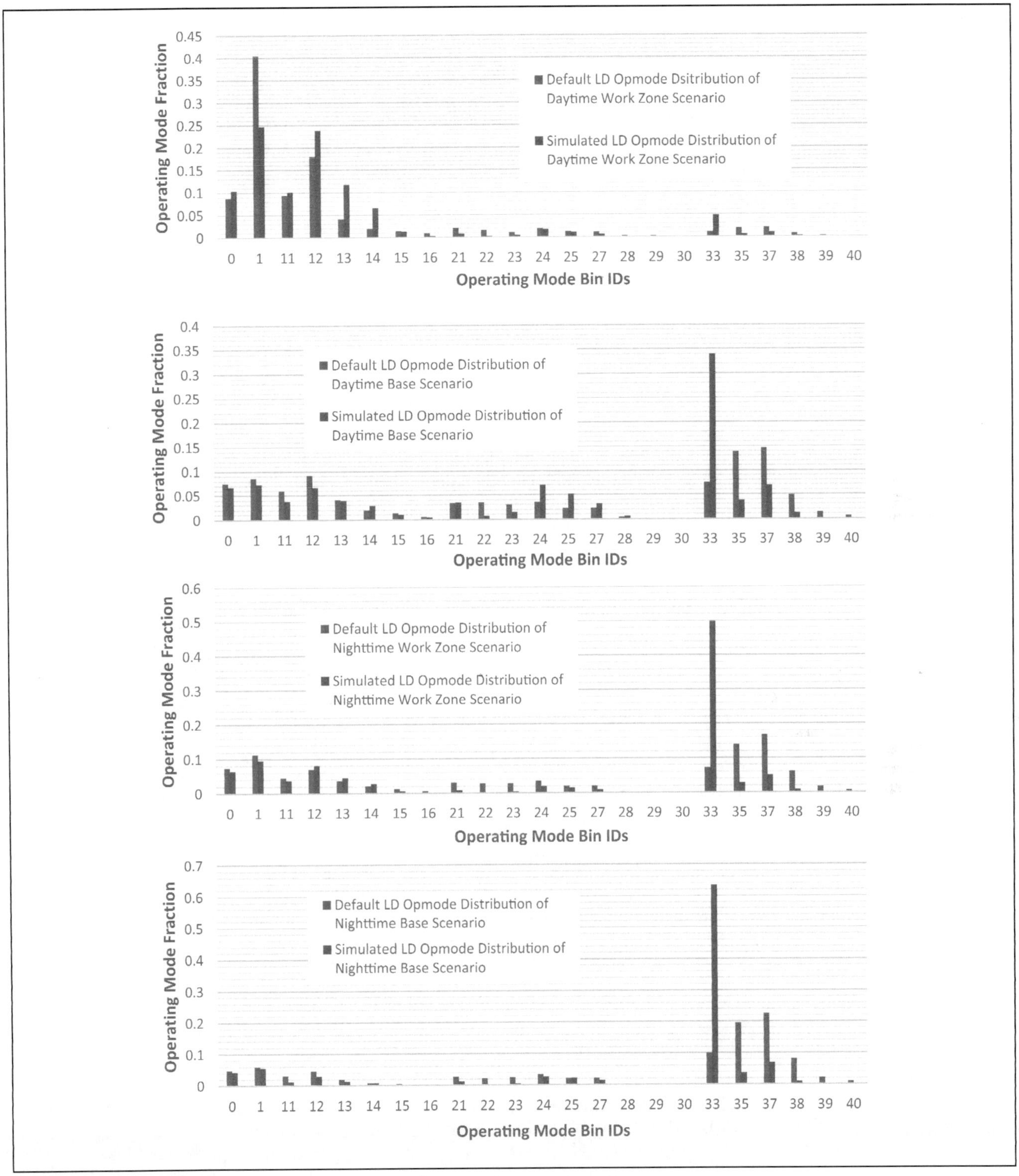

Figure 4. Fleet-mix specific operating mode distributions of all case scenarios.

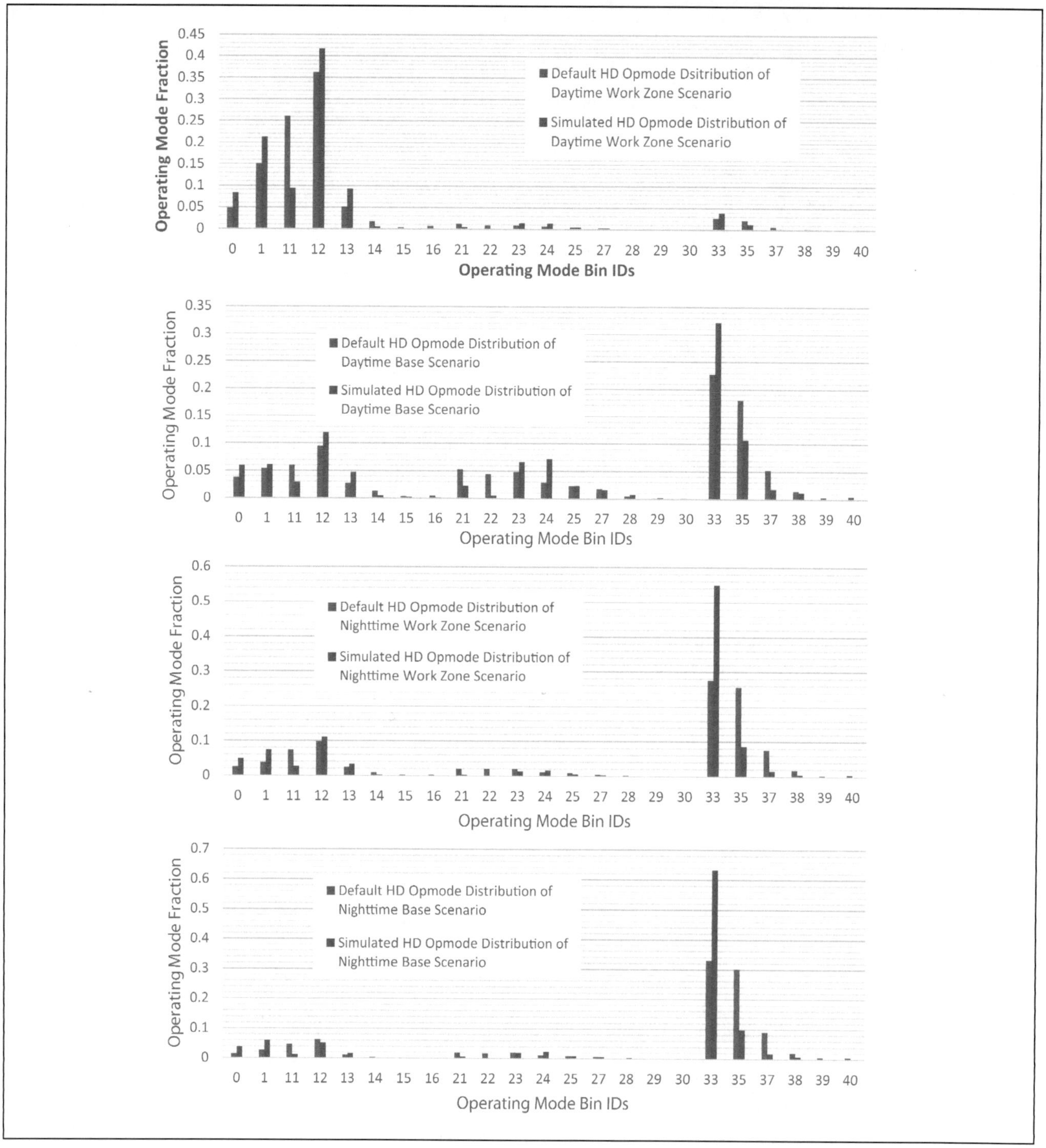

Figure 4. (continued)

Table 4. Percentage Reduction in Total Emissions for Nighttime vs. Daytime Construction

Link average speed based		VMT	Emissions THC	CO	NO_X	CO_2	PM_{10}	$PM_{2.5}$
Nighttime vs. daytime	Total	66%	87%	80%	80%	80%	83%	84%
	LDV	65%	83%	79%	71%	80%	76%	76%
	HDV	69%	89%	86%	81%	81%	84%	84%

Operating mode based		VMT	Emissions THC	CO	NO_X	CO_2	PM_{10}	$PM_{2.5}$
Nighttime vs. daytime	Total	66%	87%	86%	86%	84%	85%	85%
	LDV	65%	86.2%	85%	79%	83%	74%	74%
	HDV	69%	87.7%	86%	87%	84%	86%	86%

Table 5. Percentage Comparison between Link Average Speed-Based Emission Estimation and Operating Mode-Based Emission Estimation of All Scenarios

Total emission comparison Operating mode emission/link-based emission		Emissions THC	CO	NO_X	CO_2	PM_{10}	$PM_{2.5}$
Daytime	Work zone	97.3%	95.9%	99.1%	101.3%	98.7%	98.8%
	Base	92.3%	78.1%	81.3%	85.5%	91.6%	91.7%
Nighttime	Work zone	93.6%	70.6%	67.7%	80.9%	89.3%	89.3%
	Base	86.8%	65.0%	59.4%	72.2%	77.5%	77.5%
Total light-duty emission comparison Operating mode emission/link-based emission		Emissions THC	CO	NO_X	CO_2	PM_{10}	$PM_{2.5}$
Daytime	Work zone	97.6%	96.4%	89.3%	102.5%	87.0%	87.0%
	Base	84.4%	74.5%	71.8%	85.6%	85.0%	85.0%
Nighttime	Work zone	80.1%	65.9%	66.1%	85.4%	96.5%	96.5%
	Base	74.4%	60.3%	59.1%	77.1%	87.2%	87.2%
Total heavy-duty emission comparison Operating mode emission/link-based emission		Emissions THC	CO	NO_X	CO_2	PM_{10}	$PM_{2.5}$
Daytime	Work zone	97.1%	94.0%	100.2%	99.7%	99.7%	99.7%
	Base	97.4%	93.3%	82.8%	85.5%	92.3%	92.3%
Nighttime	Work zone	103.6%	94.2%	67.9%	74.3%	88.5%	88.5%
	Base	94.7%	84.9%	59.5%	65.2%	76.4%	76.4%

Table 6. Operating Mode-Based Emission Rates per Activity of Passenger Cars (Source Type 21) and Combination Short-Haul Truck (Source Type 61) in MOVES (gram/veh/hour)

	THC		CO		NO_X		CO_2		PM_{10}		$PM_{2.5}$	
Operating mode	LDV	HDV	LDV	HDV	LDV	HDV	LDV	HDV	LDV	HDV	LDV	HDV
0	1.59	14.5	17.6	46.9	1.19	117.7	3385.8	15929.0	0.10	4.24	0.09	3.90
1	1.12	12.1	14.5	28.6	1.09	112.3	3111.4	7845.4	0.08	4.64	0.07	4.26
11	1.21	20.0	34.0	84.8	1.90	103.7	4869.7	10527.6	0.09	7.24	0.08	6.66
12	1.46	20.9	56.1	90.2	3.66	347.1	6682.6	30634.3	0.11	11.84	0.09	10.89
13	1.99	27.7	66.6	172.9	7.72	558.5	9389.2	56111.4	0.16	29.67	0.14	27.30
14	2.53	30.5	96.0	218.6	13.49	734.2	11918.5	81881.0	0.18	38.87	0.16	35.76
15	3.09	30.7	124.8	278.9	20.58	887.6	14283.8	103549.2	0.24	56.03	0.21	51.55
16	3.92	28.5	179.1	361.3	32.58	1253.6	17385.6	142430.1	0.70	56.06	0.62	51.58
21	1.76	21.4	51.5	90.4	4.47	75.7	6549.2	8490.8	0.15	9.19	0.14	8.45
22	1.65	25.5	61.7	129.3	5.62	425.8	7533.6	39375.3	0.18	21.48	0.16	19.76
23	1.94	25.7	80.0	151.3	8.56	681.2	9256.1	65303.5	0.16	28.02	0.14	25.78
24	2.66	27.9	117.5	205.9	14.60	979.9	11905.3	94516.9	0.19	45.47	0.17	41.84
25	2.96	28.8	132.9	246.3	20.59	1207.1	15772.1	121533.2	0.25	64.76	0.23	59.58
27	4.38	30.8	212.1	322.8	33.28	1685.1	20761.1	167879.1	0.43	89.46	0.38	82.31
28	9.21	30.1	385.7	365.5	53.26	2207.5	27985.1	235031.0	1.91	139.71	1.69	128.54
29	15.07	38.7	773.7	470.0	79.25	2743.3	38321.7	302182.9	6.77	221.78	5.99	204.04
30	25.15	47.2	2388.7	574.4	104.45	3352.9	48061.3	369333.7	16.50	281.04	14.59	258.56
33	1.76	33.9	42.7	154.5	6.65	279.6	9462.6	35029.6	0.41	18.29	0.37	16.83
35	2.43	33.8	77.4	201.0	19.30	1261.0	15150.5	107117.6	0.46	31.83	0.40	29.28
37	2.97	37.6	108.1	327.2	27.88	1826.9	19718.8	167657.0	0.48	47.93	0.42	44.09
38	6.80	50.0	330.9	278.1	44.36	2404.7	25695.8	234720.8	0.92	73.31	0.81	67.45
39	9.88	64.3	413.7	357.5	64.53	3007.4	34205.6	301782.6	2.49	114.03	2.21	104.90
40	14.00	78.5	1136.0	436.9	81.28	3675.7	43560.8	368845.4	2.91	143.03	2.57	131.59

through the work zone between daytime lane-closure scenarios and nighttime lane-closure scenarios.

The link average speed-based emission estimation approach provided similar results compared with a more detailed and complex operating mode-based emission estimation approach when estimating percentage of emission reductions. However, the results from the study showed that the total emissions from link average speed-based estimation were greater than those calculated by the operating mode-based method. The reason behind the finding is the difference in operating mode distributions between MOVES default drive cycles, and real-world operating mode distributions as simulated using VISSIM in work zones. This finding highlights the potential importance of the operating mode-based approach for accurately estimate total traffic emission quantities. It also shows the advantage of using microsimulation-based vehicle activities in emission estimation for future projects where real-world data are not available. As an important step of the assessment chain of the health impact of construction sites, future implementations will be extended from this study to dispersion modeling and health studies.

Author Contributions

The authors confirm contribution to the paper as follows: study conception and design: Chaoyi Gu, Reza Farzaneh; data collection: Geza Pesti, Gabriel Valdez; analysis and interpretation of results: Chaoyi Gu, Reza Farzaneh; draft manuscript preparation: Chaoyi Gu, Andrew Birt. All authors reviewed the results and approved the final version of the manuscript.

References

1. Grundy, C., R. Steinbach, P. Edwards, J. Green, B. Armstrong, and P. Wilkinson. Effect of 20 mph Traffic Speed Zones on Road Injuries in London, 1986–2006: Controlled Interrupted Time Series Analysis. *British Medical Journal*, Vol. 339, 2009, p. b4469.
2. Wang, T., C. Kim, and J. Harvey. Energy Consumption and Greenhouse Gas Emission from Highway Work Zone Traffic in Pavement Life Cycle Assessment. Paper presented at International Symposium on Pavement Life Cycle Assessment, Davis, CA, 2014.
3. Hanson, C. S., and R. B. Noland. Greenhouse Gas Emissions from Road Construction: An Assessment of Alternative Staging Approaches. *Transportation Research Part D: Transport and Environment*, Vol. 40, 2015, pp. 97–103.
4. Huang, Y., R. Bird, and M. Bell. A Comparative Study of the Emissions by Road Maintenance Works and the Disrupted Traffic Using Life Cycle Assessment and Micro-Simulation. *Transportation Research Part D: Transport and Environment*, Vol. 14, No. 3, 2009, pp. 197–204.
5. González, V., and T. Echaveguren. Exploring the Environmental Modeling of Road Construction Operations Using Discrete-Event Simulation. *Automation in Construction*, Vol. 24, 2012, pp. 100–110.
6. Avetisyan, H., E. Miller-Hooks, S. Melanta, and B. Qi. Effects of Vehicle Technologies, Traffic Volume Changes, Incidents and Work Zones on Greenhouse Gas Emissions Production. *Transportation Research Part D: Transport and Environment*, Vol. 26, 2014, pp. 10–19.
7. Zhang, K., S. Batterman, and F. Dion. Vehicle Emissions in Congestion: Comparison of Work Zone, Rush Hour and Free-Flow Conditions. *Atmospheric Environment*, Vol. 45, No. 11, 2011, pp. 1929–1939.
8. Benz, R. J., and D. Fenno. *Emissions due to Construction Equipment and Traffic Delays – Task 6: Contribution of Vehicular Emissions Caused by Reduction in Capacity during Roadway Construction*. Texas Transportation Institute, Texas A&M University System, 1600 E Lamar Boulevard, Arlington, TX, 2001.
9. Anair, D. *Digging Up Trouble, The Health Risks of Construction Pollution in California*. Union of Concerned Scientists, Cambridge, MA, 2006. http://www.ucsusa.org/sites/default/files/legacy/assets/documents/clean_vehicles/digging-up-trouble.pdf.
10. Colbert, D. *Productivity and Safety Implications of Nighttime Construction Operations*. Independent Research Study Report. Purdue University, West Lafayette, Indiana, 2003.
11. Douglas, K., and S.-B. Park. *Selection Criteria for Using Nighttime Construction and Maintenance Operations*. Report SPR 322, Oregon State University, Corvallis, Report Number FHWA-OR-RD-03-13. OR, 2003.
12. Farzaneh, M., J. Zietsman, D.-W. Lee, J. Johnson, N. Wood, T. Ramani, and C. Gu. *Texas-Specific Drive Cycles and Idle Emissions Rates for Using with EPA's MOVES Model: Final Report*. Report FHWA/TX-14/0-6629-1. Texas A&M Transportation Institute, College Station, TX, 2014.
13. U.S. Environmental Protection Agency. *Motor Vehicle Emission Simulator. MOVES 2010 Users Guide*. EPA-420-B-09-041. U.S. Environmental Protection Agency, Assessment and Standards Division, Office of Transportation and Air Quality, Washington, D.C., 2009.

The Standing Committee on Transportation and Air Quality (ADC20) peer-reviewed this paper (18-06392).